ENTANGLEMENTS OF POWER

Practices of resistance cannot be separated from practices of domination; they are inextricably linked and produce complex entanglements of power.

Entanglements of Power is a collection of essays designed to trace thoroughly the entangled geographies of domination and resistance which are integral to all workings of power. In doing so, the contributors explore specific themes of identity formation, embodied experience, surveillance and display, colonialism and post-colonialism, representational politics, and societal transformation. The rich body of case material includes: German reunification; Jamaican Yardies on British television; Victorian sexuality and London parks; ethnicity, gender and nation in Ecuador; sport as power; and the film *Falling Down*.

Following an introduction by the editors which reframes thinking about power, domination, resistance, space and place, the book contains challenging essays from leading authors in the field which address these themes in a diversity of ways.

Joanne P. Sharp, Paul Routledge, Chris Philo and Ronan Paddison all work in the Department of Geography and Topographic Science, University of Glasgow.

CRITICAL GEOGRAPHIES

This series offers cutting edge research organised into three themes of concepts, scale and transformations. It is aimed at upper-level undergraduates and research students, and will facilitate the inter-disciplinary engagement between geography and other social sciences. It provides a forum for the innovative and vibrant debates which span the broad spectrum of this discipline.

Series Editors:

Tracey Skelton is lecturer in the International Studies department at Nottingham Trent University and **Gill Valentine** is senior lecturer in the Geography department at the University of Sheffield.

MIND AND BODY SPACES: GEOGRAPHIES OF ILLNESS, IMPAIRMENT AND DISABILITY
Edited by Ruth Butler and Hester Parr

EMBODIED GEOGRAPHIES
Edited by Elizabeth Kenworthy Teather

LEISURE GEOGRAPHIES: LEISURE/TOURISM PRACTICES AND GEOGRAPHICAL KNOWLEDGE
Edited by David Crouch

NIGHTS OUT: THE EXPERIENCE OF CLUBBING
Ben Malbon

ENTANGLEMENTS OF POWER: GEOGRAPHIES OF DOMINATION/RESISTANCE
Edited by Joanne P. Sharp, Paul Routledge, Chris Philo and Ronan Paddison

ENTANGLEMENTS OF POWER

POWER

Geographies of domination/resistance

Edited by
Joanne P. Sharp, Paul Routledge,
Chris Philo and Ronan Paddison

London and New York

First published 2000
by Routledge
2 Park Square, Milton Park, Abingdon, Oxon, OX14 4RN

Simultaneously published in the USA and Canada
by Routledge
270 Madison Ave, New York NY 10016

Routledge is an imprint of the Taylor & Francis Group

Transferred to Digital Printing 2009

Typeset in Perpetua by Keystroke, Jacaranda Lodge, Wolverhampton

British Library Cataloguing in Publication Data
A catalogue record for this book is available from the British Library

Library of Congress Cataloging in Publication Data
Entanglements of power : geographies of domination/resistance/
[edited by] Joanne P. Sharp . . . [et al.].
p. cm. — (Critical geographies)
Includes index.
1. Power (Social sciences) 2. Political science. 3. Government,
Resistance to. 4. Human geography. 5. Spatial behavior.
I. Sharp, Joanne P., 1969– . II. Series.
JC330. E58 1999
303. 3—dc21 99–18445
CIP

ISBN 0–415–18434–7 (hbk)
ISBN 0–415–18435–5 (pbk)

Publisher's Note
The publisher has gone to great lengths to ensure the quality of this reprint
but points out that some imperfections in the original may be apparent.

FOR GRAHAM SMITH

CONTENTS

CONTENTS

ILLUSTRATIONS

Figures

Tables

Figures

Maps

CONTRIBUTORS

David Atkinson lectures on cultural and political geography in the Department of Geography, University of Hull. His research interests include the connections between geopolitics and fascism, with particular reference to the Italian experience, and he is also researching the creation of spaces within both imperial and totalitarian cities.

John Bale is Professor of Sports Geography at Keele University. Since the late 1970s he has pioneered the geographical study of sports and has published widely in the fields of human geography and sports studies. He is author of *Sport, Space and the City* (1993) and *Landscapes of Modern Sport* (1994) and co-author (with Joe Sang) of *Kenyan Running: Movement Culture, Geography and Global Change* (1996).

Philip Crang lectures on cultural and economic geography in the Department of Geography, University College, London. His interests include the geographies of service, paying particular attention to the practices and experiences of service workers themselves; he is also now researching British culinary cultural geographies. He is completing a book on *Spaces of Service*.

Tim Cresswell lectures on cultural geography in the Institute of Geography and Earth Sciences at the University of Wales, Aberystwyth. His interests revolve around the geographical constitution of everyday life. He is author of *In Place/Out of Place: Geography, Ideology and Transgression* (1996) and is working on *The Tramp in America: An Exploration of Knowledge, Mobility and Marginality*.

Steve Hinchliffe is based in the Faculty of Social Sciences, Open University. His research interests focus on science, technology and nature; he has written on a diverse range of subjects, from energy debates and outdoor management training to city-natures and ontologies of disease. He is currently working on an edited collection tackling spatial theory and practice.

Philip Howell lectures on historical geography in the Department of Geography, University of Cambridge. His primary research interest in the nineteenth century covers work on the geography of popular politics, urban culture in London and New York, and the role of sexuality in the construction of public space in both Britain and its colonies.

Doreen Massey is Professor of Geography in the Faculty of Social Sciences, Open University. She is author of *Space, Place and Gender* (1994) and *Spatial Divisions of Labour: Social Structures and the Geography of Production* (2nd edn, 1995) and co-editor of *Soundings: A Journal of Politics and Culture*.

Ronan Paddison is Professor of Geography and lectures on political and urban geography in the Department of Geography and Topographic Science, University of Glasgow. He is author of *The Fragmented State: The Political Geography of Power* (1983) and more recently has written on the commodification of local power. His current research is focused on decentralisation and participation within emergent frameworks of local governance in central Scotland.

Chris Philo is Professor of Geography and lectures on social and historical geography in the Department of Geography and Topographic Science, University of Glasgow. He is interested in the geographies of mental health, the social geography of 'outsiders' and many other aspects of human geography's history and theory. He is co-editor of *Off the Map: The Social Geography of Poverty in the UK* (1995) and (with John Bale) *Body Cultures: Essays on Sport, Space and Identity by Henning Eichberg* (1998).

Sarah A. Radcliffe lectures on cultural and development geography in the Department of Geography, University of Cambridge. Her interests include social difference and political cultures in the Andes, as well as feminist geography and social theory. Her recent research has focused on the social and spatial dimensions of Ecuadorian national identity, and the intersections of gender and ethnicity. She is co-author of *Re-Making the Nation: Place, Politics and Identity in Latin America* (1996) and co-editor of *Viva! Women and Popular Protest in Latin America* (1993), both with Sallie Westwood.

Jennifer Robinson is based in the Faculty of Social Sciences, Open University. Her research has encompassed the historical geography of racial segregation in South Africa, and increasingly now concerns the changing political and urban geographies of the post-Apartheid era. She is also a contributor to debates about geography's relations to social, cultural and post-colonial theory. She is author of *The Power of Apartheid: State, Power and Space in South African Cities* (1996).

Paul Routledge lectures on cultural and political geography in the Department of Geography and Topographic Science, University of Glasgow. His current interests are in contemporary forms of popular resistance, critical geopolitics and the cultural politics of India and Nepal. He is author of *Terrains of Resistance: Nonviolent Social Movements and the Contestation of Place in India* (1993) and co-editor (with Gearoid O'Tuathail and Simon Dalby) of *The Geopolitics Reader* (1998).

Joanne P. Sharp lectures on cultural and political geography in the Department of Geography and Topographic Science, University of Glasgow. She researches popular constructions of nationalist identity, particularly the genderings of nationhood, and she is also starting to work on public art in the city. She is co-editor of *Space, Gender and Knowledge* (1997) and *A Feminist Glossary of Human Geography* (1999), both with Linda McDowell.

Tracey Skelton lectures on geography and development studies in the Department of International Studies, Nottingham Trent University. She is co-editor (with Gill Valentine) of *Cool Places: Geographies of Youth Cultures* (1998) and (with Tim Allen) of *Culture and Global Change*. She is currently working on cultural impacts of the volcanic crisis on the Caribbean island of Montserrat. She works as often as possible in the Caribbean.

Fiona M. Smith lectures on human geography in the Department of Geography, University of Dundee. Her current research addresses the gender politics of economic and urban restructuring in eastern Europe. She is co-author (with Claire Dwyer, Sarah Holloway and Nina Laurie) of *Geographies of New Femininities* (1999).

Nigel Thrift is Professor of Geography in the School of Geographical Sciences, University of Bristol. He is the co-editor (with Steve Pile) of *Mapping the Subject: Geographies of Cultural Transformation* (1995), author of *Spatial Formations* (1996) and co-author (with Andrew Leyshon) of *Money / Space*. His main current interests are in the exclusionary geographies written into software, non-representational theory, embodied 'methodologies' like dance, and the reworking of political economy.

Chris Wilbert is a post-doctoral researcher in the School of Human and Social Sciences at City University, London. He is the author of various articles on political ecology, new technology and leisure, and is co-editor (with Chris Philo) of *Animal Spaces, Beastly Places: New Geographies of Animals*.

PREFACE AND
ACKNOWLEDGEMENTS

This book is based upon the *Geographies of Domination / Resistance* mini-conference that was held at the University of Glasgow, Scotland, 19–21 September 1996, sponsored by the Department of Geography and Topographic Science. The idea for the conference emerged out of a series of discussions that Paul and Joanne had been having over the previous year concerning their respective research interests. In his work on, and participation with, various resistance movements, Paul had become increasingly interested in the various authoritarian practices that were articulated within supposedly liberatory projects. Meanwhile, in her research on gender and nationalism, Joanne had become increasingly intrigued by oppositional practices that occurred within projects of control and manipulation. It was from a desire to investigate more thoroughly such 'entangled' geographies of domination/ resistance that the conference was conceived. It was at this nascent stage that Chris and Ronan, whose research interests include the spaces of social and political power, also became involved. Moreover, it seemed like an ideal opportunity for the four of us to co-operate in a project which would express the birth of a new concern for critical human geography here in the University of Glasgow Geography Department.

The conference appeared to have been a great success – or, at the least, the four of us enjoyed it hugely – and we felt that the 'entanglements of power' theme had been explored in a sufficiently sustained fashion to warrant the preparation of a book out of the proceedings. Unfortunately, we could not go for something that included all the papers from the conference and so, reluctantly, we could select only those that most directly tackled our core theme. Nearly everybody that we approached to contribute was able to do so, and the chapters that follow are all extensive revisions of the papers initially given at the conference, revised to some extent in response to a short paper from us summarising the arguments that we have developed at greater length in Chapter 1. We hope, therefore, that there is a coherence to the present collection of chapters which will enable the sum to be more than the parts, and for the whole to stand as a new perspective on the entangled geographies of power, domination and resistance which offers both fresh conceptual and significant substantive components. We had originally envisaged arranging the chapters on a continuum from those concentrating principally on 'resistance in domination' through to those concentrating more on 'domination in resistance' (see Chapter 1), but we soon decided that this was too forced a structure, and too insensitive to the multiple entanglements identified within every chapter. As such, the chapter order amounts to nothing more than starting with those chapters containing historical materials, continuing with chapters whose materials take the reader around a variety of worldly contexts, and concluding with chapters which reflect more abstractly on the entanglings at issue (including shorter commentaries from our two conference plenary speakers). The chapters span widely across many additional themes of contemporary importance to geography and the social sciences – including those of identity, embodiment, organisation, colonialism/post-colonialism and political transformation – and, while their contents can all be read within the context of what we argue in Chapter 1, we have refrained from describing each chapter in detail. Rather, we have tried to let the chapters speak for themselves, and in so doing we believe that contributors ably articulate, albeit in very different ways, the complex web of interrelations that must be confronted in order to understand properly the spatiality of power.

Conference acknowledgements

We would very much like to thank all those who participated in the original conference. Particularly, we would like to thank Nigel Thrift and Doreen Massey, who agreed to be our plenary speakers, and all the other presenters, who were (in conference order): Suzanne Reimer, Jon Simons, Philip Howell, Tracey Skelton, Chris Wilbert, Melanie Limb, John Bale, Steve Hinchliffe, Alex Hughes, Philip Crang (paper read in his absence), Miles Ogborn (paper circulated), Pamela Shurmer-Smith, Jennifer Robinson, Dan Clayton, Lynn Stewart, David Atkinson, Sarah Radcliffe, Tim Bunnell, Emma Mawdsley, Fiona Smith, Peter Shirlow, James Sidaway, Gillian Rose, Tim Cresswell and John Silk. A special thanks also goes out to John Briggs, who, as Head of Department at the time, was able to provide the conference with valuable financial support. Further thanks are due to several of the department's postgraduate students – Oliver Valins, Helen Prescott, Lorna Philip and Mhairi Harvey – who helped enormously with the running of the conference. We would also like to thank Sandhu's Indian restaurant, who curried favour with the conference by catering a buffet for the participants. For those who attended the conference and became engaged in lively debate beyond the scheduled end of one of the days' sessions, there was perhaps no greater irony than our being requested to vacate the conference room by a security guard, and all obliging without a hint of resistance.

Book acknowledgements

First and foremost, of course, we must say a large thank you to our contributors, and in particular to thank them for their forbearance during a project which was admittedly not concluded quite as speedily as had initially been hoped. They have also been model contributors in their willingness to follow up our suggestions, and to ensure that missing references and captions and the like were tracked down with a minimum of fuss. We have been very fortunate. Similarly, we must thank Routledge, particularly Sarah Carty, Sarah Lloyd and Casey Mein, for their assistance, encouragement and patience.

More personally, Joanne would like to thank Mike Shand for help with artwork for the conference poster; Paul would like to thank Teresa Flavin for the balm of reason and the M77 crowd for the re-enchantment of resistance; Chris would like to thank Hester Parr for her support, and Eric Laurier for his many ideas and specific help at a crucial stage in the project. The four of us would also like to thank each other for being such excellent co-conspirators.

Dedication

This book is dedicated to Graham Smith, whose death earlier this year has caused much sadness and shock to his friends and colleagues in academic Human Geography and Post-Soviet Studies the world over. We personally had many connections with Graham: Ronan co-supervised his PhD work at Glasgow; Chris was taught by him at Cambridge, and was later a colleague at Sidney Sussex College, Cambridge; Joanne was also taught by him at Cambridge, and owes her own research interest in political geography to his teaching; and all four of us would have counted Graham as a valued friend. Moreover, Graham's contributions to rethinking the political geography of power comprise a very important element in the conceptual context for the writings contained in this collection. It is for these many reasons that we wish to dedicate the book to Graham's memory.

Joanne P. Sharp, Paul Routledge, Chris Philo and Ronan Paddison
Glasgow

1

ENTANGLEMENTS OF POWER

Geographies of domination/resistance

*Joanne P. Sharp, Paul Routledge, Chris Philo
and Ronan Paddison*

> Power must be analysed as something which circulates . . . And
> not only do individuals circulate between its threads; they are
> always in a position of simultaneously undergoing and exercising
> this power.
>
> (Foucault 1980: 98)

The purpose of this book is to explore fresh perspectives on the geographies of power. In particular, we wish to emphasise the myriad *entanglements* that are integral to the workings of power, stressing that there are – wound up in these entanglements – countless processes of domination and resistance which are always implicated in, and mutually constitutive of, one another. We thereby wish to retain a clear focus on the domination/resistance couplet, notwithstanding Pile's (1997: 3) remark that assumptions about this couplet now 'become questionable', our understanding is that the entanglements involved here can never escape from the endless circulations of power. At the same time, our use of the term 'entanglements', suggesting an image of knotted threads, is intended to underline the deep 'spatiality' of this spinning together of domination and resistance within power. We talk about 'entanglements' to indicate that the domination/resistance couplet is always played out in, across and through the many spaces of the world. In one sense, we obviously use the term 'entanglements' metaphorically, and we aim to introduce what we hope will be a helpful new spatial metaphor, one that brings with it other connotations, maybe a more optimistic sense of possibilities for change, than are conveyed by the more common metaphors deployed in this respect (see also Massey, Chapter 14 in this volume). At the same time our use of the term is intended to be more than just metaphorical, since it is supposed to flag the countless material spaces, places and networks which sustain, practically as well as imaginatively and symbolically, the knottings that are the subject of our attention.

J. P. SHARP, P. ROUTLEDGE, C. PHILO, R. PADDISON

Orthodox accounts of power, particularly as translated into the literatures of political and historical geography, tend to equate power straightforwardly with domination. Power thus becomes almost exclusively conceived of as the 'power to dominate' or as 'dominating power' (see Box 1.1). Similarly, orthodox accounts of resistance, particularly as translated into the literatures of social, cultural and development geography, tend to pit resistance against power or against domination (understood as a coherent oppressive force), or even to portray dominating power as so ubiquitous (as just so 'powerful') that acts of resistance appear either futile or trivial. In such accounts, moreover, matters are rarely conceived of in terms of the 'power to resist', what might be called 'resisting power' (see Box 1.2). In this introduction we briefly describe these two more orthodox strands in the understanding of power, hinting at how their simplified conceptualisations, ones which consistently strive to binarise domination and resistance, tend to become scrambled when striving to provide more grounded commentaries alert to the chaotic muddle of empirical situations. In mentioning these scramblings, we begin to signal rather different ways of thinking about the workings of power. Indeed, leading from these orthodox materials, we use the remainder of the chapter to develop an alternative perspective which concentrates squarely upon the messy and inherently spatialised entanglements of domination/resistance, as always energised and traversed by the machinations and effects of power. Something akin to this perspective on the geographies of entangled power relations is then explored substantively, so we would suggest, in the chapters that follow.

Following Foucault, we understand 'power' as having both positive and negative dimensions, operating in ways which can be repressive and progressive, constraining and facilitative, to be condemned and to be celebrated. The word power comes

---

Box 1.1

We understand *dominating power* as that power which attempts to control or coerce others, impose its will upon others, or manipulate the consent of others. These circumstances may involve domination, exploitation and subjection at the material, symbolic or psychological levels. This dominating power can be located within the realms of the state, the economy and civil society, and articulated within social, economic, political and cultural relations and institutions. Patriarchy, racism and homophobia are all faces of dominating power which attempt to discipline, silence, prohibit or repress difference or dissent. Dominating power engenders inequality and asserts the interests of a particular class, caste, race or political configuration at the expense of others.

Box 1. 2

We understand *resisting power* as that power which attempts to set up situations, groupings and actions which resist the impositions of dominating power. It can involve very small, subtle and some might say trivial moments, such as breaking wind when the king goes by, but it can also involve more developed moments when discontent translates into a form of social organisation which actively co-ordinates people, materials and practices in pursuit of specifiable transformative goals. Social movements of various sorts can be mentioned here, many of which co-ordinate everyday forms of resistance that still fall short of open confrontation, but some situations may eventually lead to violent actions. In order for all of these resistances to occur, power has to be exercised and realised, both by the leaders (in a form that can become dominating in its own right) but also in a more 'grass-roots' fashion by everyday people finding that they have the power to do and to change things.

---

from the Latin word *potere*, meaning 'to be able'. In this sense, power should not be viewed solely as an attribute of the dominant, expressed as coercion or political control, since it is also present in the ability to resist. This resistance can enable resisters to find common ground in struggle (as in the power to mobilise others), and to become em*power*ed in the very act of resistance. The implication is that power is operative in moments of both domination and resistance, and hence can be assessed in both positive and negative terms. In addition, we continue to find in Foucault valuable clues for thinking about the entangled spatialities of power, for conceiving (as in the opening quote) of the spatialised threads along and through which power circulates, entering into the worlds of individuals and groups as they are 'simultaneously undergoing and exercising this power'. As such, in this introduction we revisit some of Foucault's main contributions in this respect, but we also acknowledge criticisms which can be levelled at him for giving greater attention to the pervasiveness of an anonymous dominating power, one to which resistance appears impossible, than to the potentialities of resisting power. We also note aspects of his lesser known later work, wherein he started to uncover spaces of power within which creative acts of resistant 'self-fashioning' can perhaps gain a measure of effectivity.

Orthodox accounts I: power equals domination

While our reading of power highlights its contested nature, the theoretical foundations on which much geographical analysis has been built demonstrate the

3

significance of certain orthodox interpretations. In these, power tends to be equated with domination and coercion, rather than giving 'space' to notions of resistance. In one sense such an emphasis has been understandable, if blinkered. While there was at least a measure of consensus over the significance of power in shaping social action, the task of identifying the sources of such power was a perfectly logical one. Whether couched in terms of institutions – the state, in particular – or in terms of individuals, particularly as they operated within institutions, power relationships became understood principally in terms of the means by which dominance over others (both inside of institutions and beyond) was achieved. Surveying the developing analysis of power (over the post-war period, for example), we can argue that such a bias has been increasingly challenged with the growing recognition that, where power is more diffuse and clearly not restricted to the formal processes of governing, researchers need to take cognisance of the more nuanced ways in which the ability to shape social action takes place. It may nonetheless be useful to trace the ways in which orthodox accounts of power have foregrounded the processes of governance, and thereby the ability of rulers to govern.

Accounts of the state, both liberal and Marxist, present an image of power in terms of power as authority or domination. Liberal accounts are offered by 'pluralist' theorists in which the state is seen not only as the legitimate site of authority, but also as one which (classically) is presented as relatively neutral to different competing interests as represented by a plurality of different groups. Such a state is taken as responding to the needs of these different groups in a situation where political processes are assumed to be more or less open, where the state is construed as both responsive and neutral, and where individuals and groups are interpreted as rational actors pursuing their interests. It is understandable that such a state comes to be seen in terms of exercising legitimate power over its subjects in the interests of social order (Pringle 1999: 216). Admittedly, such pluralist accounts have been modified: 'neo-pluralism' admits to the uneven terrain over which political processes occurs, and in particular to the distinctions between individuals and groups who are relatively powerful and powerless within the overall processes which influence the actions of the state. Even so, how power is construed and located remains fundamentally unchallenged, and the centrality of the state and its ability to harness legitimacy, together with the notion that power is rooted in particular institutions, remain largely unquestioned assumptions. The overall presumption in liberal formulations continues to be that examining power equates with identifying the processes of governing.

Pluralist ideas duly became a rich source of empirical work, later sometimes termed 'surface analysis', which began to investigate both where power could be seen to reside and how it might be measured. Pioneering studies such as that by Hunter (1953) were to set the agenda, particularly for political scientists working within sub-state political arenas. The ensuing 'community power debate' (most vigorously developed among US political scientists) broached issues to do with the openness and the accessibility of the state, and also considered the extent to which power could be centralised within particular elites, a 'fact' which was measurable (in Hunter's methodology) through the use of something called 'reputational analysis'. As such, then, power not only could be centralised, but also was the means of domination, if not coercion. Such notions about power were expressed more systematically in Dahl's influential analyses of the late 1950s (Dahl 1957, 1961), which continue to enjoy some support among political scientists. His intuitive idea of power, 'that A has power over B to the extent that he or she can get B to do something that B would not otherwise do', highlights both power in terms of domination or coercion and the presumption of the unilinearity of the relationship between the two hypothetical characters. The origins of Dahl's approach lay in the need, as he argued, to achieve greater precision in the attribution of power. Yet, in empirical analysis the rigour attained by his formal model of power might be spurious:

> while considerable attention is paid to constructing a precise instrument for measuring power in terms of responses to its exercise, many less precise aspects of analysis are to be found in the model's application to the empirical analysis of the community of New Haven.
>
> (Clegg 1989: 9)

One of the defining features of much pluralist analyses has been its tendency to view power in individualistic terms, rather than adopting a more structural conception of its exercise. This was to be emphasised by Bachrach and Baratz (1962, 1963) with their notion of 'non-decision-making', wherein they were deeply critical of Dahl's one-dimensional analysis of power. While accepting the importance of observable events as outcomes of power relationships, their point was that such results also reveal the importance of events which are concealed (and often deliberately so). If power is interpretable through decision-making, then it is equally so through non-decision-making, in which there is power given by the ability to withhold from the agenda issues which might otherwise gather widespread support. Borrowing from Schattschnieder's (1960) 'mobilisation of bias' argument, Bachrach and Baratz's thesis presupposed that behind the exercise of power in any particular setting was a 'structure' which to a greater or lesser extent determined how interactions occur and outcomes arise. Clearly, such a 'hidden' source of power enabled elite groups potentially to dominate the political agenda, and, even if in their subsequent analysis (Bachrach and Baratz 1970) the thesis became somewhat diluted, it still represented a potent model of how (local) political practice could be dominated by particular interests.

Non-decision-making analysis extended the debate by showing how the study of power must take account of the unobservable rather than just the observable. In other words, power could now be read as a multidimensional concept, the 'higher' dimensions assuming an understanding of the structures within which power games take place. Such an idea of the different dimensions to power was extended by Lukes (1974, 1977), whose 'third dimension' – the first two corresponding to surface analysis and non-decision-making analysis respectively – was to underscore the reality of just how unequal power relationships could become in practice, not least because of the biases generated by the dominant ideas underpinning how society should be organised. Part of the significance of Lukes's contribution is that, when compared to the community power debate, power as such was given direct emphasis; but in addition, and as set within the terms of the 'structure/agency' debate (e.g. Giddens 1979: esp. 88–94), it identifies the role of ideology within power relationships. Subsequently, much Marxist analysis has extended the debate on power in terms of the role of dominant ideologies fashioned by hegemonic interests, thus speaking of the state being co-opted by dominant classes in order to exert control, authority and, indeed, power over others who conform or occasionally choose to resist. Marx himself once described the state as merely 'a committee for managing the common affairs of the whole bourgeoisie' (quoted in Paddison 1983: 5). In versions with a more structural Marxist bent, power simply emerges or 'condenses' from a society's underlying structure of class relations (e.g. Poulantzas 1973, 1978), but – and perhaps almost inevitably once the balance of inquiry tilts decisively in the direction of structure – power ends up being conceived of almost entirely on the side of domination.

Where so much of this endeavour had been concerned with the *distribution* of power, contemplating the practice of power identified the need for more subtle understandings. Other accounts, as in the work of Clegg (1975), have looked at the practices of power, the various ways in which organisations are able to devise the 'rules of the game' by which power is effected. Again, the emphasis is on the mechanisms by which organisations, and also individuals working within them, can achieve domination. Where so much behaviour is rule guided, the production of rules hence becomes the means of capturing advantage. Within such a 'game', the more powerful have the advantage not only of setting the rules, but also of interpreting (and reinterpreting) what is their meaning. The power of resistance in such cases becomes effectively marginalised, although not for some a totally pointless goal. Indeed, advocacy planning – and also 'advocacy geography' (Bunge 1971) – was initiated in the 1960s, and provided pointers as to how otherwise dominant sources of power could be challenged, and even subverted. Here it was deemed possible for a form of oppositional agency to work from inside institutions, and therein to use the 'rules of the game' to the potential advantage of the relatively powerless. Clegg's work demonstrated the value of distancing the analysis of power from the totalising accounts of much Marxist work, and yet it retained in-built assumptions about the location of power: that is, particularly within institutions.

More recent work by political geographers (e.g. Clark and Dear 1984; Driver 1991, 1993a; Paddison 1983; Painter 1995; G. Smith 1994) and others has shown how societies are much less neat than orthodox accounts have presumed, and that relationships in the power webs of everyday mundane political and social practices are much more entangled – with many more spaces for resistance – than most versions of liberal and Marxist thought ever imagined. For example, those who have studied the historical geography of the modern state have realised that, to paraphrase Mann (1986), societies are much messier than our theories of them. Liberal and Marxist traditions of political thought alike have tended to reduce state function to the logic of dominant political-economic systems, whereas Mann sought instead to account for the modern state as an autonomous power which does not simply support or oppose the logic of the market or of capitalism. The state has arguably gained this unique power through its control of territory, a bounded space within which it is recognised as offering legitimate rule, exercising within these bounds a monopoly of the means of violence and storing knowledge of its citizenry through a complex bureaucratic infrastructure (Mann 1986; Giddens 1981, 1985). Modern state power is not the spectacular performance of domination embodied in pre-modern despotisms, therefore, but instead is insinuated throughout the round of daily life. In modern societies, it is virtually impossible to perform much of daily life beyond the purview of the state.

Yet, and despite the apparent ubiquity of modern state power as woven into the everyday routines of its subjects, domination is in no way complete or secure. Individual citizens or groups may reject the state directly through political action, or indirectly through, for example, apathy or non-participation (and open resistance will be discussed further on pp. 8–12). In addition, resistance is woven into the heart of the state apparatus itself, in part because state power is based upon different axes of power – economic, military and informational – which do not always work in perfect consort. The bureaucratic modern state duly involves a range of different branches, interests and agents, each of which tends to operate with different agendas, support networks and constituents. Power is thereby entangled within the state, and resistances are insinuated throughout its varied apparatuses (Clark and Dear 1984). This power is also entangled spatially, since state power emerges in part from its territoriality, but this territory is never a homogenous space. There exist different layers and spaces of territoriality, the most obvious example being dual claims to legitimate representation within a piece of state territory which lead to the oppositional struggles of regionalism and minority nationalism. The limits to state power and the power of resistance thus become entangled in complicated and at times unpredictable ways, and notions of sovereignty, itself a social construction, can end up being challenged through resistances which in turn undermine the power of the state. Few examples illustrate better the complex entanglements of power, domination and resistance enacted through the state, and the different spatialities through which these processes are transmitted, than the recent British 'Poll Tax'. Given its status in fulfilling a classic function of the state, the levying of taxes, not only did rebellions *against* the 'Poll Tax' reveal clear limits to state power, but also they showed how resistance could erupt unexpectedly at different spatial scales (regional and local) and from a complex interweaving of formal and informal political arenas. Attempts to revise theories of the state so as to account for such eventualities have led to a progressively more messy account, therefore, and to an orientation that removes the state from the simple conceptual space of political-economic domination. Writers increasingly recognise the different forms of power exercised by the state, as well as the different interests that might work through these varied forms, and such differences, together with the spatial complexities of territoriality, insert new threads into thinking about the entangled geography of state power.

Orthodox accounts II: resistance movements

The other tradition in geography and related disciplines most obviously concerned with the operations of power, domination and resistance has been the literature tackling the emergence and operation of resistance movements. Although there are many different strands within this tradition, the bulk of writing concerning resistance has concentrated on social movements: groups frequently organised around opposition to the dominant forces of the state or of multinational capital. Contemporary theoretical debates concerning social movements have discussed the comparative efficacy of 'resource mobilisation theories' and 'identity-oriented theories' (e.g. see Cohen 1985). The resource mobilisation approach (e.g. McCarthy and Zald 1977; Morris and Mueller 1992; Oberschall 1973; Tilly 1978) takes as the object of its analysis the collective action between groups with opposing interests. This approach concentrates on the goals, organisation and leadership of social movements, the resources and opportunities available to them and the strategies that they employ. This perspective is concerned with movement processes over time, and is keenly interested in both the role of political parties (whether regional, nationalist or communist) in organising the disaffected and the role of the state as a mechanism of repression. As Zirakzadeh (1997) noted, the resource mobilisation paradigm includes 'indigenous community' theorists who explore how local-level social institutions (e.g. neighbourhood clubs) can provide communication and organisational resources for social movements (Adam 1987; Tarrow 1994), as well as 'political process' theorists who investigate how governmental policy-making and intra-governmental power struggles influence the strategies and tactics employed by social movements (e.g. McAdam 1982).

The identity-oriented approach, meanwhile, seeks to understand how collective actors strive to create the identities and solidarities that they defend, and it also attempts to understand how structural and cultural developments within society (such as social relations of power, domination and cultural orientations) contribute to the character and expression of a social movement. This approach criticises resource mobilisation theory because the latter is viewed as studying social movement strategies as if actors are defined by their goals, a point raised a moment ago, and not by the social and power relations in which they are situated. In addition, although identity-oriented theorists accept some claims from resource mobilisation theory, they argue that crucial to a fuller understanding of resistance are both the systems of political legitimacy that exist and the interplay between ideologies of domination and subordination. A variety of different versions of this approach can be identified, but among the most influential theorists here are Touraine (1981, 1988), who has argued that social movements frame their struggles in terms of a cultural project, their aim being the control of historicity (particular sets of cultural models that control social practices); Laclau and Mouffe (1985), who have argued that social movements constitute politics as an articulatory process, as a terrain of negotiations between hegemonic and counter-hegemonic interpretations and positions; and Melucci (1989, 1996), who has argued that, because collective action is frequently focused on cultural codes, the forms of social movements are themselves messages, operating as signs representing a symbolic challenge to dominant codes.

Despite the insights that both resource mobilisation and identity-oriented approaches provide for the study of different forms of collective action, a common tendency within their conceptualisations of conflict has been to construct a binary of opposing forces – between those of dominating power and those of resistance to it – thus posing a central dialectic of opposed forces: the rulers and the ruled (Said 1983). This tendency reinforces the idea that it is the figures and processes of domination that exclusively hold power. Such a notion has not gone unchallenged, however, notably by Tarrow (1994) when arguing that social movements are able to wield considerable power in their resistance, if only briefly. The presence of 'power in movement' is thereby seen to act as a counterforce to the dominating power of institutions like the state. Other research on the prosecution of conflict, in particular that on the strategic use of non-violent sanctions, attempts to develop a relational theory of power between these opposing (dominating and resisting) forces. 'Non-violence theory' provides an account of power based upon the principle that all power, of whatever form, depends ultimately on the tractability of those over whom the power is wielded. Writing in this tradition, Sharp (1973: 7) suggested that power is inherent in practically all social and political relationships, and he differentiated between social power (the capacity to control the behaviour of others, directly or indirectly, through action by groups which impinge upon other groups) and political power (social power that becomes exercised for political objectives, especially by government institutions and by people in support of, or maybe in opposition to, such institutions). For him, the rationale for non-violent action arises from the recognition that governments depend on people, that power is pluralistic, and that political power is fragile because it depends on many groups for reinforcement of its political sources (Sharp 1973: 8). The sources of this power (human, material and ideological resources) rely intimately on the obedience of subjects. Indeed, compliance to and enforcement of this power tend to reinforce one another, although, as Sharp pointed out, obedience is not inevitable but exists only when one has complied with, or submitted to, the command (i.e. when one has consented to obey). This means that the theory of non-violent action rests upon the withdrawal of consent, cooperation and obedience by subjects from those who purport to rule them. Through resistance by non-consenting and non-cooperation, leading into disobedience, subjects decline to supply the power-holders with the sources of their power.

Drawing on the strategy model of Clausewitz (1968), Sharp (1973) proposed a strategic approach to the practice of non-violent action – rather than a necessarily moral or principled approach – and he described three broad methods of undertaking such action: intervention, non-cooperation or protest and persuasion. The results of non-violent action produce change through one of three ways: conversion, accommodation or non-violent coercion. Where the latter operates, change is achieved against the opponent's will and without their agreement, the sources of their power having been so undercut that they no longer have control. Sharp further introduced the concept of non-violent sanctions, those forms of non-violent action that impose a cost by the 'actionist' on the opponent. Much contemporary research (e.g. Ackerman and Kruegler 1994) has emphasised the strategic elements of non-violent action, and in so doing it considers relations of power as derived from people's desire to co-operate, to give consent, and a government's ability to repress. Attention is given to how power is manifested in non-violent conflict through the methods of non-violent action, and to how non-violent sanctions are subject to the logic of strategic interaction. While such scholarship posits a relational view of power – accepting that power is not only the preserve of the dominant, but also can be wielded by resistant formations within society – it nonetheless still tends to theorise a binary of opposing (dominating and resisting) forces within the crucible of conflict.

Other social movement theorists do present the possibility of more fluid collective agency, though, and do acknowledge its interpellation within broader and more complex networks and relations (e.g. Escobar and Alvarez 1992). In consequence, such writers have to admit that there are no completely autonomous spaces from which, and within which, collective action takes place. As such, they admit that there is no necessary autonomy of such action from local, national or international relations and processes. Hence, following Escobar (1992), we may situate conflict theoretically within and between the practices of everyday life (e.g. the family, church, community) and the socio-political processes of the state and regional or national institutions. This siting of conflict in relation to both the practices of everyday life and the socio-political processes of the state is one of perpetual movement, negotiation, changing alliances and affinities, co-options and infiltrations, contingent upon the particularity of spatio-temporal conditions. Given this perpetual movement, social movements are located within a contested web of power and knowledge relations, and such a picturing opens the possibility for an entangled view of dominating and resisting powers.

In addition to social movement theorising, there is research implying that every action, no matter how apparently insignificant, is imbued with resistant intent. Such work conceptualises power as a permanently tenuous practice, and supposes that dominated groups within society are constantly challenging the hegemony of the dominant. This literature includes work on resistance practices that eschew open, direct confrontations with the dominant, notably work by Scott (1985, 1990) on the everyday forms of 'peasant' resistance (such as foot-dragging, sabotage and pilfering) and by Adas (1981) on avoidance protest (e.g. discontent articulated through flight and sectarian withdrawal). Equally significant has been the rise of the 'subaltern' literature, inaugurated by the first *Subaltern Studies* volume (Guha 1982) wherein detailed empirical inquiries into the activities of India's various 'subaltern classes and groups', ones supposedly possessing a clear 'autonomy' from the texture of British colonial power, began to spotlight an inspirational realm of both overt and more covert resistant possibilities. Related research then considers cultural expressions of resistance and transgression embedded in the rituals, attitudes and lifestyles that exist in very ordinary circumstances (e.g. Cresswell 1996; Hall and Jefferson 1976; Harlow 1987; Said 1993; Stallybrass and White 1986). While many are attracted to this literature, we recognise, with others (notably Abu-Lughod 1990: see also both Atkinson and Cresswell, Chapters 4 and 12 in this volume), the danger here of 'romanticising' resistance, and of detecting in all kinds of activities the expression of a resistant spirit refusing to knuckle under the yoke of domination. Moreover, such work also risks constructing a binary of opposing (dominating and resistant) forces, and still implies that power is the preserve of the dominant. Some of the most recent theoretical and empirical work on resistance movements has adopted Foucault's ideas about 'capillaries' of power, and thereby views power as insinuated throughout all social activity and not simply located within forces of domination (e.g. Alvarez *et al.* 1998). We certainly have much sympathy for such a Foucauldian manoeuvre, even if certain readings of Foucault can convey the impression of a dominating power so diffuse, so unspecific and yet so all-encompassingly effective that resistance against it is not only futile but impossible to achieve.

Foucauldian entanglements

As many discussions of power proceed from rather cursory engagements with Foucault's writings, it is to his work on power and space that we will now turn in order to clarify our own thinking about the entanglements of power. Foucault's so-called 'genealogical' inquiries after circa 1970 (especially Foucault 1979a, 1979b, 1985, 1986a) were all quite obviously shaped by a concern for power, and at one point he explicitly claimed that what he had always been exploring in his various substantive histories was the issue of power:

> In my studies of madness or the prison, it seemed to me that the question at the centre of everything was: what is power? And, to be more specific: how is it exercised, what exactly happens when someone exercises power over another?
>
> (Foucault 1988b: 101–102)

Although there may be few explicitly 'geographical' reviews of Foucault's writing on power (Driver 1985, 1993b; Matless 1992; Hannah 1992, 1997a, 1997b; Philo 1992), in practice a number of geographers have explicitly used his ideas to frame their own substantive research into geographies of power (e.g. both Howell and Robinson, Chapters 2 and 3 in this volume; Driver 1990, 1993a; Hannah 1993; Murray 1995; Ogborn 1993, 1995; Philo 1989; Robinson 1990, 1996) while others, including several contributions to this collection, are clearly considering such matters on what might be termed Foucauldian grounds. Much could be said about Foucault's ideas on power, but in what follows we simply suggest that, functioning more as an historian than as a theorist or philosopher (O'Farrell 1989), he provided a compelling example of how to proceed in recreating the entanglements – ones invariably associated with specific issues, societies, periods and places – integral to a highly nuanced historical geography of power, domination and resistance. We begin by considering some fairly well-known claims about his scholarship, but then touch upon some of his less-known final work where he arguably foregrounded the entangled geographies of power even more clearly.

Disciplinary geographies

Many readers look to *Discipline and Punish* (Foucault 1979a) for the clearest *theoretical* exposition of what Foucault understood by power, but it is important to appreciate the extent to which his claims here about power were rooted in a specifically *historical* narrative concerning the changing forms of power from the early modern period in Europe (the 1600s and 1700s) through to the modern period (the 1800s into the twentieth century). The former entailed a 'terrific', as in 'terror-ific', regime based on bloody retribution and the breaking of the body of the condemned (as so graphically conveyed in the gut-churning description of Damien's execution in 1757 with which Foucault commenced the book). The sovereign exercised power in a nakedly violent fashion, bullying the populace into obedience through spectacular punishments (from small mutilations to complete disembowellings) and the ambition was to control unruly hearts by this dramatic means. This was a dominating power built on elimination and display, but it was hardly an effective form of power, since it was supported by no systematic attempt to reach into the obscure corners, the hidden spaces in the alleys and forests away from the scaffold, where rebelliousness of all kinds – countless resistant thoughts and acts – could ferment. Foucault charted the gradual transition away from this highly uneven regime of terror, a transition itself marked by the eruption and disappearance of new and different practices (including that of the 'punitive city': Foucault 1979a: 113), but the key development for him was the appearance of a whole system of segregationary institutions, complete with their own internal apparatuses of capture, during the early 1800s:

> The theatre of punishment of which the eighteenth century dreamed and which would have acted essentially on the minds of the general public was replaced by the great uniform machinery of the prisons, whose network of immense buildings was to extend across France and Europe.
>
> (Foucault 1979a: 115)

Foucault noted that the rapid spread of the prison was characterised by many 'points of convergence and . . . disparities' (Foucault 1979a: 126), and he documented the countless subtle practices – notably ones to do with the micro-organisation of both time and space, including 'the art of distributions' (Foucault 1979a: 141–149) – which comprised the workings of this 'coercive, corporeal, solitary, secret model of power' (Foucault 1979a: 131) behind the high prison wall.

All of these practices were designed to act upon the person of the prisoner, to create 'docile bodies' where before there had been unruly hearts, and thereby to reform individuals with the aim of returning them to society as new additions to the productive labour force. As is well known, it was Bentham's 'Panopticon' design that Foucault took as the architectural figure *par excellence* of this novel disciplinary regime (Foucault 1979a: 200–209), a figure that enabled him to play up the role of spatial arrangements internal to an institution in facilitating a constant (threat of) inspection, of surveillance, which caught inmates within an overall field of visibility wherein their every twitch and grimace became open

13

to scrutiny (and, if necessary, rebuke). The upshot was the internalisation of discipline, the making of self-discipline, as inmates were enlisted into controlling themselves, and as the external eye in the inspection tower was replaced by the internal eye of conscience (see also Bender 1987). Bentham intended his design to be extended to other institutions such as asylums, hospitals, schools, reformatories and even manufactories, and in a sense Foucault echoed this extension by supposing that a more generalised 'panopticism' – 'the discipline-mechanism: a functional mechanism that must improve the exercise of power by making it lighter, more rapid, more effective' (Foucault 1979a: 209) – came to feature in a range of different institutional and even 'de-institutionalised' (Foucault 1979a: 211) practices. In this repect, he mapped out the contours of a widely diffused 'carceral archipelago' (Foucault 1979a: 297), as well as examining what he termed 'the swarming of disciplinary mechanisms' (Foucault 1979a: 211–212) to generate the undulating texture of a modern disciplinary society which he evidently supposed still to be with us in the second half of the twentieth century (Foucault 1979a: 308; see also Atkinson, Bale and Crang, Chapters 4, 6 and 9 in this volume).

Concentrating on what Foucault claimed about panopticism, many readers have found here the architecture of a more theoretical reflection upon the nature of power, or at least of a certain variety of 'disciplinary power' (Foucault 1979a: 221). While there is a danger of forgetting that Foucault posited this particular calculus of power as a distinctive product of a given period and place, circa 1800 in western Europe and North America (the latter when discussing the 'pentitentiary model': e.g. Foucault 1979a: 123–126), there is still some justification for finding in his words a generalisable model of how disciplinary power operates in all contexts. Accordingly, much has been made of an observation such as this:

> 'Discipline' may be identified neither with an institution nor with an apparatus; it is a type of power, a modality for its exercise, comprising a whole set of instruments, techniques, procedures, levels of application, targets; it is a 'physics' or an 'anatomy' of power.
>
> (Foucault 1979a: 215)

Foucault clearly did not regard this form of power as a simple possession of certain groups of people, something to be wheeled into action at will, but rather saw it as a 'relational' entity, as itself a 'relation', always running between two or more people (or sets of people) and carrying with it the possibility that one individual (or collectivity) will be able to achieve certain desirable effects in the thoughts and actions of another. Although this is a matter which Foucault was to explore more explicitly elsewhere (Foucault 1979b), the wider implication was that power should not be seen solely as 'repressive' (stopping things happening, blocking outcomes) but also as 'productive' (making things happen, achieving outcomes).

The further implication concerns the need to unpick the 'microphysics' of power (Dean 1994: 155–158; Gordon 1991: 3–4) which Foucault so brilliantly exposed, a task which meant establishing how 'general formulas, techniques of submitting forces and bodies, in short "political anatomy", could be operated in the most diverse political regimes, apparatuses or institutions' (Foucault 1979a: 211). Careful attention was also needed to the spatial coordinates of power, to what subsequent commentators have elaborated as the 'capillaries', 'transmissions' and 'relays' of power through the specific spatial fields wherein certain agencies strive to effect changes in the conduct of others. Taken together, all of the above amounted to a quite remarkable theoretical, historical and (we would add) thoroughly geographical picturing of power.

This is precisely not to say that Foucault's account of power in *Discipline and Punish* is beyond criticism, however, and it certainly has been criticised from a number of different angles for not being (in our vocabulary) sufficiently entangled. Many readers come away from the book with the impression that power is nothing but a sticky pall of domination, something that is always and everywhere present doing the bidding of those in authority – even if it is not all that consciously directed by them – and in the process entering into every tiny pore of the social world in an 'effective' manner (having effects contrary to what might be the preferred ways of thinking-and-acting for the people seeking to inhabit these pores). The prevailing image is arguably that of society as one vast Panopticon, or of the Panopticon as a giant lighthouse shining its beams of disciplining light into every nook and cranny, allowing nobody to escape from the 'normalising' gaze nor from the imperative that they themselves should interiorise that gaze. In this image, power, a dominating power, does take on a total(ising) cast against which no resistance is possible or all resistance is futile. Various scholars have described Foucault's account of power in this fashion, and have voiced their objections to envisioning power as something for which there appears to be no limits, no obstacles, no outside:

> [One objection] was that Foucault's representation of society as a network of ominipresent relations of subjugating power seemed to preclude the possibility of meaningful individual freedom. [Another] complaint was that Foucault's markedly bleak account of the effects of humanitarian penal reformism corresponded to an overall political philosophy of nihilism and despair.
>
> (Gordon 1991: 4)

In short, if power is so nebulous, so 'unauthored', can we have any optimism in the potential for individuals or groups to find ways, and we might also say spaces, of resisting the all-encompassing cloak seemingly spun by power in its dominating guise? Foucault portrayed the so-called 'productive' work of disciplinary power as going on anonymously, emerging from expectations, positions and discourses which coalesced from countless individual events whose particular personnel (their individual utterances) were of less import than the encompassing logic of their organisation. It is hence unsurprising that many, ourselves included, do wonder if additional concepts – ones beyond the pages of *Discipline and Punish* – are required to create fresh visions of power in which genuine possibilities for resistance, and perhaps too for a creative human agency capable of carrying these possibilities, are introduced?

Governmental geographies

It is only in his later works in and around *The History of Sexuality* (Foucault 1979b) that Foucault himself began to meet some of these objections. Dean (1994) can be our guide in this respect, along with several other writers who have discussed Foucault's admittedly somewhat inchoate concern for 'governmentality' (Gordon 1991; see also Foucault 1979c). What Dean neatly explains is how, through this concept of governmentality, Foucault effectively conjoined an interest in the 'government of the body politic', an interest which bumps up against work on both the political economy and the political geography of the state (but balks at seeking a 'theory of the state'), with a more novel concern for the 'government of the self'. As Dean puts it:

> What makes Foucault's later studies so fascinating . . . is condensed in this notion of governmentality. It defines a novel thought-space across the domains of ethics and politics, of what might be termed 'practices of the self' and 'practices of government', that weaves them together without a reduction of one to the other.
>
> (Dean 1994: 174)

The second of the domains identified by Dean, that concerning the 'practices of government', entails the complex task of governing whole populations, a task named in Foucault (1979b) as 'biopower', 'in which issues of individual sexual and reproductive conduct interconnect with issues of national policy and power' (Gordon 1991: 5). Foucault started to pose questions about how emerging technologies – notably ones invented by the natural sciences, specifically medicine and the so-called 'psy-' disciplines, but also ones emerging from the social sciences – first came to be mobilised in ensuring the constant reproduction of both a quick-witted elite and a fit and healthy working class, the latter being needed in the dual function of industrial workforce and 'manpower' of a nation-state's armed forces. Together with the term 'biopower', Foucault used the term 'pastoral power', 'a power . . . whose role is . . . constantly [to] ensure, sustain and improve the lives of each and every one' (Foucault 1988a: 67), and a power that Foucault suggested can be located at the heart of Christian notions of both Christ as 'the good shepherd' and the 'pastoral responsibilities' of priests (Foucault 1988a: 67–68). Insofar as the first volume of *The History of Sexuality* (Foucault 1979b) tackled biopower or pastoral power, it did so chiefly with reference to the installation of such power within matters of sexual reproduction in nineteenth-century western Europe, and a link was drawn between the management of populations and the specifics of how individuals were supposed to regulate their own sexual lives. Insofar as the subsequent two volumes of *The History of Sexuality* (Foucault 1985, 1986a) tackled this form of power, they turned more to the pastoral ingredients, and more narrowly still to their dynamics as installed during the times of the Ancient Greeks (vol. 2) and then the Romans (vol. 3). Furthermore, a different slant was here placed on the study of governmentality, one leading Foucault to examine more closely the situated 'arts' or 'rationality' of government – Gordon (1991: 3) suggests that the two terms were used interchangeably – and thereby to take seriously what enables power to be successfully gained, held and exerted over populations (national or perhaps, as in the Ancient world, more regional or municipal). In so doing, he considered how particular individuals, as governors, rulers or 'statesmen', could learn to function as rational political agents: how do they become familiarised with and able to mobilise a code of political conduct whose effects are inevitably ones of power?

The first of the domains identified by Dean (1994), that concerning the 'practices of the self', entails a whole terrain of concern for the ways in which individual people become instructed in, and learn to instruct themselves in, the crafting of 'the self' as a particular kind of self distinct from other kinds of self. In one of his last interviews, Foucault gave a very definite inflection to a root question concerning:

> the form of an *art of existence* or, rather, a *technique of life*. It was a question of knowing how to govern one's own life in order to give it the most beautiful form (in the eyes of others, of oneself, and of the future genera-tions for which one might serve as an example). That is *what* I have tried to reconstitute: the formation and development of a practice of self whose aim was to constitute oneself as the worker of the beauty of one's own life.
>
> (Foucault 1988c: 259, original emphasis)

With these strange words, Foucault indicated his emerging fascination with processes of 'self-fashioning', 'self-stylisation' and 'self-control', stressing that the goal was to utilise documents at his disposal in libraries and archives to discern 'the very form in which contemporaries have reflected' (Foucault 1988c: 259) upon such processes. More specifically, he wanted to tease out how particular peoples in past periods and places have sought to impose an order on their bodily regimes to ensure good physical and mental health, perhaps through reading the many 'helpful discourses' (Foucault 1986a: 101) available to these peoples on such matters. More specifically still, the focus was on the particular cast of 'ethics' or 'morality' bound up in how to deal with the self, and most notably in how to interpret, master and direct those pleasures (passions, desires, impulses) of a more sexual nature. The impression is of Foucault thinking himself into the realm of (educated) individuals in the past, following them as they consult a diversity of instruction manuals, medical guidebooks and philosophical treatises when striving to ascertain what would be the best 'self-conduct' in relation to themselves, their families, friends and others. What was at stake for these individuals was how they managed to exert power over themselves: the effecting of a power over the self; the self exerting power over the self; the supposedly 'free self' creating a supposedly 'free self'. In this vein 'Foucault affirm[ed] . . . that power is only power (rather than mere physical force of violence) when addressed to individuals who are free to act in one way or another' (Gordon 1991: 5; see also Foucault 1982), and this is very much the Foucauldian manoeuvre which underpins the chapter by Robinson (Chapter 3 in this volume). As should be obvious, such a perspective demands that human subjects be regarded as much more creative beings, much freer to choose and accorded more opportunity to translate discursive codes into embodied performances, than is the case in any of Foucault's earlier, more structural(ist) writings. In the first volume of *The History of Sexuality* (Foucault 1979b) his analysis thus pivoted around the negotiation of the sexual self in nineteenth-century western Europe, critiquing a diversity of contemporary writings – including those which would have played some part in framing the discursive wranglings around the Cremorne Gardens affair (see Howell, Chapter 2 in this volume) – and considering too the model, and indeed the key social and material space, of the 'confessional'. In the next two volumes of the history (Foucault 1985, 1986a) he reached back into antiquity, patiently excavating a diversity of Greek and Roman texts available to the elites of the ancient world, contextualising the 'story' to be told from these texts against a discussion of the extent to which, in specific meditations on the likes of the body, marriage and the 'love of boys', the ancients anticipated the ethical rigour and (sexual) austerity of the earliest Christian scholars (esp. Foucault 1986a: conclusion).

The two domains discussed here were effectively brought together in much of Foucault's later work, and were stirred together in what Dean (1994: 176) describes – a little confusingly, given the two-domain character of his initial reasoning – as 'this triple domain of self-government, the government of others, and the government of the state'. Dean identifies here two crucial continuities in Foucault's work:

First, there are the continuities between the microphysics of power (and the political technology of the body) and the concerns of the government of nations, populations and societies. Secondly, there is a continuum established between both of these and the practice of ethics as a form of government of the self . . . [Recognising these continuities] suggests a relation between the government of the self by the self, of one's own existence, and broader modalities of government, including political government.

(Dean 1994: 176)

Within this new account of power, the entity of power becomes intensely scrutinised in its many different instances, forms and levels, and many new sources of entanglement arguably swim into view. Moreover, the focus on self-fashioning individuals prizes open more of a niche for resistance than was present in *Discipline and Punish*. Individual people can now be viewed as more likely to effect resistances, since, in their relative freedom, Foucault has now granted them some leeway to pursue their own routes within, and maybe even wholly to dispute, the standard discourses, practices and technologies of dominating power. To draw upon Dean one more time, the later Foucault now 'seeks to show how it is possible to conceive of the subject as a site of independent conduct without relapsing into a humanist framework, and to address questions of *resistance to relations of power and domination*, and problems of freedom and autonomy more generally' (Dean 1994: 175, added emphasis, although note that Dean risks equating power and domination in a manner which we are questioning). In his later writings, Foucault duly offered a fresh take on the relations between domination and resistance, as set within the wider operations of power, albeit one predicated on putting humans as humans, creative and creating, into the picture more forcefully (and optimistically) than ever before.[1] It is possible that recognising these dimensions of late-Foucauldian thought might cause Thrift to pause in his critique of Foucault's failure to give any 'space' for 'something else' to happen (Thrift, Chapter 13 in this volume) and maybe as well to qualify his claim that for Foucault 'the body's functions and movements are shaped by discourse without considering the other, *expressive side* of its existence' (Thrift 1997: 137, original emphasis). While agreeing that Foucault's emphasis has remained chiefly on the discourses which 'programme' conduct, we would propose that in his later work something of a gap *does* open up between discourse and practice, between scripted invocations of what embodied selves should be like and the particular performances of self that individuals fabricate in their everyday lives. In this gap, or perhaps constituting the gap, are many of the entanglements of interest to this book.

J. P. SHARP, P. ROUTLEDGE, C. PHILO, R. PADDISON

The entanglements of power

Leading from our critique of orthodox accounts of power, and also from our more detailed engagement with the Foucauldian corpus, let us now begin to state more directly what we understand by the entanglements of power. The first thing to notice is that we do accept the broadly Foucauldian reading of power as a thoroughly entangled bundle of exchanges dispersed 'everywhere' through society, as comprising a 'micro-physical' or 'capillary' geography of linkages, intensities and frictions, and as thereby not being straightforwardly in the 'service' of any one set of peoples, institutions or movements. The implication is that power 'should not be understood as "blocs of institutional structures, with pre-established, fixed tasks (to dominate, to manipulate), or as mechanisms for imposing order from the top downwards, but rather as a social relation diffused through all spaces"' (Alvarez *et al.* 1998: 11, citing Garcia Canclini 1988: 474). More schematically, we propose that power be conceptualised as an amalgam of *forces*, *practices*, *processes* and *relations*, all of which spin out along the precarious threads of society and space (see Box 1.3). Such a perspective then leads us to suppose that neither dominating power nor resisting power are total, but rather are both fragmentary, uneven and inconsistent to varying degrees, and this realisation prompts us to deploy the Foucauldian dyad of 'domination/resistance' (as in the title of both the book and this chapter). Such a formulation acknowledges that domination and resistance cannot exist independently of each other, but neither can they be reducible to one other: they are thoroughly hybrid phenomena, the one always containing the seeds of the other, the one always bearing at least a trace of the other that contaminates or subverts it. No moment of domination, in whatever form, is completely free of relations of resistance, and likewise no moment of resistance, in whatever form, is entirely segregated from relations of domination: the one is alway present in the constitution of the other. As Vaneigem (1983) reminds us, there is no end to subversion, whether of practices which we might categorise as domination or of those we might label as resistance. The familiar Taoist image of the Yin-Yang symbol helps us to visualise our claims about the entanglements of domination and resistance: although the symbol implies a dynamic balance of opposed forces, it implies there to be no complete separation between two seemingly opposed practices, in that the one will always contain at least the seed of the other.

Accepting a Foucauldian position, one which envisages the play of domination/resistance caught within ever-knotted relations of power, does not necessarily commit us to a kind of intellectual and political quietism: a shrug of the shoulders in the face of power's apparent all-pervasiveness, elusiveness and complexity. Intellectually, and as Alvarez *et al.* (1998: 11, citing Garcia Canclini 1988: 474) insist in a more specific context: a 'decentred view of power and politics . . . should not divert our attention from how social movements interact with political society and the state', and 'must not lead us to ignore how power sediments itself

Box 1.3

In order to break down the components of power, we suggest that it involves the following: the *forces of power*, which involve the use of power over others (as in coercion, persuasion, influence) or the power to act effectively by individuals or organised groups within a particular situation; the *practices of power*, which involve the use and/or application of strategic and/or tactical knowledges within a particular situation; the *processes of power*, which involve the particular methods of doing particular actions over time, which may develop, change, or adapt during the course of events; and the *relations of power*, which involve the myriad social, economic, political, and cultural connections and networks within and between groups, institutions and organisations.[2]

and concentrates itself in social institutions and agents'. Instead, there is much work to be done in conducting patient and detailed research documenting precisely how the entanglements of power arise, what they look like and how they operate, which will also mean continuing to look closely at the specifics of power – its forces, practices, processes and relations – together with its sedimentation, and its mechanisms and effectivities, in certain 'social institutions and agents' (and, we would add, spaces). A stress on these entanglements also does not mean that we cannot tell the difference between domination and resistance, in which case we agree with Said (1983, 1993) when he points out that, fashionable cultural theories notwithstanding, there *are* still the oppressed of the world. There are evidently still acts of brutal domination occurring, and also instances of spirited resistance, and we do not wish to obscure a recognition of such acts and instances. Politically, we must not be led to suppose that resistance cannot be effected against the optics and grip of dominating power, and we are sympathetic to Konrad's (1984) notion of an 'anti-politics' which opposes any species of (political) power that is exerted over people against their will.

Nonetheless, the simple purity that moments of domination and resistance were once thought to embody does need to be queried, and for analytical and strategic purposes it is arguably more useful to think of domination and resistance as occupying a continuum: one running between two idealised poles which might (if a little glibly) be characterised as *resistance in domination* and *domination in resistance*. Accordingly, Haynes and Prakash (1991) speculate that any inquiry into moments of domination should take into account the ways in which the subjectivities of the dominant are constrained, modified and conditioned by power relations, thus acknowledging that dominating power is constantly fractured by the struggles of

21

the subordinate. But, so they continue, moments of resistance are also constantly conditioned by the structures of dominating social and political power, hinting that resisting power is constantly in danger of replicating the structures of the dominant. This means that, to echo what we said above, 'neither domination nor resistance is autonomous' (Haynes and Prakash 1991: 3; see also Moore 1997: 92): resistance turns up in domination and domination turns up in resistance. To be cognisant of such entanglements is to be able to enact resistance while being critical and vigilant to its internal oppressions, and it is to be able to confront, negotiate and enter into dialogue with the manifestations of dominating power from a sensitivity to the 'feeling space' of domination. It is within this latter space that we can then maybe locate pockets of resistant thought and action to be engaged, nurtured and mobilised.

Starting at one end of the continuum just noted, even in the most totalitarian of societies, where it might be assumed that the power to dominate is all pervasive, there exists what Havel (1985) has termed the 'power of the powerless'. Within the cultural and political context of 'communist' Czechoslovakia, for instance, this entailed challenges to the system's ideological manipulation 'from below', involving dissident political and artistic expression in myriad forms. Such challenges sought to extend the 'space' available within Czechoslovakian society for autonomous action beyond the control and discipline of a state political culture. Resistance sometimes arose within the strata of an *intelligentsia* who ostensibly occupied a social position on the side of domination (see also G. Smith's work on the roles of the 'ethnoregional' *intelligentsias* in the old Soviet Russian republics: G. Smith 1985, 1989). More generally, it is possible to identify what Guattari (1996) terms 'soft subversions' occurring within regimes of domination, whereby subtle, hidden or indeed confrontational forms of resistance may appear, thus fracturing the facade of totalising power. In one respect, the state, as a particular manifestation of dominating power, may uphold certain democratic rights within a society that are being challenged by (anti-democratic) resistances. Examples here would include the Sandanista government's struggle against the Contra rebels in Nicaragua during the 1980s, and also the upholding of reproductive rights by various state legislatures in the United States in the face of anti-abortion oppositions that have entailed the fire-bombing of reproductive heath clinics. In another respect, the state may fail to function as anything approaching a coherent and smooth-running engine of domination (and see our earlier remarks about the state). All manner of resistances tend to surface in a state's own internal operations, less from the principled oppositions of key individuals and groups within their make-up (although this can happen), and more from the simple difficulties of co-ordinating spawling politico-legal-administrative machineries. To use Paddison's (1983) evocative phrase, we need to think in terms of 'the fragmented state', one where attempts to hold together policies across national, regional and local domains consistently encounter 'frictions', as too do policies which require the co-operation of different central ministries, departments and inspectorates (see also Driver 1993a). Jurisidictional wrangling, parochial jealousy and downright incompetence all contribute a variety of 'soft subversion' to the state's dominating power, and many other examples of dominating power encountering what can be conceptualised as 'internal' resistances can doubtless be cited. Indeed, Radcliffe (Chapter 7 in this volume) identifies the resistances allowed by the ambiguities of nationalism and heritage, while Skelton (Chapter 8) identifies the resistances allowed by contradictions within the media machine.

Moving to the other end of the continuum, it can be argued that even the seemingly most overt and successful occasions of resistance may be scarred by lines of power which reinforce, rather than dismantle, certain forms of domination. One particular aspect of this process occurs when the state insinuates itself into resistance (even revolutionary) movements through *agents provocateurs*, who then 'teleguide' the actions of these movements to the state's own agenda. In this respect Sanguinetti (1982) discusses how the Brigaddo Rosso was infiltrated by Italian state agents who teleguided the would-be revolutionary cells to conduct bombings and other atrocities, and in the process, arguably, to conduct the assassination of Aldo Moro (then Prime Minister of Italy) to further a deeper state agenda against Moro's accommodation with leftist unions. Another aspect of this process arises in the form of 'minor' reversals within resistance practices, such as occurs with the creation of internal hierarchies, the silencing of dissent, peer pressure and even violence; or in how various forces of hegemony are internalised, reproduced, echoed and traced within such practices. Social movements hence frequently suppress their own internal heterogeneities and sub-groups in the interests of some broader strategy. Rubin (1998) recounts how the Coalition of Workers, Peasants, and Students of the Isthmus in southern Mexico, while struggling against the Mexican state, continued to enact violence toward, and exclusion of, women, as well as more widely repressing internal democracy within its practices. Similarly, the US women's movement has been criticised for its predominantly bourgeois nature, and for excluding lesbians and women of colour (de Lauretis 1990; Mohanty 1991).

Certain resistances are themselves a reproduction or extension of dominating power, rather than a challenge to it. For example, we could point to the ways in which Operation Rescue in the United States forms part of a broader attack upon women's reproductive, civil and economic rights within the country (Faludi 1991). Alternatively, we could turn to the South African-backed RENAMO guerrillas in Mozambique, whose war against peasant communities formed part of the Apartheid regime's campaign of destabilisation against the front-line states. Registering such instances of domination in resistance also problematises in a critical way the agency of 'political' subjects, as in the case of striking General Electric workers in the United States during the mid-1980s whose jobs involved producing the Vulcan 2 gun for the El Salvadorean military to use against Salvadorean peasants. Resistance in one place may therefore be complicit with domination in another, which begins to introduce the entangled geographies about which we shall speak further. To reflect more abstractly on this issue, we would propose that individuals and groups are commonly subject to contradictory consciousness, supporting some aspects of the social order while opposing others. As Mitchell (1990) notes, even resistance practices are partially conditioned by hegemony, since their logics of enactment accept, at least in part, the larger structures of political and economic power: there is no autonomous consciousness or completely self-determining subject who thoroughly escapes the effects of hegemonic practices. Moore (1997) states that the notion of resisting subjects being autonomous of dominating power is based upon supposing a subjectivity or selfhood which pre-exists, and is maintained against, an objective, material world. It also depends upon a corresponding conception of power as an objective force that must somehow penetrate this non-material subjectivity, which means that resistance becomes the intentional thwarting of 'external' forces from an imagined space of autonomy conceived as somehow outside of power (a point to which we shall return).

Entangled geographies of power

Having elaborated our sense of power's many entanglements, let us now elaborate the claim that these entanglements are thoroughly spatial and are indeed themselves inherently geographical. The term 'entanglements' is meant to conjure up the threadings, knottings and weavings of power, thus deploying a metaphor full of spatial imagery to convey the complexity of what we see in the workings of power, domination and resistance. Yet, our use of the term is also meant to be *more* than metaphorical, and is intended to signal that relations of power are really, crucially and unavoidably spun out across and through the material spaces of the world. It is within such spaces that assemblages of people, activities, technologies, institutions, ideas and dreams all come together, circulate, convene and reconvene – it cannot but be so – and it is only as a consequence of the spatial entangling together of all of these elements that relations of power are established. In this view, the entanglements are a precondition for the appearance of power, and in a sense we might say that such entanglements are precisely what releases power, enables power, permits power to 'do its business'. We are even tempted to speak of the *power of entanglements*, thereby accenting the 'power', or capacity, of entanglements to make power happen, to set in train the relational encounters which are always replete with the effects of power, even of highly uneven contests between dominating and resisting power. We acknowledge that we shift here to a perspective which regards power very much as an *effect* of the entanglements, emerging from the spatial assemblages rather than somehow pre-existing them in disembodied but coherent units, and we also acknowledge that such a perspective echoes claims made by Latourian 'actor-network theorists' about how many properties previously taken as pre-given to the world – notably that of 'agency' – are better viewed as effects or accomplishments of specific worldly assemblages (or networks of varying spatial extension: see also both Hinchliffe and Wilbert, Chapters 10 and 11 in this volume). Once couched in this way, and while recognising objections that may be raised to such a formulation, the geographies become vital: far from being incidental outcomes of power, they become regarded, in their ever-changing specifics, as absolutely central to the constitution of power relations.

There are all manner of theoretical lenses on space, and also place and networks, which could be drawn into the picture, and the claims just made, deceptively simple as they may appear, owe much to nearly three decades of hard labour by human geographers and others striving to conceptualise the articulations of society and space (what Soja (1980) memorably termed the 'socio-spatial dialectic'). The argument is hence that the operations of power, domination and resistance must be seen as integrally rolled up in these many articulations of society and space. Massey (1993), for instance, argues that place is a heterogenous social construct, a dynamic locus of community, in which case she also echoes Young's (1990) claim that community, and therefore place, may involve a variety of exclusions (manifested in local displays of sexism, racism, ethnic chauvinism and class devaluation) as well as inclusions (ones joining people together in the fostering of an energising 'community spirit'). This means that place can imply control and surveillance as much as community and free expression and, through the effects of collective memory, imaginaries and institutions, it can be understood as solidifying configurations of 'social relations', 'material practices', elements in 'discourse' and forms of 'power'. Places are thereby constructed and experienced as both material ecological artefacts and intricate networks of social relations, being the focus of the imaginary, of beliefs, desires and discursive activity, filled with symbolic and representational meanings. They are also the products of institutionalised social and political-economic power. For Harvey (1996), the construction of a secure place – through these myriad and interrelated processes – is thus a fundamental moment in an ongoing political power struggle. It is the very certainty, the foundational truths and beliefs, of right-wing politics and religious parties that offers such safety and attractiveness to individuals in the face of dissolving boundaries and frontiers in a time of heightened uncertainty. There is a danger of allowing the right the ability to claim for itself the solid ground of certainty and clarity, and to represent others, including left theorists, as occupying much less tenable positions on fluid ground. Clearly, materiality, representation and imagination within the social practices of everyday life are interrelated, and forms of domination and resistance draw upon all of these realms in particular, context-dependent configurations, but all nested within and productive of the play of 'power' (Lefebvre 1991).

Different social groups endow space, and of course place, with amalgams of different meanings, uses and values. Such differences can give rise to various tensions and conflicts within society over the uses of space for individual and social purposes, and over the domination of space by the state and other forms of dominating social (and class) power:

> Socio-political contradictions are realised spatially. The contradictions of space thus make the contradictions of social relations operative. In other words, spatial contradictions 'express' conflicts between socio-political interests and forces; it is only in space that such conflicts come effectively into play, and in doing so they become contradictions of space.
>
> (Lefebvre 1991: 365)

Within such contradictions of space, particular places frequently become sites of contestation where the social structures and relations of power, domination and resistance are interwoven. These sites may involve alternative constructions of place, or 'counter spaces', as Lefebvre (1991: 382) calls them, which become inserted into spatial reality 'against power and the arrogance of power' (we take him here to mean dominating power). In a related vein, for Foucault, space is 'where discourses of power and knowledge are transformed into actual relations of power' (Wright and Rabinow 1982: 14). The abstract forces of domination and resistance become tangible relations of power-creating subjects, identities and knowledges, but only when enunciated in particular places at particular times. As many geographers have noted, much social science has assumed the 'dead' nature of the spatial in contrast to the dynamic dimension of history, the inevitability of change over time (Massey 1993; Soja 1989, 1993). Yet prominent figures such as Foucault and Lefebvre have suggested that there is no hope of revolutionary politics and activism having any effect without there being a change in the spatial organ-isation of society: the material and hence discursive spaces that make up modern life, the spatial connections that exist between, and creating, individuals and groups, therefore need to be altered. Dominant spatial organisations have to be transcended in the creation of both new subjectivities and new possibilities for relations of power. 'The subject experiences space as an obstacle', opined Lefebvre (1991: 57), insisting upon the problems inhering in the dominant spatialities encircling many people's lives, and reflecting upon the possibilities for resistant spatialities full of new possibilities for social existence.

It might be argued that there is a danger in what is too often the metaphorical nature of space-talk about 'margins', 'decentrings' and so on, without a concomitant concern with the related issue of the material manifestation of spaces wherein domination and resistance are outworked (see also N. Smith and Katz 1993). It is a concern voiced by Lefebvre (1991: 5), albeit in a more regimented sense than we would like to adopt here, when suggesting that 'the philosophico-epistemological notion of space is fetishised and the mental realm comes to envelop the social and physical ones'. This is not to suggest that the epistemological and mental realms of space are separable from the social and physical ones, but nor do they neatly collapse one into another. Lefebvre also raised the issue that, although Foucault stimulated so much interest in the spatiality of social life, his use of space remains an elusive one: Foucault never explains what space it is that he is referring to (see also Philo 1992), nor how it bridges the gap between the theoretical realm and the practical one, between mental and social, between the space of the philosophers and the space of the people who deal with material things (Lefebvre 1991: 4). Both Harvey (1989, 1993, 1996) and Castells (1983) have shown how space is constitutive of different power relations, and how it is manifested within practices of resistance, either in geographically circumscribed communities (resistance within localities: Castells) or in spatialised communities (such as class: Harvey). As Castells notes, multiple axes of power coalesce to create political identities, and, moreover, different power relations become spatialised in different ways. This means that the different scales and articulations of these relations (from the global to the local) are constituted by the very spatialisations through which struggles take place.

Mohanty (1991: 2: see also F. Smith, Chapter 5 in this volume) usefully argues as follows: 'we occupy . . . a world that is definable only in relational terms, a world traversed with intersecting lines of [dominating] power and resistance . . . of gender, color, class, sexuality, and nation'. We duly understand the geography of domination/resistance as a contingent and continuous bundle of relations; a geography that enacts a contested encounter within and between dominant and resistant practices which are themselves hybrid, rather than binary, and which are contingent upon and enmeshed within social networks, communication processes and economic relations. And this is also a geography that comprises 'webs of tension' (Rose 1996) across space, hybrid assemblages of knowledges, vocabularies, judgements, technologies; a geography suspended in an interwoven web of specific practices, processes, subjectivities and spatialities of enactment, expression and materialisation. More schematically, our aim in this book is to investigate the spatiality of the processes of dominating and resisting power, forging an understanding of the many ways in which space constitutes the active medium through which the relations of domination/resistance can be discerned and assessed. We are arguing for an ambiguous, entangled view of power, raising questions concerning: the spatiality of power relations inherent in domination/resistance; the spatial relations and constraints imbued in the processes of domination/resistance; and the impact of such a spatial perspective on both our understanding of society and our reflections on social-cultural theory. In order to make progress along these lines, though, it is necessary to pursue more grounded inquiries into the practices of domination/resistance whereby specific spaces, places or 'sites' are created, claimed, defended and used (strategically or tactically). Such 'spatial practices' may entail a strategic mobility that involves the tactical interactions and communication relays within social collectivities, and also between them and their opponents and allies (Routledge 1996, 1997a, 1997b). These practices may also effect movements of territorialisation and/or deterritorialisation: the former implies the temporary or permanent occupation of space, while the latter involves a movement across space, fostering an imminent dispersion (Deleuze and Guattari 1987). The relationship between such practices and the sites wherein they are articulated are mutually constitutive, albeit in different ways. Hence, while the strategic mobilities of domination/resistance may constitute particular places as sites of action, the material, symbolic and imaginary characteristics of these places will also influence the exact effects of such practices. As a multiplicity, the precise blend of these relations is always unique. Indeed, they are endowed with varying degrees of strategic force, movement and meaning according to the particular spatial, cultural and historical contexts of a given conflict. What this also implies, following our earlier discussion, is that it may be appropriate to offer a preliminary distinction between *resistance within spaces of domination* and *domination within spaces of resistance*.

If we consider the first of these distinctions, a salient example arose during the conflict in Northern Ireland when practices of domination, as a clear manifestation of state power, not only were imbued with a counterforce to practices of resistance but also engendered their own forms of resistance. For instance, the prison of Long Kesh (also known as the H-Blocks and The Maze) represented a particularly brutal regime of confinement, and indeed of death (during 1981 ten Republican prisoners, led by Bobby Sands, died while on hunger strike to demand political status in Long Kesh). However, the caging of Republican prisoners together for control and surveillance – in 1974 there were twenty cages, each containing a hundred prisoners (Bennett 1998) – enabled an attenuated form of their family, friend and activist networks, as had existed outside, to be reinscribed within the prison. Moreover, the length of incarceration enabled younger Republican activists, including Gerry Adams, to rethink the political strategy and political organisation of the IRA (Bennett 1998). The particular spatiality of domination within Long Kesh therefore enabled an articulatory and strategic space of Republican resistance to be created. In a further entanglement, the gendered character of prison life became reinscribed in the Republican communities outside, as ex-prisoners' clubs became sites witnessing the marginalisation of women (Dowler 1998). Linked to this example, the classic work of Goffman (1961) explains how prisons, mental asylums, monasteries and other such 'total institutions' install a regime of domination which superficially appears comprehensive, but which in practice is always subverted by inmates who find means of eluding, even if in tiny ways, the demands of the institution (its attempt to impose a singular 'institutional identity' upon them). As Goffman explains, this is a thoroughly spatialised resistance, since it involves inmates slipping out of 'surveillance space' and into 'free space' (under the stairs, in the toilets, behind the bicycle sheds) supposedly beyond the reach of the institution's dominating power. The situation is arguably even more entangled, however, since to some extent these other spaces are knowingly left to inmates as part of a 'geography of license' to which the authorities turn a 'blind eye', the idea being that permitting such inmates these small moments of resistance will defuse the build up of resentments which might then prompt a more serious threat to institutional control.

If we consider the second of these distinctions, we find that in a specific situation such as the Nepali revolution of 1990, practices of resistance, as a clear articulation of subaltern power, not only were imbued with a counterforce to autocratic domination but also engendered their own forms of domination. For example, the Movement for the Restoration of Democracy organised a series of blackout protests in the capital, Kathmandu, in order to register a symbolic protest against the country's autocratic regime and to enable the movement to challenge more physically the government's dusk to dawn curfew on the streets. In the space of resistance that constituted the discursive relay for this particular tactic, the rooftops of the capital's houses from which people hailed each other, dominating practices were also entangled since residents were threatened by activists with having their windows broken if they did not comply with movement demands to participate in the blackouts. In another example, as movement swarms took control of certain parts of the city, so some of these 'liberated spaces' became sites of execution as movement cadres performed revolutionary justice by executing suspected *agents provocateurs* (Routledge 1997a).

We can also make the broader observation that subject positions are cross-cut by multiple matrices of power, even within what hooks (1991) terms the 'home-places' of resistance, the spaces of self-dignity and solidarity in which, and from which, resistance can be conceptualised and organised. Even such seemingly safe homeplaces may become sites of difference and distance, ones in which separation, limitation and violence can all too often be present (Kirby 1995: 208). Following Foucault (1986b), we might go on to think about 'heterotopias' as real sites that are 'other' to the norms of conventional space and time, resistant counter-sites perhaps in which other real sites of culture are simultaneously represented, contested and inverted (Foucault 1986b: 24). But such 'performed spaces' still contain physical and social boundaries where resources are available to some and not to others (Pile and Thrift 1995: 374), and as such they remain ambiguous in character and may better be conceptualised as 'third spaces' (Bhabha 1994). That is, they are sites where resistance is never a complete, unfractured practice, but rather are ones where practices of resistance always become entwined in some manner with practices of domination such as marginalisation, segregation or imposed exile. These hybrid, interstitial spaces are thereby sutured and structured by intersecting and interwoven spatialities of power, whether this is primarily power from above (dominating), power from below (resisting) or even power from within (spiritual) (Comaroff 1985; Ong 1987; Routledge and Simons 1995).

Slipping outside of power?

Before concluding, and letting the chapters that follow 'speak for themselves', we want finally to revisit Pile's (1997: 3) claim that 'assumptions about the domination/resistance couplet become questionable'. This claim is made from within a conceptual framework which contextualises an excellent collection of papers addressing the 'geographies of resistance' (Pile and Keith 1997), in the course of which Pile provides a particularly compelling introduction to the many spatialities which are implicated in the resistances of given peoples, situations and events. At the same time, he argues that resistance should be at least in part conceptually 'uncoupled' from domination, not being seen as merely a 'mirror image' or effect of domination, which means that domination and resistance should cease being viewed as locked in 'some perpetual dance of control' where they are always moving together, 'neither able to let go' (Pile 1997: 2). For him, this also means that 'geographies of resistance do not necessarily (or even ever) mirror geographies of domination, as an upside-down or back-to-front or face-down map of the world' (Pile 1997: 2); and he goes further to state that inserting 'geography' into the picture forces us to appreciate the extent to which domination and resistance end up being quite literally 'dislocated'. Instances of resistance need to be understood in the specificities of when and where they erupt, with detailed attention paid to how the resistant identities of the peoples involved are pieced together in the process (not arriving fully formed), and also with a sense of how 'alternative spatialities' of resistance are forged whose contours elude those laid down by the dominating forces of control. Moreover, Pile (1997: 3) then declares that 'it is no longer enough to begin stories of resistance with stories of so-called power', not least because 'resistant political subjectivities are constituted through positions taken up not only in relation to authority . . . but . . . through experiences which are not so quickly labelled "power", such as desire and anger, capacity and ability, happiness and fear, dreaming and forgetting'. This account is also spatialised, as when suggesting that 'resistance may invoke spatialities that lay beyond "power"' (Pile 1997: 5), and he insists that the 'material effects of power are everywhere . . . but wherever we look power is open to gaps, tears, inconsistencies, ambivalences, possibilities for inversion, mimicry, parody and so on; open, that is, to more than one geography of resistance' (Pile 1997: 27). These observations signal a deeper presumption that resistance can somehow slip outside of power, a dream that resistance is possible because all sorts of other peoples can also dream of evading power, imagining that they can begin 'throwing away imposed maps, unfolding new spaces, making alternative places, creating new geographies of resistance' (Pile 1997: 30).

We are certainly attracted to Pile's arguments, and to the parallel ones of Thrift (1997; Chapter 13 in this volume) about uncovering embodied geographies of practices, competencies, skills, performance, play and fantasy whose dynamics might be cast as resistant acts somehow lying beyond the normal compass of power. Yet we also have reservations, chiefly because at root Pile (if not Thrift) *dis*entangles what we regard as always unremittingly *en*tangled. Ostensibly, he is wishing to disentangle domination and resistance, a point to which we will return presently, but the initial problem for us is that he tends to equate power quite straightforwardly with dominating power (see Box 1.1). Despite noting that power can be 'conceived of as oppression or authority or capacity or even *resistance*' (Pile 1997: 2, added emphasis), the logic of his subsequent explicit statements is to pit resistance against power, to effect a fundamental disentanglement of power and resistance, which then sets up the possibility of imagining resistances that slip outside of power. Throughout his paper the extensive 'cartographies' of power are referenced, as are 'the fixed grids of the latitudes and longitudes of power relations' (Pile 1997: 27) alongside 'the spatial technologies of power configurations' (Pile 1997: 30), and the imagery is always of resistance finding ways to evade these rigidities through the fluidity of fleeting moments and geographies. From such comments it seems that his take on power *has* to mean dominating power, the power inhering in hegemonic institutions, ideas and identities as they strive to 'territorialise' the world, to fix it, to grid it, to survey and to discipline it. For us, he operates with an opposition, that between power and resistance, which it is precisely the objective of the current volume to question. Since we conceive of power as a complicated nexus of forces, practices, processes and relations (see Box 1.3) which cannot but run through *all* instances of domination and resistance, as irreducibly present in any occasion of resistance (even in the foot-dragging) as in any act of domination, it makes no sense for us to separate power and resistance. Resistance involves power, it requires it, releases it and generates effects of power, just as much as does domination; and it is only because there is power in resistance that we can be as optimistic as Pile in supposing that resistance will happen.

In the details of his own essay, Pile arguably does not achieve the clear-cut separation of power and resistance that his more programmatic remarks imply.

This is also true of how he treats the domination/resistance couplet, whose initial 'dislocation' (Pile 1997: 2) quickly becomes transposed into the more complex figure of '*dis-located* interaction' (Pile 1997: 16, original emphasis), indicating that domination and resistance are still meeting, mediating one another, even if doing so in a messy, disjointed fashion which may entail a spatial distancing between exactly where domination occurs (e.g. on the street, in the dictator's proclamations) and where it is then resisted (e.g. on the roof, in the rebel's heart). Pile draws upon Fanon's ([1959] 1965) diagnosis of the Algerian struggle against French colonisation, and shows how the encounter of coloniser and colonised was deeply fractured, uneven and contingent, but still clearly pivoting around the 'play' of domination and resistance meeting, making and remaking one another across an ensemble of different spaces. Building on Fanon's thoughts about the internal conflicts of Algerian women striving to overcome inbred senses of Muslim womanhood so as to serve the needs of the 'revolution', Pile discusses how 'psychic resistance' (understood in a Freudian sense of 'repression') can work counter to 'political resistance'. Political resistance 'seeks to overthrow the perceived dangers in the practices of the powerful, while psychic resistance seeks unconsciously to maintain the repression of traumatic or potentially dangerous memories, feelings or impulses' (Pile 1997: 24). All the time that the latter are repressed, it is unlikely that the passions necessary to energise the former will be released. Pile duly provides a stunning example of domination in resistance, indicating how totally entangled the two become in the power relations crystallising between coloniser and colonised, and offering materials which square well with our own favoured understanding of power, domination and resistance.

In addition, several contributors to the Pile and Keith (1997) volume seemingly prefer a vision akin to ours, notably Moore (1997: 92) when writing as follows: 'Instead of conceiving of a space of subalternity, insurgency and resistance "outside" of power, domination or hegemony, the challenge becomes to understand their mutual imbrication'. Similarly, in her evocation of the entangled 'economies of meaning and power' found in Philippines *a-go-go* bars, key sites of an international sex tourism, Law (1997) speculates about:

> supposedly powerless bar women, who have their own sights/sites of power, meaning and identity. It can also be conveyed that bar women are capable of positioning themselves in multiple and intersecting relations of power – which include, but are not exclusive to, their encounters with men – and that it is in these spaces that subjectivities, capable of resistance, are forged.
>
> (Law 1997: 108)

The Abu-Lughod notion of resistance being a 'diagnostic' of power relations is mentioned by Law (1997: 112: see also Cresswell, Chapter 12 in this volume), and

Pile (1997: 3) himself appears to agree with such a notion in noting that 'resistance becomes a mode through which the symptoms of different power relations are diagnosed'. What is more, in the concluding essay to *Geographies of Resistance*, Keith (1997: 283) echoes Moore in asserting that 'mobilisations and resistance occur within and are in part defined by territorialised structures of governmentality', an assertion informed by his broader argument that geographies of resistance should be explored in relation to 'the spaces through which political subjects are rendered visible' (Keith 1997: 282). This is no dragging apart of domination and resistance, but rather an attempt to see how the latter grows out of 'spaces' left by the former, as fostered in reaction to his worry that too 'often in social analysis the heroic celebration of the transgressive moment erases the subject creation of the forces that are being transgressed' (Keith 1997: 282). Keith goes on to borrow from Foucault's later work concerning the varied fields of governmentality, those which specify what is the possible 'conduct of conduct' (Keith 1997: 286) for individuals and groups endeavouring to create themselves as agents with political effectivity, and as such his take on matters of power, domination, resistance and their geographies enters fully into those Foucauldian entanglements already debated (and see also Note 1). Something happens, then, in the progress through the *Geographies of Resistance* collection from Pile's introduction to Keith's conclusion: something which *re*entangles what Pile had initially sought to *dis*entangle. And it is precisely the task of a (spatialised) *re*entangling that now continues into our volume, so we would hope, and into the substantive chapters which follow.

Notes

1 To illustrate the abstract claims made in the main text, we shall consider Part 3, Chapter 2 of *The Care of the Self* (Foucault 1986a), entitled 'The political game'. Foucault here rehearsed the argument sometimes made about a supposed 'retreat' into matters of the self by Roman elites throughout Europe in the third century: an alleged withdrawal from public, civic and political life into private, personal and psychological life, wherein self-directed philosophising became a 'shelter from the storm' (Foucault 1986a: 82) for the authorities of city-states losing their autonomy in the face of imperial entrenchment. But this was not an argument which he accepted, in part because he reckoned the political and urban geographies of the era to entail more than simply a collapsing system of city-states:

> Rather than imagining a reduction or cessation of political activities through the effects of a centralised imperialism, one should think in terms of the organisation of a complex space. Much vaster, much more discontinuous, much less closed than must have been the case for the small city-states, it was also more flexible, more differentiated, less rigidly hierarchised than would be the authoritarian and bureaucratic Empire that people would attempt to organise after the great crisis of the third century. It was a space in which the centres of power were multiple; in which the activities, the tensions, the conflicts were numerous.
>
> (Foucault 1986a: 82)

The story was not of urban decline, then, but one of 'municipalisation' in which local powers were actually enhanced so as 'to stimulate the political life of the cities' (Foucault 1986a: 83). Local municipal elites thereby took on responsibilities for governing municipal and regional populations on behalf of the empire, an act of government requiring them to secure the loyalty of such populations in more subtle ways than had been the case when they had been more obviously autonomous rulers. In consequence their power was 'relativised' (Foucault 1986a: 87–88), since the 'political life' to which the elites now dedicated themselves – as framed by the polycentric political geography of the age – could be enacted much less as a sovereign exertion of power and more as a negotiation of power through the crafted performances of the elite officers themselves. Foucault put matters like this:

> Short of being a prince himself, one exercises power within a network in which one occupies a key position. In a certain way, one is always the ruler and the ruled. Aristotle, in the *Politics*, also evoked this game, but in the form of an alternation or rotation: one is now the ruler, now the ruled. On the other hand, in the fact that a man [*sic*] is one and the other at the same time, through an interplay of directions sent and received, of checks, of appeals of decisions taken, Aristides sees the principles of good government . . . Anyone who exercises power has to place himself in a field of complex relations where he occupies a transition point.
>
> (Foucault 1986a: 88–89)

Notwithstanding the unquestioned masculinism of this passage, it brilliantly expounded the entanglements of power integral to the situation, time and place under study. Third-century imperial cities, albeit not functioning as the autonomous entities of old, were key 'transition points' between the exertions of imperial power from above and the impositions of municipal power on the masses below. From these sites it was possible for the municipal elites to effect a measure of resistance to the dominating power from Rome, but in turn they themselves experienced resistances rebounding back upon them from the localities (through 'checks' and 'appeals of decisions').

The crucial further aspect is the new countenance which Foucault identified to the exercise of power by the city senators, administrators and military officers. This was one which demanded a new political rationality dependent upon an enlarged sense of the role to be played by the self in municipal government, an increased appreciation of the need for skilled self-possession in the individual (inter)actions of these highly visible elite figures. It was this circumstance, so Foucault reasoned, which accounted for the supposed 'retreat' from public to private life mentioned earlier:

> if one wishes to understand the interest that was directed in these elites to personal ethics, to the morality of everyday conduct, private life and pleasure, it is not that pertinent to speak of decadence, frustration and sullen retreat. Instead, one should see in this interest the search for a new way of conceiving the relationship that one ought to have with one's status, one's functions, one's activities, and one's obligations.
>
> (Foucault 1986a: 84)

Foucault thus delimited a new form of 'political game' being played by the municipal elites, one no longer based on 'a system of signs denoting power over others' (Foucault

1986a: 85), but rather on the cultivation of an ethical relationship with oneself which would also be the underpinning of an ethical, reciprocated and probably successful relationship with others. It was an ethics 'concerned to define the principle of a relation to self that will make it possible to set the forms and conditions in which political action, participation in the offices of power . . . will be possible or not possible' (Foucault 1986a: 86). In other words, and once more noting the geopolitical basis to what was occurring:

> In a political space where the political structure of the city and the laws with which it is endowed have unquestionably lost some of their importance, although they have not ceased to exist for all that, and where the decisive elements reside more and more in men [*sic*], in their decisions, in the manner in which they bring their authority to bear, in the wisdom they manifest in the interplay of equilibria and transactions, it appears that the art of governing oneself becomes a crucial political factor.
>
> (Foucault 1986a: 89)

The heart of the matter clearly involved an interweaving of the ethical and the political, a fusing of the government of the (local) body politic with the government of the (elite individual) self. Although specific, with this example Foucault illustrated several more threads in the entanglements of power. To be sure, these look like delicate threads, less ones weaving the pattern of larger structures and more ones knotted up in particular individuals, but they do show the importance of a deliberate, self-conscious 'modelling of political work' (Foucault 1986a: 91) which can presumably be done by both more elite figures and ones of lesser status (it is not *only* elites who can creatively 'model' themselves in the manner meant by Foucault). Through this example, he thereby hinted at processes central to the authorship of power, which he duly depicted not as anonymous but rather as constituted by people quite intentionally, even if they cannot avoid drawing upon a wider terrain of 'helpful discourses' to do so. In the process he offered the materials for a revision of his previous inquiries into power, and he also opened up a much larger gap for resistance, for things to happen differently because they are made to do so by the people involved.

2 In the sociological sense, tactics form the specific tools for the articulation of broader political strategies. Such tactics can be deployed by both dominating powers and resisting powers, albeit in potentially different ways (e.g. Ackerman and Kreuger 1994; Chaliand 1982; Clausewitz 1968; Sharp 1973; Sun tzu 1988). Moreover, as effects of power, both tactics and strategies are inherently spatial in their operation. In contrast, de Certeau (1984: 35–36) defines strategy as

> the calculation . . . of power relationships that become possible as soon as a subject with will and power (a business, an army . . .) can be isolated. It postulates a *place* that can be delimited as its *own* and serve as a basis from which relations with an *exteriority* composed of targets and threats (. . . competitors, enemies . . .) can be managed. [In contrast,] a tactic is a calculated action determined by the absence of a proper locus . . . The space of the tactic is the space of the other. Thus it may play on and with a terrain imposed on it and organised by the law of a foreign power . . . [I]t is a man[oeuvre] 'within the enemy's field of vision' . . . and within enemy territory . . . [A] tactic is the art of the weak . . .

[A] tactic is determined by the *absence of power* just as a strategy is organised by the postulation of power.

(de Certeau 1984: 36–38)

Strategies, for de Certeau (1984: 38–39), 'privilege spatial relationships', while tactics pin their hopes on 'a clever utilisation of time'.

We find this conceptualisation problematic on at least three grounds. First, de Certeau equates power solely with the forces of domination, and juxtaposes this power against the practice of resistance or the arts of the powerless, thus falling into a trap which we are critiquing throughout the chapter. Second, he regards strategies only as expressing the workings of power, a thoroughly dominating power, and thereby positions resistance as somehow without power, again in contrast with our own claims here. Third, he locates dominating power within the realm of the spatial, but locates resistance practices within the realm of the temporal, thus setting up another opposition which we would question. Simply put, de Certeau disempowers and despatialises resistance, and inscribes too stark a binary relationship between domination and resistance (see also Massey, Chapter 14 in this volume).

References

Abu-Lughod, L. (1990) 'The romance of resistance: tracing transformations of power through Bedouin women', *American Ethnologist* 17: 41–55.

Ackerman, P. and Kreugler, C. (1994) *Strategic Nonviolent Conflict*, Westport, CT: Praeger.

Adam, B. D. (1987) *The Rise of a Gay and Lesbian Movement*, Boston, MA: Twayne.

Adas, M. (1981) 'From avoidance to confrontation: peasant protest in precolonial and colonial Southeast Asia', *Comparative Studies in Society and History* 23: 217–247.

Alvarez, S. E., Dagnino, E. and Escobar, A. (eds) (1998) *Cultures of Politics, Politics of Cultures*, Oxford: Westview.

Bachrach, P. and Baratz, M. S. (1962) 'Two faces of power', *American Science Review* 56: 947–952.

Bachrach, P. and Baratz, M. S. (1963) 'Decisions and non-decisions: an analytical framework', *American Political Science Review* 57: 641–651.

Bachrach, P. and Baratz, M. S. (1970) *Power and Poverty: Theory and Practice*, Oxford: Oxford University Press.

Bender, J. (1987) *Imagining the Penitentiary: Fiction and the Architecture of Mind in Eighteenth-Century England*, Chicago: University of Chicago Press.

Bennett, R. (1998) 'Farewell to hell', *Guardian* G2, 3 September: 2–3, 6.

Bhabha, H. (1994) *The Location of Culture*, London: Routledge.

Bunge, W. (1971) *Fitzgerald: The Geography of a Revolution*, Cambridge, MA: Schenkman.

Castells, M. (1983) *The City and the Grassroots*, Berkeley: University of California Press.

Castells, M. (1997) *The Power of Identity*, Oxford: Blackwell.

Chaliand, G. (1982) *Guerilla Strategies*, Berkeley: University of California Press.

Clark, G. and Dear, M. (1984) *State Apparatus: Structures and Language of Legitimacy*, Boston, MA: Allen and Unwin.

Clausewitz, C. V. (1968) *On War*, New York: Penguin.

Clegg, S. R. (1975) *Power, Rule and Domination: A Critical and Empirical Understanding of Power in Sociological Theory and Organisational Life*, London: Routledge and Kegan Paul.

Clegg, S. R. (1989) *Frameworks of Power*, London: Sage.

Cohen, J. L. (1985) 'Strategy and identity: new theoretical paradigms and contemporary social movements', *Social Research* 52: 663–716.

Comaroff, J. (1985) *Body of Power, Spirit of Resistance: The Culture and History of a South African People*, Chicago: University of Chicago Press.

Cresswell, T. (1996) *In Place/Out of Place: Geography, Ideology and Transgression*, Minneapolis: University of Minnesota Press.

Dahl, R. A. (1957) 'The concept of power', *Behavioural Science* 2: 201–215.

Dahl, R. A. (1961) *Who Governs?*, New Haven, CT: Yale University Press.

Dean, M. (1994) *Critical and Effective Histories: Foucault's Methods and Historical Sociology*, London: Routledge.

de Certeau, M. (1984) *The Practice of Everyday Life*, Berkeley: University of California Press.

de Lauretis, T. (1990) 'Eccentric subjects: feminist theory and historical consciousness', *Feminist Studies* 16: 115–150.

Deleuze, G. and Guattari, F. (1987) *A Thousand Plateaus: Capitalism and Schizophrenia*, Minneapolis: University of Minnesota Press.

Dowler, L. (1998) 'Till death do us part: deconstructing the masculine spaces of the Irish hunger strike', paper presented at the Association of American Geographers' annual meeting, Boston, MA.

Driver, F. (1985) 'Power, space and the body: a critical assessment of Foucault's *Discipline and Punish*', *Environment and Planning D: Society and Space* 3: 425–446.

Driver, F. (1990) 'Discipline without frontiers? Representations of the Mettray Reformatory Colony in Britain, 1840–1880', *Journal of Historical Sociology* 3: 272–293.

Driver, F. (1991) 'Political geography and state formation', *Progress in Human Geography* 15: 268–280.

Driver, F. (1993a) *Power and Pauperism: The Workhouse System, 1834–1884*, Cambridge: Cambridge University Press.

Driver, F. (1993b) 'Bodies in space: Foucault's account of disciplinary power', in C. Jones and R. Porter (eds) *Reassessing Foucault: Power, Medicine and the Body*, London: Routledge.

Escobar, A. (1992) 'Imagining a post-development era? Critical thought, development and social movements', *Social Text* 31/32: 20–56.

Escobar, A. and Alvarez, S. E. (eds) (1992) *The Making of Social Movements in Latin America*, Boulder, CO: Westview.

Faludi, S. (1991) *Backlash: The Undeclared War Against American Women*, New York: Doubleday.

Fanon, F. ([1959] 1965) *A Dying Colonialism*, New York: Grove.

Foucault, M. (1979a) *Discipline and Punish: The Birth of the Prison*, Harmondsworth: Penguin.

Foucault, M. (1979b) *The History of Sexuality, Volume 1: An Introduction*, London: Allen Lane.

Foucault, M. (1979c) 'Governmentality', *Ideology and Consciousness* 6: 5–21

Foucault, M. (1980) 'Two lectures', in C. Gordon (ed.) *Michel Foucault: Power/Knowledge – Selected Interviews and Other Writings, 1972–1977*, Brighton: Harvester.

Foucault, M. (1982) 'Afterword: the subject and power', in H. L. Dreyfus and P. Rabinow, *Michel Foucault: Beyond Structuralism and Hermeneutics*, Brighton: Harvester.

Foucault, M. (1985) *The History of Sexuality, Volume 2: The Uses of Pleasure*, Harmondsworth: Penguin.

Foucault, M. (1986a) *The History of Sexuality, Volume 3: The Care of the Self*, Harmondsworth: Penguin.

Foucault, M. (1986b) 'Of other spaces', *Diacritics* 16: 22–27.

Foucault, M. (1988a) 'Politics and reason', in L. D. Kritzman (ed.) *Michel Foucault: Politics, Philosophy, Culture – Interviews and Other Writings, 1977–1984*, London: Routledge.

Foucault, M. (1988b) 'On power', in L. D. Kritzman (ed.) *Michel Foucault: Politics, Philosophy, Culture – Interviews and Other Writings, 1977–1984*, London: Routledge.

Foucault, M. (1988c) 'The concern for truth', in L. D. Kritzman (ed.) *Michel Foucault: Politics, Philosophy, Culture – Interviews and Other Writings, 1977–1984*, London: Routledge.

Garcia Canclini, N. (1988) 'Culture and power: the state of research', *Media, Culture and Society* 10: 467–497.

Giddens, A. (1979) *Central Problems in Social Theory: Action, Structure and Contradiction in Social Analysis*, London: Macmillan.

Giddens, A. (1981) *A Contemporary Critique of Historical Materialism, Volume 1: Power, Property and the State*, London: Macmillan.

Giddens, A. (1985) *A Contemporary Critique of Historical Materialism, Volume 2: The Nation-State and Violence*, London: Macmillan.

Goffman, E. (1961) *Asylums: Essays on the Social Situations of Mental Patients*, Harmondsworth: Penguin.

Gordon, C. (1991) 'Governmental rationality: an introduction', in G. Burchell, C. Gordon and P. Miller (eds) *The Foucault Effect: Studies in Governmentality*, London: Harvester Wheatsheaf.

Guattari, F. (1996) *Soft Subversions*, New York: Semiotext(e).

Guha, R. (ed.) (1982) *Subaltern Studies*, vol. 1, Delhi: Oxford University Press.

Hall, S. and Jefferson, T. (eds) (1976) *Resistance through Rituals: Youth Sub-Cultures in Post-War Britain*, London: Routledge and Kegan Paul.

Hannah, M. G. (1992) 'Foucault deinstitutionalised: spatial prerequisites for modern social control', Unpublished PhD thesis, Department of Geography, University of Pennsylvania.

Hannah, M. G. (1993) 'Space and social control in the administration of the Oglala Lakota ('Sioux'), 1871–1879', *Journal of Historical Geography* 19: 412–432.

Hannah, M. G. (1997a) 'Imperfect panopticism: envisioning the construction of normal lives', in G. Benko and U. Strohmayer (eds) *Space and Social Theory: Interpreting Modernity and Postmodernity*, Oxford: Blackwell.

Hannah, M. G. (1997b) 'Space and the structuring of disciplinary power: an interpretive review', *Geografiska Annaler* 79B: 171–180.

Harlow, B. (1987) *Resistance Literature*, London: Methuen.

Harvey, D. (1989) *The Condition of Postmodernity: An Enquiry into the Origins of Cultural Change*, Oxford: Blackwell.

Harvey, D. (1993) 'Class relations, social justice and the politics of difference', in M. Keith and S. Pile (eds) *Place and the Politics of Identity*, London: Routledge.

Harvey, D. (1996) *Justice, Nature and the Geography of Difference*, Oxford: Blackwell.

Havel, V. (1985) *The Power of the Powerless*, Armonk, NY: M.E. Sharpe.

Haynes, D. and Prakash, G. (1991) *Contesting Power: Resistance and Everyday Social Relations in South Asia*, Berkeley: University of California Press.

hooks, b. (1991) *Yearning: Race, Gender and Cultural Politics*, Boston, MA: South End Press.

Hunter, F. (1953) *Community Power Structure*, Chapel Hill: University of North Carolina Press.

Keith, M. (1997) 'Conclusion: a changing space and a time for change', in S. Pile and M. Keith (eds) *Geographies of Resistance*, London: Routledge.

Keith, M. and Pile, S. (eds) (1993) *Place and the Politics of Identity*, London: Routledge.

Kirby, K. (1995) *Indifferent Boundaries: Exploring the Space of the Subject*, New York: Guilford.

Konrad, G. (1984) *Antipolitics*, New York: Henry Holt.

Laclau, E. and Mouffe, C. (1985) *Hegemony and Socialist Strategy*, London: Verso.

Law, L. (1997) 'Dancing on the bar: sex, money and the uneasy politics of third space', in S. Pile and M. Keith (eds) *Geographies of Resistance*, London: Routledge.

Lefebvre, H. (1991) *The Production of Space*, Cambridge, MA: Blackwell.

Lukes, S. (1974) *Power: A Radical View*, London: Macmillan.

Lukes, S. (1977) 'Power and structure', in S. Lukes, *Essays in Social Theory*, London: Macmillan.

McAdam, D. (1982) *Political Process and the Development of Black Insurgency, 1930–1970*, Chicago: University of Chicago Press.

McCarthy, J. D. and Zald, M. N. (eds) (1977) 'Resource mobilization and social movements: a partial theory', *American Journal of Sociology* 82 (May): 33–47.

Mann, M. (1986) *The Sources of Social Power, Volume 1: A History of Power from the Beginning to AD 1760*, Cambridge: Cambridge University Press.

Massey, D. (1993) 'Politics and space/time', in M. Keith and S. Pile (eds) *Place and the Politics of Identity*, London: Routledge.

Matless, D. (1992) 'An occasion for geography: landscape representation and Foucault's corpus', *Environment and Planning D: Society and Space* 10: 41–56.

Melucci, A. (1989) *Nomads of the Present*, London: Radius.

Melucci, A. (1996) *Challenging Codes*, London: Verso.

Mitchell, T. (1990) 'Everyday metaphors of power', Theory and Society 19: 545–577.

Mohanty, C. T. (1991) 'Cartographies of struggle: Third World women and the politics of feminism', in C. T. Mohanty, A. Russo and L. Torres (eds) *Third World Women and the Politics of Feminism*, Bloomington: Indiana University Press.

Moore, D. S. (1997) 'Remapping resistance: "ground for struggle" and the politics of place', in S. Pile and M. Keith (eds) *Geographies of Resistance*, London: Routledge.

Morris, A. D. and Mueller, C. M. (1992) *Frontiers in Social Movement Theory*, New Haven, CT: Yale University Press.

Murray, M. (1995) 'Correction at Cabrini-Green: a socio-spatial exercise of power', *Environment and Planning D: Society and Space* 13: 311–328.

Oberschall, A. (1973) *Social Conflicts and Social Movements*, Englewood Cliffs, NJ: Prentice Hall.

O'Farrell, C. (1989) *Foucault: Historian or Philosopher?*, London: Macmillan.

Ogborn, M. (1993) 'Law and discipline in nineteenth-century English state formation: the Contagious Diseases Acts of 1864, 1866 and 1869', *Journal of Historical Sociology* 6: 28–55

Ogborn, M. (1995) 'Discipline, government and law: separate confinement in the prisons of England and Wales, 1830–1877', *Transactions of the Institute of British Geographers* 20: 295–311.

Ong, A. (1987) *Spirits of Resistance and Capitalist Discipline*, Albany, NY: SUNY Press.

Paddison, R. (1983) *The Fragmented State: The Political Geography of Power*, Oxford: Blackwell.

Painter, J. (1995) *Politics, Geography and Political Geography: A Critical Perspective*, London: Edward Arnold.

Philo, C. (1989) '"Enough to drive one mad": the organisation of space in nineteenth-century lunatic asylums', in J. Wolch and M. Dear (eds) *The Power of Geography: How Territory Shapes Social Life*, London: Unwin Hyman.

Philo, C. (1992) 'Foucault's geography', *Environment and Planning D: Society and Space* 10: 137–162.

Pile, S. (1997) 'Introduction: opposition, political identities and spaces of resistance', in S. Pile and M. Keith (eds) *Geographies of Resistance*, London: Routledge.

Pile, S. and Keith, M. (eds) (1997) *Geographies of Resistance*, London: Routledge.

Pile, S. and Thrift, N. (1995) 'Conclusions: spacing the subject', in S. Pile and N. Thrift (eds) *Mapping the Subject: Geographies of Cultural Transformation*, London: Routledge.

Poulantzas, N. (1973) *Political Power and Social Classes*, London: New Left Books.

Poulantzas, N. (1978) *State, Power, Socialism*, London: New Left Books.

Pringle, R. (1999) 'Power', in L. McDowell and J. Sharp (eds) *A Feminist Glossary of Human Geography*, London: Edward Arnold.

Robinson, J. (1990) 'A perfect system of control? Territory and administration in early South African locations', *Environment and Planning D: Society and Space* 8: 135–162.

Robinson, J. (1996) *The Power of Apartheid: State, Power and Space in South African Cities*, Oxford: Pergamon.

Rose, N. (1996) 'Identity, genealogy, history', in S. Hall and P. du Gay (eds) *Questions of Cultural Identity*, London: Sage.

Routledge, P. (1996) 'Critical geopolitics and terrains of resistance', *Political Geography* 15: 509–531.

Routledge, P. (1997a) 'A spatiality of resistances: theory and practice in Nepal's revolution of 1990', in S. Pile and M. Keith (eds) *Geographies of Resistance*, London: Routledge.

Routledge, P. (1997b) 'Space, mobility and collective action: India's Naxalite movement', *Environment and Planning A* 29: 2,165–2,189.

Routledge, P. and Simons, J. (1995) 'Embodying spirits of resistance', *Environment and Planning D: Society and Space* 13: 471–498.

Rubin, J. (1998) 'Ambiguity and contradiction in a radical popular movement', in S. E. Alvarez, E. Dagnino and A. Escobar (eds) *Cultures of Politics, Politics of Cultures*, Oxford: Westview.

Said, E. (1983) *The World, the Text, and the Critic*, Cambridge, MA: Harvard University Press.

Said, E. (1993) *Culture and Imperialism*, London: Vintage.

Sanguinetti, G. (1982) *On Terrorism and the State*, London: Chronos.

Schattschnieder, E. E. (1960) *The Semi-Sovereign People*, Hinsdale, IL: Dryden.

Scott, J. C. (1985) *Weapons of the Weak: Everyday Forms of Peasant Resistance*, New Haven, CT: Yale University Press.

Scott, J. C. (1990) *Domination and the Arts of Resistance: Hidden Transcripts*, New Haven, CT: Yale University Press.

Sharp, G. (1973) *The Politics of Nonviolent Action*, Boston, MA: Porter Sargent.

Smith, G. (1985) 'Ethnic nationalism in the Soviet Union: territory, cleavage and control', *Environment and Planning C: Government and Policy* 3: 49–73.

Smith, G. (1989) 'Ethnoregional societies, "developed socialism" and the Soviet ethnic intelligentsia', in R. J. Johnston and E. Kofman (eds) *Nationalism, Self-Determination and Political Geography*, London: Croom Helm.

Smith, G. (1994) 'Political theory and human geography', in D. Gregory, R. Martin and G. Smith (eds) *Human Geography: Society, Space and Social Science*, London: Macmillan.

Smith, N. and Katz, C. (1993) 'Grounding metaphor: towards a spatialized politics', in M. Keith and S. Pile (eds) *Place and the Politics of Identity*, London: Routledge.

Soja, E. (1980) 'The socio-spatial dialectic', *Annals of the Association of American Geographers* 70: 207–225.

Soja, E. (1989) *Postmodern Geographies: The Reassertion of Space in Critical Social Theory*, London: Verso.

Soja, E. (1993) 'Postmodern geographies and the critique of historicism', in J. P. Jones III, W. Natter and T. Schatzki (eds) *Postmodern Contentions: Epochs, Politics, Space*, New York: Guilford.

Stallybrass, P. and White, A. (1986) *The Politics and Poetics of Transgression*, London: Methuen.

Sun tzu (1988) *The Art of War*, Boston, MA: Shambala.

Tarrow, S. (1994) *Power in Movement: Social Movements, Collective Action and Politics*, New York: Cambridge University Press.

Thrift, N. (1997) 'The still point: resistance, expressive embodiment and dance', in S. Pile and M. Keith (eds) *Geographies of Resistance*, London: Routledge.

Tilly, C. (1978) *From Mobilization to Revolution*, Reading, MA: Addison-Wesley.

Touraine, A. (1981) *The Voice and the Eye: An Analysis of Social Movements*, Cambridge: Cambridge University Press.

Touraine, A. (1988) *The Return of the Actor*, Minneapolis: University of Minnesota Press.

Vaneigem, R. (1983) *The Revolution of Everyday Life*, London: Rebel Press.

Wright, G. and Rabinow, P. (1982) 'Spatialisation of power: a discussion of the work of Michel Foucault', *Skyline* March: 14–20

Young, I. M. (1990) 'The ideal of community and the politics of difference', in L. Nicholson (ed.) *Feminism/Postmodernism*, New York: Routledge.

Zirakzadeh, C. E. (1997) *Social Movements in Politics*, London: Longman.

2

VICTORIAN SEXUALITY AND THE MORALISATION OF CREMORNE GARDENS

Philip Howell

Domination and resistance in the understanding of Victorian sexuality

Our understanding of Victorian sexuality has been constructed around the paradigm of domination and resistance.[1] In an historiographical sense we can point to the well-established picture of a dominant, repressive, ultimately middle-class sexual moralism, and of the pockets of resistance to this sexual orthodoxy (e.g. Cominos 1963; Cott 1978; Harrison 1977; Pearl 1980; Pearsall 1971; Vicinus 1972). In this historiography, the model of repression, conceived of as a form of domination, has been understandably most prominent – 'Sexuality was not freed from domination nor the urge toward domination', as Cominos (1972: 170) put it – but the awareness of heterodox sexual cultures, conceived of as some form of resistance to the dominant sexual morality, is clearly evident as well. Explorations of the 'double standard' of morality, the Victorian 'Underworld' or the 'other Victorians', point out the dualistic and hierarchical structure of a great deal of writing about Victorian sexuality (e.g. Chesney 1972; B. Harrison 1967; Marcus 1966; Peckham 1975; Thomas 1959). In a theoretical and epistemological sense too we can refer to the way that Victorian sexuality has been understood through a comparison to our own, apparently more liberal, culture of sex: that is to say, the contrast between a 'repressive' Victorian and a modern, 'anti-repressive', moral code. Here the trope of domination and resistance, bolstered by analogues from psychology and psychoanalysis, is sometimes deployed in the service of a self-congratulatory narrative which draws a sharp dividing line between 'Victorian' and enlightened 'modern' mores. Most forcefully, it was Foucault (1980) who pointed out the epistemological politics of this contemporary claim to sexual enlightenment, arguing that the modern discourse of sex, with its 'repressive hypothesis' concerning Victorian sexual culture, conceals its own repressive, normalising force in its rhetorical claims to sexual and sensual liberation. The dualism of dominance and resistance is built in this way into the very concept of sexuality itself: in its epistemology and theory as well as its history and chronology. Indeed, notwithstanding the Victorians' own persuasive claims to have invented the term, and the clear continuities between the Victorians' sexual culture and our own, 'sexuality' has for many been most easily formulated as simply 'whatever it was that the middle-class Victorian mind attempted to hide, evade, repress, deny' (Miller and Adams 1996: 1).

Recognising that it is not only the derided adjective 'Victorian' which was mobilised to support a self-congratulatory image of modernity, but also the concept of sexuality itself, helped to establish the significance of the domination/ resistance dualism at the same time as of course it promised to undermine it. Foucault's history of sexuality took as its starting point the need to discard the 'repressive hypothesis' through which our understanding of the nature of the Victorian domination of sex is bound up with 'our' resistance to the culture of the Victorian middle classes. Foucault's work, as it has been debated and developed by others, is certainly the most influential attempt to redefine sexuality, principally towards seeing it as a social construction and a technology of power rather than as an innate essence subject to periodic bouts of repression. There can be no deny-ing that Foucault's most valuable points – that sex is discursively produced, that all 'sexual knowledge' is a social construction, and that there is a fundamental recursiveness in our own knowledge about the sexuality of the past – have trans-formed the way in which we think about sexuality in history. Bolstered too by several strands of substantive revisionism (e.g. Barret-Ducrocq 1991; Gay 1984, 1986; Mason 1994a, 1994b; Seidman 1990; Stearns and Stearns 1985) our under-standing of Victorian sexuality is considerably more sophisticated. The simpler narratives of domination and resistance have now, thankfully, more of the air of historiography than history.

Yet while it is certainly true that the general picture of Victorian sexuality has been revised, and in parts even revolutionised by recent work, the dualism of domination and resistance is still very much in force. Historians and sociologists of sex still fall back on the established narrative of domination and resistance, albeit refigured in innovative and often informative ways.[2] While Foucault has been immensely influential, his rhetorical demolition of the 'repressive hypothesis' is less thorough than it seems. Part of the problem is that Foucault notably fails to specify the different natures and significance of the various sexual discourses whose proliferation he charts, and in a related way fails to properly connect beliefs about sex with sexual practices (Porter and Hall 1995: 9; Mason 1994a: 172–173). This leaves the nature of sexuality as a technology of power curiously vague and insubstantial, and paradoxically allows the discursive nature and disciplinary deployment of sexuality to assume an air of total dominance. The focus of much work has been indeed on the deployment of sexuality, not in a merely discursive but in a solidly institutional sense, in the service of social order: sexuality remains a central element in the domination of bodies and their pleasures. There is a certain irony in this reproduction of a version of the repressive hypothesis, of course, in its extension of the eminently Victorian project of putting sex into discourse. Foucault himself is disarmingly straightforward about the deployment of a historically bourgeois sexuality on a doggedly resistant working class (Foucault 1980: 120–122). What Foucault's work has not done is to dislodge the static pairing of a dominant bourgeois repressive culture and heroic resistance of individuals, groups or bodies. The Foucault effect in this field has been if anything to bolster rather than undercut this structuring dualism.[3]

Consider the problem of prostitution in the Victorian age. The (female) prostitute is a key figure in understanding Victorian sexuality, and is established as the most prominent construct of the discourse of dangerous sexuality (see Mort 1987; Nead 1988). Here, however, we face the immediate and acknowledged danger of viewing prostitutes as either passive victims or successful 'working girls'. The whole discussion of prostitution has tended to revolve around this question, with its debts to the structuring dualism of domination and resistance (e.g. Finnegan 1979; Hill 1993; Hobson 1987; Mahood 1990). To take just one example, Walkowitz's (1980a) analysis of the campaign leading to the repeal of the British Contagious Diseases Acts in 1886 is a detailed examination of how sexual ideology became embedded in policy and practice. Here we see how a 'science of sexuality' and a 'technology of power' worked to uncover new areas of illicit sexual activity, legitimising state intervention in the lives of the poor and creating an outcast group; and also how this 'technology of power' generated an impressive social and political resistance, not least among the prostitutes themselves. It is a 'painstaking documentation of a feminist movement of resistance' (Wood 1982: 62), indebted to Foucault but developing what were only theoretical points into historical and political realities (see Walkowitz 1980b). Irreplaceable as this work is, it is not immune from criticism that the stark dualism of domination/resistance is in many ways quite straightforwardly reproduced. Wood (1982: 72–74) notes that Walkowitz's reference to the way in which the discursive construction of a reticent female sexuality had repressive political effects is itself too closely bound up with the repressive hypothesis, as is her analysis of the state as a repressive power. Similar critiques may in fact be made of several other works on prostitution and the social construction of sexuality (e.g. Mahood 1990; Nead 1988). Resistance in these works tends to be marginal and rhetorical; domination by discursive and disciplinary powers seems all but complete. Anderson (1993) is particularly critical of what she sees as a generalised and unexamined notion of agency and resistance in these texts, which reproduces the ideological character and presuppositions of the discursive and practical emphasis on the 'fallenness' of public women.

Let me not give the impression that such work should be dismissed; on the contrary, these are key works in any consideration of Victorian sexuality. What should be recognised is that the nature of domination and resistance remains problematic and provisional; it is overdetermined by contemporary political concerns. This is arguably inescapable, but it can and does result in some real confusion. We still know too little about the diversity and variety of Victorian sexual ideologies and cultures, about their relative strengths and weaknesses, about their degrees of overlap and shared assumptions, and about the nature of their mobilisation in social contests and conflicts, to allocate domination and resistance such a privileged status. And the question to be asked here is therefore: to what degree does the continuing use of the trope of domination and resistance hinder or help in answering questions about Victorian sexuality?

I want to try to answer this question by looking at the struggle around the Cremorne pleasure-gardens in Chelsea, an especially notorious public space which was the focus over several decades for the ire of social purity campaigners shocked at the apparent licentiousness and intemperate behaviour of its patrons and outraged at the seemingly symbiotic relationship between Cremorne and the prostitutes who were said to frequent its gardens and gates. Cremorne will serve here to illustrate the attempted moralisation and purification of urban public space in the later nineteenth century, through the use of the dominant sexual moralist arguments which were the currency of social reformers. But it will also illuminate the determined and multifaceted resistance to this moralisation, which was able to draw on several alternative strains of sexual culture in counterargument. I shall go on to describe these elements of Cremorne's story in turn, before sketching out the tangled geographies in which these elements were necessarily embedded and the difficulties involved in explicating its historical geography through such categories. Ultimately, indeed, I want to warn against hypostasising the dualism of domination and resistance, and move to the foreground the complex geographies of Victorian sexuality.

Cremorne and the Chelsea Dogberries[4]

Almost from its inception, London's Cremorne Gardens had achieved a reputation which was inseparable from heterodox sexual behaviour, in both its casual and commercial forms. The precursor of the Gardens, an inoffensive rural sporting arena, opened in 1832. But in 1843, under the brief ownership of Renton Nicholson, Cremorne was quickly transformed into a pleasure-garden associated with immodest sexuality. Cremorne's entertainments as they developed in the next thirty years were in themselves by and large inoffensive. Patrons entered the twelve acres of gardens from either the King's Road or the River Thames and enjoyed the 'monster pagoda' with its orchestra, surrounded by the extensive dancing platform and the numerous refreshment booths, supper tables and side-shows. These pleasures were not sophisticated; rather than a pleasure-garden in the sense of Vauxhall and Ranelagh, it is probably best thought of as an amusement park, devotedly demotic and popular where the older pleasure-gardens attempted to be exclusive and aristocratic in tone. But Nicholson was already well known for his scurrilous journal, *The Town*, the paper of choice of the capital's would-be men about town, and for presiding over the obscene Judge and Jury trials in Maiden Lane and the Strand. Nicholson's reputation, his championing of worldly sexuality and the 'sporting' male culture he represented, was quickly and indelibly trans-ferred to the Gardens, remaining long after his involvement with the Gardens had ceased. In fact 'Cremorne' became no less than a by-word, or even a brand-name, for male sexual licence.[5]

It was an association with prostitution that was really damaging to the Gardens' reputation. William Acton, the well-known 'expert' on the 'social evil', famously treated Cremorne as a notorious haunt of prostitutes, the 'demure immorality in silk and fine linen' who claimed the Gardens at night as the respectable world vacated them (Acton 1972: 16–17). Cremorne had to fight for years against the accusations that it was, through the presence of prostitutes, a dangerous blight on the entire neighbourhood. Canon John Cromwell, the principal of a training college a few yards down the King's Road from Cremorne, was its chief tormentor, claiming that the streets near Cremorne were 'infested with prostitutes' who represented a danger to the boys in his care, to his family, and to the respectable men and women of the neighbourhood (*Times* 10 October 1874: 5). The Chelsea vestry agreed wholeheartedly and considered the Gardens to be 'the greatest curse of anything in the parish' (*Chelsea News and General Advertiser* 7 October 1871: 4). For them, the Gardens formed a gross public nuisance, collecting together 'crowds of noisy, disorderly inhabitants, and forming an accumulation of vice and immorality which shocked the respectable inhabitants'; they noted that 'Gentlemen were afraid to allow their female servants to go beyond the doors after seven o'clock at night' (see *Reynolds* 15 October 1871: 5). To these men, and the constituency they represented, Cremorne was palpably immoral, 'a pestilential hot-bed, mis-named "a garden," where the very flowers seemed to drop their heads in shame, and the trees threw more sombre shadows than is their wont to veil in darkness that which dare not brave the light' (*Church Opinion*, quoted in *Chelsea News and General Advertiser* 21 October 1871: 5). Year after year the vestry opposed the license to the Gardens on the grounds that Cremorne had become 'a special resort of prostitutes' (*Chelsea Vestry Supplemental Reports* 18 September 1877: 344).

Now these complaints were clearly not wholly fantastical, though the term 'prostitute' was typically applied with abandon. But for morality campaigners there was little room for compromise. To suggestions that Cremorne was a safety-valve for the exuberance of London's pleasure-seekers, its opponents retorted that this argument was

> nothing more than a direct plea for vice and shameless extravagance. Healthy athletic exercise, well directed study and labour, intellectual recreation, such are the true safety-valves for youth, but not late hours, bad company, excess of intoxicating drink and wild excitement. It is no sin for a man or a woman to spend some portion of the day and some portion of their substance in rational refreshing amusement and recreation, but it is a sin, for man or women to pay one single coin, to spend one minute, to obtain the spurious and guilty delight offered by such places of amusement.
>
> (*Church Opinion*, quoted in *Chelsea News and General Advertiser* 21 October 1871: 5)

Throughout a quarter of a century, the opponents of Cremorne marshalled arguments of this kind, arguments which were steeped in the evangelical morality and domestic ideology that dominated sexual moralism towards the end of the nineteenth century. Over and over again, these charges were repeated in neighbourhood petitions, legal arguments and public correspondence. This was a long drawn out dispute, but the end for the Gardens came quickly. In 1877, the owner John Baum was forced to resort to suing a local Baptist minister over a piece of libellous doggerel entitled 'the Horrors of Cremorne'. The minister in his defence called a Cremorne waiter and a woman from a reformatory who both traced their downfall to the Gardens. Although Baum won the case, he was awarded just a farthing damages without costs and he did not reapply for the licence. The moralists had finally succeeded in their aim of eradicating the pernicious spot of Cremorne. Its remains were covered over, in time, with a power station; later still a small, wholly inoffensive public park restored Cremorne's name and its gates but nothing of its inglorious reputation.

In its bare outlines, then, the end of Cremorne Gardens seems to confirm the effectiveness of Victorian sexual moralism in 'purifying public space' (Bland 1995); its history has largely been understood in terms of the growing dominance of middle-class sexual ideology. The closure of the Gardens is only one incident in the international campaign against urban immorality, of course. Backed by evangelical Christianity and middle-class ideological prescriptions, social purity reformers targeted zones of vice, removed prostitutes from the streets, and cleaned up public amusements (see Boyer 1978; Gilfoyle 1991; Mumford 1997; Pivar 1973). The previously unrestrained public space of the pleasure-gardens and other popular venues was remade by urban reformers in a process through which Victorian sexual moralism helped directly to transform urban geography. This unrestrained and disordered public – whose commingling of men and women, respectable and unrespectable, was precisely its most disturbing and threatening element – was marked out for replacement by a moral public, in a morally homogeneous city, where classes could meet and learn from each other, and where men and women might meet in a civilised and domesticated public world.[6]

Despite the temptation to read the closure of Cremorne as a straightforward case of the bourgeoisie extending their dominance over urban geography, in the face of popular and working-class resistance, we must exercise a great deal of caution here. The evidence that we have as to the social constituencies that lay behind the expressions of support for and antagonism to Cremorne suggests a rather more complex sociology and an entangled geography. Using an analysis of the five thousand-odd signatories of petitions for and against the renewal of Cremorne's licence, it is possible to sketch out these constituencies.[7] What they reveal says something of the strength of popular resistance to the moralists, collecting four times the number of signatures than the opponents of Cremorne, but also about the rough similarities between the two opposed camps (Tables 2.1 and 2.2). In terms of age, sex, marital status, residential geography and occupation, these two camps were in fact surprisingly similar. A greater proportion of men signed in support of the Gardens than in opposition to them, and these petitioners tended to be slightly younger, but overall there is little evidence here that the Gardens were the focus of very clearly delineated groups. Neither does the social geography of the petitioners reveal any simple sense of one community pitted against another (Figure 2.1). The main indication of social differences dividing the two camps comes from the fact that the rateable value of the homes of supporters is lower than that of opponents, as we would expect, yet even here the difference is not startling.

Table 2.1 Summary breakdown of information on the signatories for and against the renewal of Cremorne's licence, 1846–1877

	Con		*Pro*	
Total	874		4,160	
Male	721	82%	3,801	91%
Female	153	18%	359	9%
Unmarried	21	12%	34	8%
Married	118	68%	355	82%
Widowed	35	20%	45	10%
Average age in 1877	45		42	
Rateable value of residence in 1876	£38		£29	

Table 2.2 Occupational breakdown of signatories for and against the continuation of Cremorne's licence, 1846–1877

	Con	Pro		Con	Pro
Annuitant and independent means	8	2	Doctor and physician	2	
			Domestic servant	1	5
Artist, photographer, sculptor	3	3	Draper	7	2
			Dressmaker and milliner	3	5
Baker	3	6	Dyer	1	
Barmaid		1	Eating house keeper	1	
Bird dealer		1	Engine driver	1	
Boat builder		1	Engine fitter		1
Bookseller	1		Engineer		4
Bootdealer	2	3	Excavator		1
Boot and shoe maker	9	16	Fishmonger	1	3
Brass founder		1	Foreman	1	1
Brewer's servant		1	French polisher		1
Bricklayer	1	8	Fruiterer	1	2
Brushmaker		1	Furniture packer		1
Builder	3	5	Furniture salesman		1
Butcher and poulterer	1	14	Gardener		5
Cab driver		1	Gasfitter	2	2
Cabinet maker		2	Gaslighter		2
Carman	1	4	General dealer		3
Carpenter and joiner	15	19	Glass manufacturer	1	
Carter	1		Glove dyer	1	
Carver and gilder	2	3	Goldsmith	1	1
Cellarman		1	Greengrocer		13
Charwomen		4	Grocer	3	7
Cheesemonger		1	Hairdresser		6
Chemist	1		Harness maker	1	
China and glass ware	1	1	Herbalist		1
Clergyman and missionary	3		Horseshoe nail maker		1
Clerk	2	3	Hosier		1
Club servant		1	House agent		1
Coach painter	1	2	House keeper		1
Coachman and omnibus employee	2	7	Iron worker	1	3
			Ironer	1	
Coal dealer	2	1	Labourer	2	11
Coffee house keeper		3	Launderer	3	10
Confectioner		3	Letter carrier and messenger		4
Cook	1		Licensed victualler and refreshment house keeper		8
Cooper		1			
Corn chandler	1		Lodging house keeper	1	1
Dairyman	1	2	Machine puller		1
Dealer in toys		1	Marine store dealer		1
Decorator		3	Mason		4

Table 2.2 continued

	Con	Pro		Con	Pro
Moulder		1	Shopkeeper	1	4
Museum attendant	1	1	Signwriter		2
Musical instrument maker and piano tuner	1	3	Smith and farrier		4
			Staymaker		1
Newsagent and stationer	5	5	Surgeon and dentist	3	2
Oilman	2	2	Tailor and haberdasher	5	9
Painter	2	15	Ticket collector		1
Paper hanger		1	Tobacconist		3
Paper stainer	1	1	Undertaker	1	
Pitman		1	Upholsterer	2	1
Plasterer	2	6	Van proprietor		1
Plumber	1	5	Waiter	5	
Policeman		3	Wardrobe dealer		2
Printer and compositor	2	4	Warehouseman	1	
Professor of music		1	Watchmaker and jeweller		3
Publican and beer retailer		3	Waterman		1
Rate collector	1		Wheelwright	1	
Sawmaker		1	Wholesale rag merchant		1
Sawyer	1	1	Wine merchant	1	1
Scalesmaker		1	Woodcutter		1
Sergeant major		1	Zinc cutter		1

There are many difficulties in using petitions as proxies for social communities and ideological camps, but this information is useful as a guide to the cultural geographies which often remain opaque when discussing sexual cultures. It is difficult to argue that the signatories for the Gardens were directly representing or even condoning the kind of sexual licence associated with Cremorne – the Gardens were a major employer and economic force in this neighbourhood, and many of the tradesmen and artisans had a great deal to lose if the Gardens were closed. Nor, too, should the petitioners against the Gardens be taken to represent a unitary sexual moralism – no doubt the Gardens did represent a nuisance and a brake on property values quite irrespective of any sexual connotations. But, still, these petitioners were prepared to put their names to strongly worded arguments that centred on accusations of sexual impropriety and they demand our attention insofar as we have so little reliable evidence of ordinary people's views on sexuality and its public expression. Again, if anything, this analysis tends to count against the argument that a distinctively bourgeois sexual moralism was imposed upon a

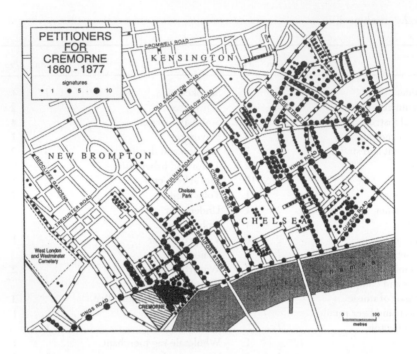

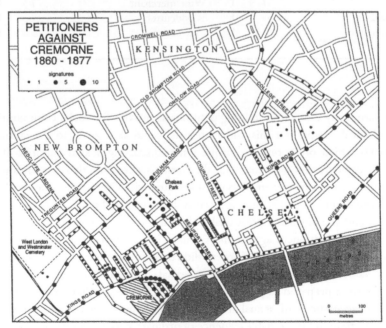

Figure 2.1 Residences of petitioners for and against Cremorne, 1860–1877
Source: GLRO, Middlesex Sessions for Music and Dancing Licences (see note 7)

resistant working people and their institutions. There were for instance several policemen willing to sign up here in the defence of Cremorne, alongside the publicans, carmen and carpenters whom one might expect to form the natural constituency of support for the Gardens. While the antagonists of Cremorne included representatives of the clergy and the professional classes, it numbered also many of the respectable artisans of the vicinity.

This, then, is a more tangled social and cultural geography than our expectations might suggest, primed as we are by the idea of domination and resistance that I have argued is at the heart of our understanding of Victorian sexuality. Perhaps the most interesting element of Cremorne's historical geography is the equally entangled nature of the arguments deployed in its defence, ranging as they do from claims to complete innocence to robust and worldly defences of sexual licence. The extent of the resistance to the sexual moralism of Canon Cromwell and the Chelsea vestrymen is rather impressive; there was a remarkable degree of popular support for the licence not just in Chelsea but in London as a whole. But it is the fact that Cremorne's supporters were able to mount a defence of the Gardens which was as various as the public it attracted which is precisely the significant point here. Cremorne's owners could insist on the one hand that the Gardens catered to the respectable: it was 'a place where the husband can bring his wife, the mother her children for a day's or evening's enjoyment, without the risk of annoyance' (Cremorne programme 1866). The proprietors instituted Sunday promenades, flower-shows and charitable fêtes, as they worked hard to defend themselves from accusations of immorality, and to project an image of Cremorne as a place for respectable rational recreation. The Gardens, supporters insisted, were 'the habitual resort of large numbers of the respectable portion of the inhabitants and their families . . . no expense has been spared to render the Gardens attractive and to supply amusements of an unobjectionable character' (Petition in support of the licence, GLRO, MR/LMD 19/2).

However, there were firmly pragmatic arguments that stressed Cremorne's role as a 'safety-valve . . . for the exuberant gaiety of an immense metropolis; which else, pent in dingy holes and corners, would turn into rank, sodden vice' (*Weekly Dispatch* 4 October 1857). Defenders of the Gardens acknowledged and accepted some degree of impropriety, but dismissed this as more or less harmless and certainly unavoidable in an institution of its kind. In this vein, *Punch* preferred to accentuate the positive and turn a blind eye to the alleged presence of vice:

> [Mr. Punch] states, without the least hesitation, that he concurs with several of his friends, members of the Royal Family, that MR. SIMPSON's gardens are very delightful ones, and for a daylight visit, a place to which a Bishop may go without risk of a speck upon [his] Apron . . . the behaviour of the visitors is exceedingly exemplary, far better, especially

as regards the dancers, than that of many of the attendants at similar Parisian places, to which Paterfamilias, once away from the respectability of Bloomsbury Square, hurries, and very often takes Materfamilias, and thinks he has rather done a knowing thing than not. And whether all the said visitors may take with them 'all the Virtues under Heaven,' . . . we do not exactly know . . . *Mr. Punch* begs to state, with equal distinctness, that he knows, and desires to know nothing of the Gardens after the evening's *programme* is over. They may, after midnight, be as orderly as before. He has no evidence before him. Decent people walk off before tomorrow walks in. And so they ought. . . . We have nothing to say to anybody who stays at Cremorne or anywhere else, at unseemly hours, except that he ought to be ashamed of himself.

(Punch 24 October 1857: 175)

This defence of Cremorne, which was characteristic of many of the popular papers, was combined with the vilification and lampooning of Canon Cromwell and the Gardens' opponents. They ridiculed their attempts to cleanse the neighbourhood of prostitutes (Figure 2.2) and satirised the obsession with regulation by picturing a Cremorne 'as it will be under future regulations' (Figure 2.3). This lampooning could have a serious and political side, and the popular papers were never very far away from articulating a populist political line which condemned alike evangelical and philanthropic sanctimony and upper-class patronage. The ex-Chartist paper *Reynolds*, for instance, railed against the 'drivelling Dogberries' who were willing to deny the people of one of their favourite places of amusement, and on the occasion of an 'aristocratic fête' at Cremorne, when the Gardens were hired to a private aristocratic party, *Reynolds* took the opportunity to rub in its denunciation of hypocrisy and the presumptuous encroachment on the people's amusements (*Reynolds* 4 July 1858: 12; 18 July 1858: 4); in particular, it railed against the fact that aristocratic women were allowed to 'act the part of *Violetta* and her sisters safe from the intrusion of professional Cyprians and plebeian pleasure-hunters' (*Reynolds* 4 July 1858: 12). This denunciation of hypocrisy and the acceptance of the existence of prostitutes – even a pointed defence of them – could thus be combined with a political analysis of the great social evil. *Reynolds*, as a paper which took aim at middle-class Christian Pharisees, countered charges of immorality by advocating radical political solutions to the social problem of prostitution. This represents the most radical contemporary analysis of the meaning of prostitution and of the philanthropic efforts to counter it, and of all the forms of 'resistance' to bourgeois moralism to be found in this story, this is perhaps the closest to that most valued in our own day. Here it can stand for the genuine strength and cogency of the resistance to the bourgeois moralisation of Cremorne.

A NEW OCCUPATION FOR REVEREND PRECEPTORS.—WHAT WE MAY EXPECT TO SEE IN THE NEIGHBOURHOOD OF CREMORNE.

"THERE SHALL BE AT LEAST ONE PLACE SECURE FROM THE INROADS OF WOMEN."

Figure 2.2 'A new occupation for reverend preceptors'
Source: *The Days' Doings* 11 November 1871

Classic moralism and the ambiguities of resistance

Nevertheless, it is the variety and admixture of arguments in support of Cremorne that is significant, not their value as a unitary resistance to bourgeois moralism. If we can detect a single discursive idiom here, in fact, it is that of the 'classic moralism' identified by Mason. As 'the age-old anti-sensual, but anti-idealistic, secular code' (Mason 1994b: 57) whose worldly approach was founded upon what it took to be an empirical and pragmatic line, such a sexual attitude could at the

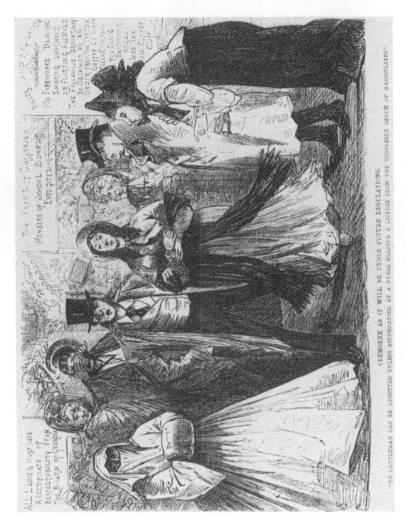

Figure 2.3 'Cremorne as it will be under future regulations'
Source: The Days' Doings 4 November 1871

same time deplore the consequences of immorality and accept a degree of sexual transgression. It is not the pro-sensual advocacy we tend to imagine when we think of resistance to bourgeois sexual morality, but a sort of pragmatic moralism concerned not to suppress licentiousness but to manage it as best possible. The idealist anti-sensualism which reached its apogee in the social purity crusades was ridiculed for its high-minded and wrong-headed cant. Supporters of Cremorne thus typically took the line that popular public amusements were never going to be entirely respectable, but that they provided more or less innocent fun for the great majority of their patrons. That they were at the mercy of sanctimonious kill-joys like Cromwell was a shame and an outrage.

This is nicely illustrated by an account in *The Days' Doings* (27 May 1871) of what was perhaps the bachanalian highpoint of the Cremorne season – 'that foulest of Saturnalia' as the clergymen had it (*Church Opinion*, 21 October 1871) – the Derby Night at Cremorne. The narrator of the piece, initially reluctant, is dragged to the Gardens by his friends on this special night, to confront a scene of unbridled sensual excess (Figure 2.4):

> the Cremorne of to-night is not the normal Cremorne. There is a leaven of what is carnivalesque in it; there is an extra fastness in the fun; an unusual spirit of gaiety abroad in the Gardens. How crowded it is, to be sure! . . . Every seat is full, every alcove; and up in the over-hanging balconies Lesbias, with beaming eyes, are looking down upon the scene, and enjoying it thoroughly; while at times they will fill the brimming beaker and dally with the attendant swains. . . . Cremorne is acting as a focus, so to speak, and centralizing all the scattered rays of revelry in the metropolis. . . . the Derby Day is not an ordinary day; nor is the Derby night as other nights; and Cremorne, reinforced by contingents from the giddy 'west,' is preparing to make a night of it. Preparing, did I say. It *is* making a night of it. Don't tell me that we English have not notion of fun – that we are slow, elephantine, dull, crass, stupid, heavy, and the rest of it. I ask you to look around. Is there no gaiety, no sprightliness, no *abandon*, no *chic*, no grace, no merriment here? Isn't it a gay scene, I ask you? Isn't there something almost indescribably exciting in the whole or part of the place? Did you ever see so many pretty girls before? What a chance for studying dresses; for seeing ravishing toilettes; for wondering at astonishing *chevetures*. Verily, these ladies of the 'half-world' know how to make a sensation.

This account, with its worldly acceptance of the presence of prostitutes and unashamed enjoyment of the crowded public entertainment, is notably different from Acton's experience. But lest we be too caught up in this apparently sensual

Figure 2.4 'The Derby Night at Cremorne: keeping it up'
Source: The Days' Doings 27 May 1871

enjoyment of Cremorne, note that there is in this description a defence of Cremorne that strikes a perhaps unexpectedly moralistic tone:

> But, speaking seriously, can anyone affect ignorance of the free and good-natured amusement at Cremorne? None but very simple people indeed could be found to believe that it was kept up mainly for the purpose of Cromwellism and Puritan recreation. Cremorne is no worse now than it was at any time during the long succession of years in which the license was regularly granted with hardly any objection being raised to it. What was tolerated then as a necessary evil, of which the suppression would be worse than the toleration, is now to be suppressed. But does any-one suppose that the vice against which this crusade is directed will disappear with the disappearance of Cremorne? It will be driven into worse places: and while our moral zealots are congratulating themselves on having abolished a place of resort which certainly had its good points about it, hundreds will spring up in the dark to work unmitigated evil.

Ritchie (1858: 192–199) catches this tone too in one of his urban sketches. In his description of Cremorne he emphasises that his contemporaries shy away from an accurate and graphic account of Cremorne, of both its pleasures and its dangers. 'English people only see what they wish to see,' he adds, 'this is just another of the "frightful hypocrisies of our domestic life".' In fact, of the morality of the place, Ritchie affects to be little interested – some he says will be tempted by the immoral delights of Cremorne and gather moral strength, while others may be ruined. As for the possibility of moralising places like Cremorne, Ritchie simply adds that 'Places are what we make them'. 'I fear', he goes on to say,

> there are many blackguards at Cremorne; the women most of them are undoubtedly hetaerae, and yet what a place it is for fun! How jolly are all you meet! How innocent are all the amusements. . . . There is no harm in Cremorne, if man is born merely to enjoy himself – to eat, drink, be merry, and die.

Yet there is a lurking moralism here too, though certainly quite different from that of the increasingly effective sexual moralism of social reformers. Cremorne is, Ritchie notes, nevertheless a dangerous place, a resort of prostitutes and of beautiful and bewitching women who provide dangerous temptations for young men precisely because they are well dressed and well behaved. As a result, he adds, 'For one man that is ruined in a gin-shop there are twenty that are ruined at Cremorne':

amidst mirth and music, foaming up in the sparkling wine, looking out of dark blue eyes, reddening the freshest cheeks, and nestling in the richest curls, there lurks the great enemy of God and man. Young man, such an enemy you cannot resist; your only refuge is in flight. Ah, you think that face fair as you ask its owner to drink with you; it would have been fairer had she never gone to Cremorne. A father loved her as the apple of his eye; she was the sole daughter of his home and heart, and here she comes night after night to drink and dance; a few years hence and you shall meet her drinking and cursing in the lowest gin-palaces of St. Giles's, and the gay fast fellows around you now will be digging gold in Australia, or it may be walking the streets in rags, or it may be dying in London hospitals of lingering disease, or, which is worse than all, it may be living on year after year with all that is divine in man utterly blotted out and destroyed.

This is I think the unmistakable tone of classic moralism – knowing, worldly, male, prurient yet moral, amused yet sententious, above all quietly pragmatic.[8]

This attitude represents a toleration of the coexistence of the respectable and the unrespectable in public spaces like Cremorne. What Acton and others found so disconcerting and demoralising was precisely the 'elbowing of vice and virtue' in London's public spaces, the amalgamation of prostitutes, married women and sweethearts. Cremorne was like this, a commingling of public and private space which served to subvert the bourgeois's ideal world of public and private spheres readily demarcating vice from virtue. Prostitution was the central issue because in itself the open sale of sex blurred those distinctions, linking the respectable with the disreputable in an unacceptable way for Victorian reformers. But for Cremorne's defenders, despite their acceptance of prostitution as an unavoidable temptation, the Gardens were a deeply moral, not an immoral, space, given the classic moralist attitude which accepted the mixing of respectability and its other. This ability to be two-faced, as it were, represents not hypocrisy but an ability to be both prurient and respectable, so that Cremorne was accepted as a kind of polyvalent public space catering to many tastes. Unworried as it was by strict ideological consistency, classic moralism was lenient on prostitution and sexual licence without giving up on the often starkly moralistic claims to the safeguarding of young women's purity from adulterers and exploiters. Insofar as it fed into a sexual culture, attitudes like this reconfigured rather than refuted conceptions of respectability and unrespectability, and valorised an urban geography marked by moral diversity and a degree of tolerance.[9]

The fact, however, that this worldly and knowing appreciation of Cremorne is centred around the *male* enjoyment of sex, which I have not much emphasised so far, clearly runs counter to any unproblematic valorisation of such 'resistance' to dominant anti-sensual attitudes. These attitudes and this culture owe much in their popular and working-class expression to the 'sporting' male sexual ideology which developed in the early and mid-nineteenth century, rooted in an urban public culture and especially in the places of urban entertainment, and constituting a male subculture whose gender, class and race prescriptions were as oppressive as any variant of middle-class sexual moralism (see Cohen 1992; Gilfoyle 1992; Gorn 1987; Srebnick 1995). In this culture a positive view of male heterosexual indulgence prevailed, one that challenged and confused class divides in its embracing of male camaraderie and aggressive assertiveness. 'Sporting' culture defined its influential form of popular masculinity through men's access to an urban space for which women, as sexual objects, acted as geographical markers. The battle over 'Cremorne' was in part therefore a struggle over the rights of the capital's *flâneurs* to this 'sporting' public and its spaces of sexual opportunity.

Synonymous to this sporting construction of 'public' space, and more transparent perhaps as to the specificity of its sexualised cultural geography, is that of the 'town' itself. In its 'sporting' definition, the 'town' indicated a unified body composed of different social strata, high and low, who shared and negotiated what constituted urban culture (Buckley 1984). In this explicitly demotic and theatrical symbolic geography, male sexual prerogatives were asserted and celebrated. Thus, 'men about town' mixed with and sought out women 'on the town', 'public women' whose dubious reputation derived from their very presence in urban public space. While it promoted an attitude that allowed men and women to mix in heterosocial public space, and the respectable to mix with the unrespectable, 'sporting' culture did so on terms set by the larger gender culture of which it was a part (see Nord 1995; Ryan 1990; Stansell 1986). If it valorised and promoted an urban public geography marked by sexual diversity, it nevertheless insisted on the prerogatives set by what it accepted as innate male sexual urges. Such attitudes centred on male prerogatives; and in plain words, at its margins were violence, rape and sexual murder. Even regarded in its best light, sporting culture was exclusive, aggressive and predatory. It is impossible for us therefore to read the 'resistance' to the closure of Cremorne as any kind of unproblematically pro-sensual advocacy. As a public space, the mixed nature of Cremorne was certainly welcome. Whereas the middle-class readings of working-class behaviour tended to regard participation in mixed public spaces like Cremorne as evidence in itself of immorality, both classic moralist attitudes and urban 'sporting' culture allowed working people to mix with the morally lax without stigma. What little evidence we have about women's attitudes to this kind of public space suggests that many ordinary women took great pleasure in leisure spaces like Cremorne, which were part of 'a world of desire and pleasure where, for a time at least, no attempt was made to count the cost' (Barret-Ducrocq 1991: 83); women were certainly often victims, but they took pleasure in the city, knew and accepted its moral codes, and valued such public spaces as Cremorne. Nevertheless, women's sexuality was still regarded by most moralists as dangerously alluring, and the figure of the prostitute was still its dominant representation. More generally, sexuality was still coded in spatial terms, and a conservative, restrictive, sexually moralistic attitude still largely prevailed. It is as important for us to recognise the danger of over-reading evidence of resistance to middle-class sexual moralism, and of valorising it un-problematically (Mason 1994a: 167–168), as it is to acknowledge the nature and strength of that resistance. Indeed, we need to recognise the complex admixture and entanglement of sexual cultures and their geographies.

Conclusion

I hope to have indicated through a historical geography of the closure of Cremorne Gardens how limiting is the dualism of domination and resistance. While we need to recognise the ability of Cremorne's supporters to mobilise considerable practical and ideological support for the Gardens in order to resist the middle-class reformation of public space, we need to be wary of coding this as a unitary, unproblematic resistance to the dominant forms of sexual moralism. For Cremorne's support was plural in both practical and ideological ways. It is impossible to posit a straightforward rejection of middle-class sexual moralism by these supporters; rather, their actions and arguments drew on several strands of counterargument which ranged from claims to respectability endorsing such sexual moralism to a worldly tolerance that satirised it. Characteristically, in fact, Cremorne's supporters seem to have mixed such attitudes, in a manner that is analogous to the ways in which in its public space commingled the respectable and unrespectable. Cremorne's supporters could be genuinely sincere about the Gardens' respectability as well as cynically disingenuous about the presence of prostitutes and sexually licentious behaviour, without contradiction or hypocrisy. In this, I have noted, such a multifaceted defence reflects the attitudes and prescriptions of classic moralism and the sexual culture it endorsed. We need in place of problematic valorisations like domination and resistance, to connect these sexual cultures with specific cultural geographies. While high moralists feared the eroticisation of the public sphere, with its encouragements to egoism and the loss of self-control, and regarded public sexual expression as a mark of social breakdown, alternative sexual cultures clearly regarded urban space as a theatre of male sexual opportunity, revelling in its very 'public' nature. The concept of the 'town', with its prescriptions of gender and sexuality integral to its valorisation of urban life, may stand for this alternative construction of public space, and the defence of Cremorne is one example of a moral urban geography constructed along lines antithetical to those of bourgeois moral reformers. But as such Cremorne represents a culturally overdetermined space, and we should rightly hesitate to endorse its defence with the normative significance of 'resistance'.

Indeed, we might conclude that narratives and topographies of repression and resistance conspire against recognising the complexity and significance of such sexual geographies.

Notes

1 I have treated 'Victorian sexuality' here in an Anglo-American sense, with the United States as both the most useful comparison and as a society which shared many aspects of nineteenth-century British sexual culture; this will explain the prevalence of American references in this chapter.

2 Mason's dual volumes on Victorian sexual behaviour (1994a) and sexual attitudes (1994b) are brilliantly detailed, for instance, and the best accounts of their kind. Yet the complex synchronic and diachronic patterns of his narrative are difficult to reduce to simple formulae; and he largely abandons the attempt to contextualise the changing meanings of sexuality within larger social, cultural and political developments. One particularly interesting result comes where he considers the viability of the notion of sexual excess and argues the need for repression: 'We should not be inhibited by inhibitions: at all times and all places constraints and prohibitions have been woven into the fabric of human sexuality' (Mason 1994a: 5). In failing to put these constraints and prohibitions – which amount to the dominance exerted by sexual cultures – properly into the broad and dynamic context of historicised systems of class, gender and race, however, it is difficult to push forward his normative concerns any further than this.

3 The ranging of 'essentialist' versus 'constructionist' views of sexuality might well be taken as a political and rhetorical rehearsing of the dominance/resistance dualism: '"Essentialism," that is, too readily figures as yet another hypostasized image of Victorian power implacably opposed to human capacity for self-definition, and to the vitality of human signification': see Miller and Adams (1996: 6).

4 References to Cremorne from the point of view of social purity and moral reform are common, but made usually in passing and without analysis. The best general account of Cremorne is still Wroth (1907). The bulk of the source material for this chapter comes from the clippings book compiled by Wroth and deposited in the British Library (1800. c. 9), from the Greater London Record Office (hereafter GLRO) and from local newspapers.

5 As confirmation of the sexual reputation of Cremorne, see *The Cremorne: A Magazine of Wit, Facetiae, Parody, Graphic Tales of Love &c.* (1882 though dated 1851), a journal of hand-coloured pornography dedicated to the 'Votaries of Venus' and the 'worshippers of Priapus'. Showing that the name Cremorne had achieved an international brand recognition for promiscuous male sexuality there was a concert saloon in New York of extremely low reputation – virtually a brothel – named the Cremorne (Gilfoyle, 1992). Even now a title search on 'Cremorne' reveals some modern pornography, purporting to be Victorian confessions, entitled *Cremorne Gardens, The Ecstasies of Cremorne* and *The Temptations of Cremorne* (all 1992).

6 For the encouragement of a moral public, note the development of public parks in the second half of the nineteenth century; indeed, when the Royal Horticultural Society's new central London garden at Kensington Gore was opened in 1861 by Prince Albert, it was described by one commentator as a 'moral Cremorne' (Elliott 1986: 142).

7 These data come from the petitions held in the GLRO, Middlesex Sessions for Music and Dancing Licenses, MR/LMD/19/2–14 and 20/1–10. These petitions date from

1846 to 1877. These data have been cross-referenced, where possible, with the censuses and rate books; the resulting dataset amounts to about one-tenth of the 5,000 or so signatories, large enough to provide a reasonable sample here, though it is likely to be biased to those individuals residentially stable enough to appear in censuses and ratebooks over several years.

8 It is not very surprising that the defence of Cremorne should take this form, for the classic moralistic attitude is exactly that of Baron Nicholson and his popular journal the *Town* (1837–1842). Notably, even in the late 1830s, the *Town* took issue with threatening sabbatarians aiming to close public amusements. Magistrates were accused of bowing to hypocrites, sanctimonious sinners, the lowest of London's prigs.

9 There are important overlaps here with the argument of Bailey (1994), concerning the 'knowingness' of Victorian popular culture.

References

Acton, W. (1972) *Prostitution, Considered in its Moral, Social, and Sanitary Aspects, in London and Other Large Cities and Garrison Towns*, London: Frank Cass.

Anderson, A. (1993) *Tainted Souls and Painted Faces: The Rhetoric of Fallenness in Victorian Culture*, Ithaca, NY: Cornell University Press.

Bailey, P. (1994) 'Conspiracies of meaning: music-hall and the knowingness of popular culture', *Past and Present* 144: 138–170.

Barret-Ducrocq, R. (1991) *Love in the Time of Victoria*, London: Verso.

Bland, L. (1995) *Banishing the Beast: English Feminism and Sexual Morality 1885–1914*, Harmondsworth: Penguin.

Boyer, P. (1978) *Urban Masses and Moral Order in America, 1820–1920*, Cambridge, MA: Harvard University Press.

Buckley, P. G. (1984) 'To the Opera House: culture and society in New York City, 1820–1860', PhD dissertation, State University of New York at Stonybrook.

Chesney, K. (1972) *The Victorian Underworld*, Harmondsworth: Penguin.

Cohen, P. C. (1992) 'Unregulated youth: masculinity and murder in the 1830s city', *Radical History Review* 52: 33–52.

Cominos, P. T. (1963) 'Late-Victorian respectability and the social system', *International Journal of Social History* 8: 18–48.

Cominos, P. T. (1972) 'Innocent femina sensualis in unconscious conflict', in M. Vicinus (ed.) *Suffer and Be Still: Women in the Victorian Age*, Bloomington: Indiana University Press.

Cott, N. (1978) 'Passionless: an interpretation of Victorian sexual ideology, 1790–1850', *Signs* 4: 219–236.

The Cremorne: A Magazine of Wit, Facetiae, Parody, Graphic Tales of Love &c. (1882 though dated 1851).

Elliott, B. (1986) *Victorian Gardens*, London: B. T. Batsford.

Finnegan, F. (1979) *Poverty and Prostitution: A Study of Victorian Prostitutes in York*, Cambridge: Cambridge University Press.

Foucault, M. (1980) *The History of Sexuality, Volume I: An Introduction*, New York: Vintage.

Gay, P. (1984) *The Bourgeois Experience: Victoria to Freud, Volume I: Education of the Senses*, Oxford: Oxford University Press.

Gay, P. (1986) *The Bourgeois Experience: Victoria to Freud, Vol. II: The Tender Passion*, Oxford: Oxford University Press.

Gilfoyle, T. J. (1991) 'Policing of sexuality', in W. R. Taylor (ed.) *Inventing Times Square: Commerce and Culture at the Crossroads of the World, 1880–1939*, New York: Russell Sage Foundation.

Gilfoyle, T. J. (1992) *City of Eros: New York City, Prostitution and the Commercialization of Sex, 1790–1920*, New York: Norton.

Gorn, E. J. (1987) '"Good-bye boys, I die a true American": homicide, nativism, and working-class culture in antebellum New York City', *Journal of American History* 74: 388–410.

Harrison, B. (1967) 'Underneath the Victorians', *Victorian Studies* 10: 239–262

Harrison, F. (1977) *The Dark Angel: Aspects of Victorian Sexuality*, London: Sheldon Press.

Hill, M. W. (1993) *Their Sisters' Keepers: Prostitution in New York City, 1830–1870*, Berkeley: University of California Press.

Hobson, B. M. (1987) *Uneasy Virtue: The Politics of Prostitution and the American Reform Tradition*, New York: Basic Books.

Mahood, L. (1990) *The Magdalenes: Prostitution in the Nineteenth Century*, London: Routledge.

Marcus, S. (1966) *The Other Victorians: A Study of Sexuality and Pornography in Mid-Nineteenth-Century England*, London: Weidenfeld and Nicolson.

Mason, M. (1994a) *The Making of Victorian Sexuality*, Oxford: Oxford University Press.

Mason, M. (1994b) *The Making of Victorian Sexual Attitudes*, Oxford: Oxford University Press.

Miller, A. H. and Adams, J. E. (eds) (1996) *Sexualities in Victorian Britain*, Bloomington: Indiana University Press.

Mort, F. (1987) *Dangerous Sexualities: Medico-Moral Politics in England since 1830*, London: Routledge.

Mumford, K. J. (1997) *Interzones: Black/White Sex Districts in Chicago and New York in the Early Twentieth Century*, New York: Columbia University Press.

Nead, L. (1988) *Myths of Feminism: Representations of Women in Victorian Britain*, Oxford: Blackwell.

Nord, D. E. (1995) *Walking the City Streets: Women, Representation, and the City*, Ithaca, NY: Cornell University Press.

Pearl, C. (1980) *The Girl with the Swansdown Seat: An Informal Report on Some Aspects of Mid-Victorian Morality*, London: Robin Clark.

Pearsall, R. (1971) *The Worm in the Bud: The World of Victorian Sexuality*, Harmondsworth: Penguin.

Peckham, M. (1975) 'Victorian counterculture', *Victorian Studies* 18: 257–276.

Pivar, D. J. (1973) *Purity Crusade: Sexual Morality and Social Control, 1868–1900*, Westport, CT: Greenwood.

Porter, R. and Hall, L. (1995) *The Facts of Life: the Creation of Sexual Knowledge in Britain 1650–1950*, New Haven, CT: Yale University Press.

Ritchie, J. E. (1858) *The Night Side of London*, 2nd edn, London: William Tweedie.

Ryan, M. P. (1990) *Women in Public: Between Banners and Ballots, 1825–1880*, Baltimore, MD: Johns Hopkins University Press.

Seidman, S. (1990) 'The power of desire and the danger of pleasure: Victorian sexuality reconsidered', *Journal of Social History* 24: 47–68.

Srebnick, A. G. (1995) *The Mysterious Death of Mary Rogers: Sex and Culture in Nineteenth-Century New York*, New York: Oxford University Press.

Stansell, C. (1986) *City of Women: Sex and Class in New York, 1789–1860*, New York: Alfred A. Knopf.

Stearns, C. Z. and Stearns, P. N. (1985) 'Victorian sexuality: can historians do it better?', *Journal of Social History* 18: 625–634.

Thomas, K. (1959) 'The double standard', *Journal of the History of Ideas* 20: 195–216.

Vicinus, M. (ed.) (1972) *Suffer and Be Still: Women in the Victorian Age*, Bloomington: Indiana University Press

Walkowitz, J. R. (1980a) *Prostitution and Victorian Society: Women, Class, and the State*, Cambridge: Cambridge University Press.

Walkowitz, J. R. (1980b) 'The politics of prostitution', in C. R. Stimpson and E. S. Person (eds) *Women: Sex and Sexuality*, Chicago: University of Chicago Press.

Wood, N. (1982) 'Prostitution and feminism in nineteenth-century Britain', *M/F: A Feminist Journal* 7: 61–77.

Wroth, W. (1907) *Cremorne and the Later London Gardens*, London: Elliot Stock.

POWER AS FRIENDSHIP

Spatiality, femininity and 'noisy' surveillance

Jennifer Robinson

Women housing managers who followed in the traditions of the Victorian reformer, Octavia Hill, had as their guiding principle the idea of forming friendships with the tenants. They were inspired by the possibility that as women they would find it easy to establish such friendships with women tenants and as rent collectors they would have frequent opportunities for interaction. Through these friendly interactions they hoped to improve the tenants and at the same time to enhance the management of housing for the poor. Their practices, although closely related to many different techniques of surveillance and government which have been incorporated into the modern state, bring into view some relatively unexplored aspects of the exercise of power.[1] I suggest that the experiences of these women, as well as their emphasis on femininity, encourage us to consider the extent to which descriptions of power may have relied upon a rather masculine vocabulary, restricting our understanding of how power operates. More substantially, the idea of power as friendship draws our attention to the spatialities implicit in accounts of power relations. This is most apparent when we think specifically about how surveillance operates.

Surveillance is understood to be a key technique in the exercise of power, and was a central aspect of the activities of the women housing managers I am interested in. The women walked from one house to the next, talking to the tenants, collecting rents, observing the exterior and interior of the houses, as well as activities going on and the people inside and outside the house. Comparing this with the simplified version of surveillance captured in the idea of the Panopticon, where a subject internalises an ever-present but invisible gaze enabled through a clever architectural design, I am struck by how 'noisy' this form of surveillance is. Not only are there speaking subjects and moving bodies involved (rather than the invisible gaze and the silence of separated bodies implied by the Panopticon) but also, as we consider how surveillance operates to enable the transformation of subjects, we realise that the surveillant gaze experiences 'interference' in a whole variety of different ways. The originator of a particular surveillant gaze brings her own interpretation to the principles of observation involved, and is always located in particular contexts which will shape the content of the gaze. And the multiple mis-recognitions which accompany the journey of the gaze from the gazer to the subject and back make the content and internalisation of power far from predetermined – as the Panoptical tale seems to imply.

The visual regime of surveillance which is assumed to underpin the exercise of power also carries with it an implied spatiality of power. Compared to the distanced and unidirectional view of power relations which the visual regime of the Panopticon suggests, the experiences and techniques of management of the women housing managers who I shall be discussing bring into view a very different spatiality to the exercise of power. The idea that friendship could form the basis for management and surveillance emphasises mutuality and an active role for the subjects of power. Power, then, need not be thought of as simply objectifying, intrusive and imposing. In many circumstances modern tactics of power operate by enticing subjects to participate in forms of self-surveillance. Subjects are drawn in to certain forms of behaviour and thinking because their own concerns or interests are met. The interiorisation of power does not happen in a straightforward manner, though, and the transformative effects of surveillance do not happen entirely on the terms of the powerful. Power can be understood as a mutual, although rarely equal, relationship (rather than simply a technology) in which active subjects (both the 'dominated' and 'dominator') participate. Thinking about the spatiality and masculinity of our conceptions of power in the light of the Octavia Hill managers' experiences will highlight the ways in which power already incorporates elements of the dominated subject's concerns. A substantial conclusion of this chapter, then, will be to suggest that power is always contaminated with resistance.

I shall first describe the form of management and surveillance involved in the Octavia Hill tradition of housing management, especially the idea that friendship could form the basis for management and surveillance. The second section addresses the visual regime of surveillance which enables the exercise of certain forms of disciplinary power. Refiguring the surveillant gaze as unsuccessful in its objectifying and transformative effects, I argue that it is precisely in the resultant and necessary failures of objectifying surveillance that the successes of the exercise of power, as well as the possibilities for resistance, reside.

Octavia Hill Women Housing Managers: friendship in housing management

Torn between the intrusiveness of inspection and the need to manage property and people efficiently, a group of women housing managers in the late nineteenth and early twentieth centuries promoted a system of management based on 'friendship' and the possibilities which the common domestic concerns of women managers and women tenants presented for establishing relationships of trust and confidence. The work of the Octavia Hill Women Housing Managers (OHWHMs) spanned the first half of the twentieth century and came in the wake of Octavia Hill's nineteenth-century initiatives in the field of housing management in London. These were part of a much wider movement of women involved in charitable work during the late nineteenth century. Women engaged in philanthropy at this time were often drawn by Christian motivations and were strongly influenced by contemporary assessments of appropriate activities for middle-class women. Generally restricted to the domestic sphere, evangelical Christianity motivated women to venture into the public realm, but usually in a manner thought proper to their particular skills and sensibilities. Caring and nurturing, spreading love and domestic skills all figured large in the various visiting activities with which women's involvement in charity came to be associated. Visiting societies proliferated through the century and more or less systematically undertook to visit the poor in their homes (as well as in the streets, workhouses, prisons, drinking dens and brothels) to dispense, in various proportions, the gospel, Christian charity, domestic training and moral reform (Prochaska 1980).

Among these women reformers, Octavia Hill was a prominent figure. The activities of such women played an important part in influencing the authorities' interventions in relation to the poor – despite their confinement in the process to roles that seemed to reinforce contemporary notions of domestic femininity. It has been argued that this explosion of female philanthropy contributed in the long term to the transformation of women's position in society and even fed into the emergence of the feminist movement (Rendall 1985). However, many individual women continued to promote a domestic(ated) femininity despite their own increasingly public roles, resolving the contradictions which this entailed in a variety of different ways (Lewis 1991). For Octavia Hill, her growing involvement in housing management, district visiting, general charity work and in advising government was accompanied by a reticence in committee work and an assertion of the priority of the home and family in women's lives.

Like many other Victorian reformers in the 1870s (Stedman-Jones 1984), Octavia Hill felt that the presence of the rich among the poor was itself a positive force for good. The growing geographical segregation of the poor, though, was separating them from the good influence of the wealthier classes. As she noted in one of her lectures:

We live upon the labours of the poor in districts far from our homes. Our fathers and brothers may have chambers, factories, offices right down among them. We are content to draw our wealth from these. Does this imply no duty? Is the whole duty fulfilled when the head of a firm draws a cheque for donations to the local charity, and are the gentle ministrations of the ladies of the family to be confined to the few pampered poor near their house? It is our withdrawal from the less pleasant neighbourhood to build for ourselves substantial villas with pleasant gardens, which has left these tracts what they are . . . Is not our very presence a help to them? I have known courts nearly purified from very gross forms of evil merely by the constant presence of those who abhorred them . . . Mere intercourse between rich and poor, if we can secure it without corrupting gifts, would civilise the poor more than anything. See then that you do not put your lives so far from those great companies of the poor which stretch for acres in the south and east of London, that you fail to hear each other speak.

(Hill 1877: 171, 97–98)

What Octavia Hill thought of as the 'corrupting' gifts of the innocent district visitor were to be replaced in her system with a careful, observant relationship between the Charity Organisation Society and groups of visitors whereby, in her words, a 'continuous watchfulness' aided by the intimate observation of face-to-face contact in the homes of the poor would enable each individual case to be assessed and appropriate direction given to those seeking to help them.

While she claimed never to give alms to any person (although she seems to have made many a loan), Octavia Hill based her interactions with the poor upon the hopes of building friendships and providing those things which no person could be expected necessarily to provide for themselves – decorations, country trips, music, flowers – as well as assistance for those seeking employment or training. The former, she suggested, would bring joy to the poor (Hill 1877: 96), the latter independence. Perhaps her early experiences of charity work among young children when her own family had encountered hard times set a pattern for her relationships with the poor in later life. The dependencies induced by gifts would, she felt, mean that friendship with the poor was impossible. She suggested that very often inexperienced visitors did not 'think of them as *people*, but as *poor* people' (Hill 1877: 49, original emphasis). The work of assisting the poor, she argued, required the face-to-face and intimate contact which the district visiting system had constructed – and the mutual friendship which a refusal to engage in indiscriminate alms giving would enable. For, she noted, the poor also have much to teach 'us': lessons 'of patience, vigour, and content, which are of great value' (Hill 1877: 60).

Her experiments in managing housing drew directly on the traditions of women visitors – as a later commentator noted, 'There may still be a few of the comparatively small number of persons who know of her work who look upon it as an attempt to insinuate a District Visitor under the disguise of a rent collector' (Gibbon, in Hill 1921: 6). However, adopting the position of landlady was the basis for her attempts at establishing a mutual, respectful relationship with the poor, rather than one which rested upon the guilt of the visitor and the dependencies of those being visited. Undeniably, the relationships which she, her fellow workers, and those who followed in her traditions described were part of a system of inspection and moral regulation. Their work was part of the elaboration of the technologies of surveillance which sought to create responsible, self-disciplining citizens.

The site of moral inspection here was the home – the technology and profession which she has been credited with founding, that of housing management (and more indirectly, social work). However, the techniques of surveillance and management deployed by these women emphasise aspects of surveillance which are frequently overlooked. Their procedures were grounded in principles of friendship and mutuality: rent collectors and managers were to concentrate on building friend-ships with the poor, and dispensing 'joy' rather than charity, rendering themselves able, at times of crisis, to be of assistance in a substantial and self-sustaining way. She recognised that any visitors were likely to meet with dismissal from at least some of those they tried to befriend. Weekly rent collection provided a basis for visiting and potentially influencing all tenants – even those who were resistant to outside intervention. It nonetheless remained an important principle of the women rent collectors that they entered premises only at the express will of the tenant. A neighbourly respect for rights of tenants in their own home was to be maintained, and this respect was reinforced by the mutuality of the landlady–tenant relationship. As she asked in one of her letters to her 'fellow-workers':

> Why did you not take up district-visiting, where, if any family did not welcome you, you could just stay away? Because you preferred a work where duty was continuous and distinct and where it was mutual . . . it implies your determination not simply to do kindnesses with liberal hand, popular as that would be, but to meet the poor on grounds where they too have duties to you.
>
> (Hill [1879] 1921: 68–69)

This relationship might have been exaggerated at times by Hill and her followers, but it reminds us that relations of power are constituted in situations of (uneven) mutuality. Here the everday resistances of tenants in the face of poor management (exemplified, for example, in refusals to maintain property, tardiness with rent payments, lack of concern for public areas) were accommodated in the reliable fulfilment of the responsibilities and duties of the landlord. The right of entry to homes for inspection and for efforts at moral instruction was negotiated, and from

Hill's accounts, seldom imposed. The spatiality of inspection, then, cannot be imagined as one of unidirectional or panoptic-like surveillance. The practice of negotiation and mutual responsibility highlights the agency and power of those who were being inspected, as well as the difficulties associated with achieving the inspector's gaze. For many of the young women, the engagements with poor tenants were stressful and difficult. Ensuring the moral and social reform of the poor tenant and the slum-dweller demanded a subtle and negotiated access to their world – and the techniques for securing that access and visibility had to be acquired during a period of training. This is not to suggest that the class-based moral authority of lady rent collectors, or their possession of coercive sanctions such as the right to eviction, should be ignored. But a distinctive set of ideas about management had emerged from the particular historical context of Octavia Hill's initiatives, based upon her philanthropic practices and encompassing mutual responsibility and friendship.

Octavia Hill's system of housing management, then, rested upon the idea that women rent collectors could become friends with women tenants, and develop relations of trust which would enable tenants to be convinced of the values of both moral and domestic improvements. This central relationship between the manager and the tenant was conceived of as a form of negotiation, in two ways. The first was that the housing manager was seen to be in a position of negotiation between the landlord and the tenant. She was to 'recognise and give effect to the principle of the mutual responsibility of the landlord–tenant relationship' (South African Society of Women Housing Managers (SASWHM) 1959) – to ensure, directly following Octavia Hill, that landlords met the requirements of maintenance and service provision, in return for tenants' prompt payment of rent and clean and orderly behaviour. The *quid pro quo* involved here suggests a healthy appreciation on the part of Octavia Hill of the power of the tenant. Tenants were of course able to damage property, make it filthy and disturb neighbours, refuse to pay rents, respond angrily and agressively to landlords' representatives, and generally confound or subvert the efforts of these lady visitors to maintain the property or to improve themselves. It is possible to read the evolution of this system of responsible management as an outcome of the power of tenants, incorporating their resistances into the form of supervision.

The mutuality of the relationship is also apparent in the rules followed by the women for inspecting properties. The visit to collect rent was the basis for the interaction between the manager and the tenant. The manager was expected not only to inspect the property for needed repairs and to ensure that it was being properly maintained, but also to inspect the tenants in terms of their financial, social, domestic and moral circumstances. Access to property was always to be secured on the basis of the tenant's permission, and never to be forced. The women managers' techniques for doing this were dependent upon a negotiated access to the property and to the confidences of the women involved. Octavia Hill's nineteenth-century project of attempting to improve the morals of the labouring classes was writ large in the much more substantial twentieth-century investments in public housing and related efforts to rehabilitate slum-dwellers and to turn unsatisfactory tenants into responsible citizens; these technologies of negotiated surveillance persisted for some time (see Brion 1995 for a history of the Octavia Hill Women Housing Managers in Britain). An image of surveillance as intrusive, objectifying and imposed is not appropriate in this example – and perhaps in general.

Negotiated power and apartheid urban government

These ideas continued to influence housing managers long after Octavia Hill's death, and extended far beyond their origins in nineteenth-century London. My initial contact with Octavia Hill Women Housing Managers took place in the archives of the Port Elizabeth Municipality in South Africa. In the late 1930s a group of women councillors encouraged the appointment of a woman manager from England to oversee large sub-economic public housing developments for Coloured and white residents. It has taken me some time to find out who Octavia Hill was, and to trace something of the history I have written here. Women trained as Octavia Hill Women Housing Managers travelled around the world and worked in various capacities in countries as diverse as the USA, Hong Kong, Tanzania, Australia and South Africa. In South Africa, the Cape Town Municipality employed the first Octavia Hill Women Housing Manager in 1934 on the advice of the National Council of Women of South Africa and the particular enthusiasm of Mrs Muriel Jones. The South African government then sponsored a scheme to train women housing managers. About seventy or so women were trained and then employed in larger urban areas around the country through to the 1960s and beyond (for more detail see Robinson 1998). In some ways, implementing the Octavia Hill system in South Africa represented a test for the idea of managing through friendship. Objectors to the employment of Octavia Hill managers in South Africa pointed out some of the reasons why it might be expected to fail, especially the vast social, cultural and linguistic gulf which they saw existing between middle-class, white, British managers and poor, predominantly Coloured, South African tenants.[2] It could also be said that the South African example represents a test case for my claims that the exercise of power involved negotiation with and the active particpation of subjects. Imagining that racial domination and apartheid operated through forms of power other than cynical authoritarianism might seem inappropriate. However, the organisation of urban space was central to the maintenance of racial domination in South Africa (Robinson 1996), and the ways in which public housing was designed and managed played an important role in constituting and shaping racially differentiated subjects/citizens. That local government in South Africa drew at least to some extent on a technology of power built on friendship is surprising and brings to our attention how seldom we think of power relations as involving negotiation, trust and mutual interactions.[3]

The implementation of this system of management in South African cities therefore highlights a number of different issues about the exercise of power, especially the translation of the model of power as friendship into a politically charged, raced and culturally differentiated environment. Although there was a great deal of concern about the problem of language and of cultural differences between the managers, especially those who came from the UK, and their tenants, the records consulted give little indication that any difficulties of the sort were experienced by the managers. In their public correspondence, the women continued to assert that the Octavia Hill model of mutuality and friendship enabled them to solve difficulties with tenants, and to help them to become better citizens and to access the assistance they required in times of hardship. It was assumed that this set of management techniques, based on an emotional relationship of some kind between the manager and the tenant, was able to overcome the cultural and linguistic distances encountered in some of the areas they were administering. For the managers, then, it would seem that this way of thinking of their relations with the tenants captured something of their experiences. The difficulties of managing tenants on the Octavia Hill model across divides of cultural and linguistic difference were raised frequently when the appointment of Octavia Hill Women Housing Managers was discussed by councillors and officials. The women managers were (until 1956 when Coloured women were trained) white, middle-class women; only a few came from Afrikaans-speaking backgrounds (in Cape Town most of the tenants were Afrikaans-speaking). In the formal documentation and in discussions with some of the women housing managers they had no hesitation in describing their activities as conforming to the standard model of 'friendship' together with detailed knowledge of tenants and their situation. Most important to the commentators was the need for an awareness of the particularities of the South African situation – something they felt the British women would not have. In Cape Town, most of the tenants managed by the Octavia Hill Women Housing Managers were Coloured, whereas in Pretoria and Johannesburg, most of the work was done among white (Afrikaans) tenants. Some discussions were evidently held concerning the extension of the scheme to African areas: clearly it was assumed that the technical expertise of the managers including the 'visit' to tenants and the application interview, as well as numerous aspects of accounting, public health and property maintenance, would enable them to implement their methods anywhere.

The first Octavia Hill manager appointed in South Africa, Margaret Hurst, responded to concerns as to how the close relationships between manager and tenant could be forged across linguistic, racial and cultural boundaries, by suggesting that a

> trained manager, under contract to learn Afrikaans and with bilingual assistants, is the nearest expedient. Conditions in South Africa are different from conditions overseas, but England in 1937 is also very different from England in 1860 when Miss Octavia Hill began her work. I have seen a London-trained Manager do admirable work on a Liverpool Estate, where tenants were independent Lancashire men, seamen of all sorts, low class Irish, Polish Jews, Chinese etc.[4]

Margaret Hurst was insistent that the work of Octavia Hill managers was 'adaptable to very different circumstances e.g. London in 1860 and Capetown in 1936.'[5] Indeed, in Octavia Hill's Victorian London, it is arguable that the cultural divides between rich and poor were possibly as significant (and frequently as racialised) as those across which South African managers were operating.

On her arrival in Cape Town in 1934, as a result of the lobbying of the Cape Town City Council by the National Council of Women of South Africa, Margaret Hurst was placed in charge of a variety of council housing schemes, 950 tenancies in all, of which 900 were for 'non-Europeans'.[6] As a response to various aspects of urban disorder (physical, social, political), the South African state encouraged and subsidised often quite vast sub-economic council housing schemes over time. Weekly inspections were held of most properties except the ones let to poor Europeans where monthly inspections were considered adequate. As in Octavia Hill schemes elsewhere, the basic technique of management was this individual visit to tenants: 'The individual visits to tenants result in a personal and friendly relationship . . . mutual responsibility is emphasised . . . cleanliness and orderliness are insisted upon. The visiting also gives an opportunity for noticing sub-letting, which is then dealt with without delay.'[7] However, the traditional Octavia Hill link between rent collection and the visit was not possible in Cape Town until after the Second World War. The use of rent collection as the point of interaction between the managers and the tenants was replaced by weekly visits of inspection. There was already a pre-existing system of rent collection by caretakers under the auspices of the City Treasurer, and because 'of conditions amongst coloured tenants' some areas were considered too dangerous for the women to undertake rent collection.[8] The visit, then, was for some time much more overtly a routine inspection than that imagined by Octavia Hill, where the inspection side of the visit was partly hidden by the function of rent collection. The Octavia Hill Women Housing Manager, though, was involved in dealing with those in arrears and attempting to provide access to relief agencies or alternative accommodation. Her efforts at referral to agencies (such as social work, health and charity) at times required 'much persuasion'; nonetheless she noted that 'every encouragement is given to the tenants to look on the supervisor and her staff as friends'. Here 'the inquisitive, interfering and patronising attitude of the "investigator" is entirely avoided.'[9]

Despite the absence of rent collection as a cover for inspection, it was felt that 'during her weekly visits a Manager acquires a knowledge of the character and failings of each tenant which is of great value in maintaining peace and harmony on the estates.'[10] Initially, and for most of the housing offices around the country, maintaining these visits was made difficult by the shortage of staff. Margaret Hurst (and managers who followed her) suggested that her obligation to train student managers would assist in meeting the demands for house-to-house visiting: 'I cannot establish the personal contact with tenants, which is the basis of the Octavia Hill system, when I have over 1000 families to manage and a good deal of general administrative work.'[11] She was soon to be assisted by another trained British woman, and by local students she was training herself.

Aside from these problems of understaffing, one might expect that in the context of racial discrimination in which the managers worked, stereotypes of the communities would undermine their ambitions of friendly management. The Octavia Hill Women Housing Managers practised their techniques in the face of such discrimination, and a prominent Octavia Hill manager felt the need to comment that:

> It is important that the Housing Manager recognise the dignity of every human being, no matter how wretched; and in our country with its multi-racial problems, these values are perhaps more apposite than anywhere else in the world.
>
> (SASWHM 1959)

Nonetheless, they also proclaimed that:

> The ultimate goal of trained management is to train the tenants of sub-economic housing schemes in citizenship. We feel that particularly at this juncture, with the Government pursuing a policy of apartheid, the need for fully trained housing managers is urgent.[12]

Admittedly this latter claim was made in an effort to encourage the incoming (and hardly anglophile) Apartheid government to support their training programme, but the linkages between the management of housing and the implementation of racial segregation were significant. The slum-dwellers who they were eager to train were to become citizens of a very particular sort – differentiated by race – and were expected to live up to the standards of the public housing provided: racially differentiated standards. The practice of friendly management was thus strongly racially inflected. However, as I shall argue in the following sections, the stereo-typing (or objectification) involved in both this racialised manoeuvre and in the wider project of observing and categorising tenants (subjects) is unstable. The success of the project of improving tenants depended, I suggest, on exactly the humanising interactions which makes the Octavia Hill system of management seem so very out of place in the South African context.

The next section investigates these interactions, and in so doing interrogates the exercise of surveillance, contrasting its description in the ideal (silent) terms of the imagined Panopticon, with its 'noisy' form, as practised by the Octavia Hill Women Housing Managers. The conclusion considers the implications of this form of management for our interpretations of power relations in general, and for the exercise of surveillance in particular. As we have already noted, the relationships and strategies involved in this system of housing management suggest a spatiality to power relations which emphasises processes of negotiation and mutuality, rather than imposition and undirectionality (see also Stenson 1993 on social workers). This results from focusing on the *relations*, rather than on the *technologies* of power. In extending this argument through an assessment of the dynamics of surveillance, I suggest that it is the moments when the surveillant gaze fails, as well as the necessary moments of self-subjectification, which (somewhat ironically) ensure its success. Both are also moments when power is least secure in its achievements, and when resistance is most able to be inscribed in the form of power. Resistance could be understood, then, as central to the exercise of power.

The spatiality of the surveillant gaze

I never knew what changed these people, but change they did.
(D. C. K. 1956)

Surveillance describes a variety of different processes, including the acquisition and accumulation of information, the observation of individuals and their activities and bodies and, where appropriate, the interiorisation of discipline. Central to all of these aspects of surveillance is the inspection, the look or the gaze. Here I want to explore the detailed dynamics of the 'look' that is at the heart of many forms of surveillant power relations. I have already suggested that the dynamics of relations between the Octavia Hill Women Housing Managers and the tenants involved mutual interations and an inscribing of resistance within the form of power. Interrogating more closely the way in which the surveillant gaze works will enable me to propose that, while this mutuality might have been characteristic of this 'friendly' form of power, it is not unique. The conclusion will suggest that most power relations take this form.

77

What I want to do here is two-fold. First, I want to set out and then disturb a basic Foucauldian account of the surveillant gaze. Second, by using the example of the Octavia Hill Women Housing Managers, I want to demonstrate how the look at the heart of surveillance is both 'noisy' and yet successful only in its necessary failures, thereby inscribing the possibilities of resistance at the heart of its operation.

Visibility is not the only technique involved in the exercise of disciplinary power, but it is an important one, and in the telling of Foucault's account of its emergence, the role of surveillance has been emphasised through his use of the Panopticon model to exemplify the operation of modern forms of power. The Panopticon, in contrast with sovereign or violent forms of power, exemplified the less costly exercise of power upon which the organisation of modern society has come to depend. As Foucault comments:

> There is no need for arms, physical violence, material constraints. Just a gaze. An inspecting gaze, a gaze which each individual under its weight will end by interiorising to the point that he is his own overseer, each individual thus exercising this surveillance over, and against, himself. A superb formula, power exercised continuously and for what turns out to be a minimal cost.
>
> (Foucault 1980: 155)

This exercise of power is principally dependent upon the effect of being (or potentially) being observed. The assumption is that the gaze of the overseer is imagined to be ever present, and internalised by the prisoner who then conforms to the demands of power. In the Panopticon, and in many other contexts, it is the organisation of space which enables this internalisation of power (see Foucault 1979: 200–204). On Foucault's account, the gaze is constituted through this physical space as always possible, and yet not necessarily dependent upon any particular embodied source. The spread of the disciplinary achievements assumed to be consequent upon these dynamics involves an expanded understanding of the gaze as perpetuated in popular opinion or public surveillance (see Burchell et al. 1991). However, it is important to consider that the various behaviours, beliefs and norms involved in this process had to be made known at some stage to those they were to influence. How did the lonely prisoner in the Panopticon know what was forbidden, know how to police himself? The content and aims of the surveillant gaze, its intentions and demands, needed to be made apparent. That the form of modern power came to involve internalised self-regulation and mutual (neighbourly, perhaps even democratic) surveillance should not hide from us that the content of what was expected had to be created and learnt somewhere and in particular circumstances. The gaze, then, has a particular history and is always located, specific and, importantly for my discussion here – and contrary to the Panopticon model – frequently embodied (see also Crossley 1993).

Transforming family sexual and domestic life, for example, involved the education and transformation of the population – at schools, churches, through popular knowledge, personal instruction or texts. The successful 'disciplining' of citizens in many societies has made much of this instruction obsolete – and partly explains why this tradition of housing management I am discussing died out.[13] Tactics for training, disciplining, educating and transforming subjects – or constituting a form of subjectivity enabling choice and self-transformation – evolved in a wide range of practical and discursive domains. Many of these came to centre on the family, children and mothers, and directly informed the strategies of the women housing managers as well as the battery of social workers and psychological professionals who more regularly attract our attention in this regard.

The internalisation of modern disciplinary power required a wide variety of situated historical processes, such as that of housing management, medical institutions, educational training, or missionary instruction. This has caused us to reflect upon the particular dynamics and different spatialities of the mechanisms involved in each of these settings – following Foucault's claim that space was crucial to the effectiveness of the Panopticon. The different spatialities of the built environments conducive to observation, calculation, instruction, healing, management and self-monitoring have attracted a lot of attention (see e.g. Mitchell 1988; Philo 1989; Driver 1990; Thomas 1990; Crush 1993; Ogborn 1993). However, the spatialities and particular visual economies of the surveillant relationships involved have slipped from view. For a variety of reasons, we have failed to attend to the spatiality of surveillance as an embodied social relationship. It is assumed that surveillance involves visibility, observation, recording and documentation, classification and distributions, discipline and instruction; all drawing upon a common (if unarticulated) understanding of a form of objectifying visuality which is commonly associated with masculinity (see e.g. Mulvey 1989; G. Rose 1995; Kern 1996; Nash 1996). We could consider, rather, that the surveillant gaze might take a different form in different contexts and that it might not be generally successful or ubiquitous. The look of inspection, I would suggest, is both more complicated (see Kern 1996) and more varied than is assumed in the Panoptical model and in much writing inspired by this Foucauldian account. Centrally, I suggest that this is the case because the gaze is never sufficient in and of itself to instigate change in a subject.

In its simplist(ic) Panoptical form, surveillance involves the gaze, coming from somewhere, or from everywhere and nowhere, focused upon and constituting a subject who, in being aware of the (possibility of the) gaze and for a variety of (unspecified) reasons, internalises its effectiveness, ensuring that now the gaze is perpetuated from within, invoking a process of self-subjectification. There are a number of reasons why this image of surveillance is problematic. First, the gaze which subjects imagine falling upon themselves is not going to be the gaze as it might be intended by overseers: the gaze itself is going to be displaced, if not fundamentally (re-)constituted by the subject. If the gaze is specifically embodied – as it frequently is – then gazing subjects will themselves be constitutive of the particularity of the look. In these circumstances the locatedness of gazing subjects, their context and history, will be shaping its content, meaning and the way in which it might be put into practice. Even where the gaze is coming from no one in particular, the local constitution of social respectability and norms will make the gaze geographically and socially specific.

In similar vein, a subject apprehended and constituted by the gaze is always likely to be elsewhere than the gaze(r) expects, and is anyway only ever apprehensible in her exteriority or her communications: ultimately her interior choices and reflections – her reasons for complying with or refusing the intentions of the gazer – must remain mysterious. While all these dynamics disturb the smooth operation of surveillance – constitute a form of 'interference' – the relationships involved in surveillant looking are already rendered complicated by the ways in which the subjects are involved, through processes of identification, in constituting one another's identity. A housing manager, interacting with a tenant, is placed through her own fantasy life in a complex relationship with the individual tenant, and with the socially meaningful categories which she associates with either tenants in general or the specific group of tenants she perceives the individual to belong to.

Many of the women involved in housing management were drawn to the profession out of concern for the poor, out of a desire to care for the less fortunate, to improve the world, to do something worthwhile with their lives. Their own internal and emotional dynamics were closely tied to the people they interacted with daily – the tenants were already an intimate part of their own sense of identity. In a racialised environment, the categorisations of tenants and the interpersonal dynamics were also mediated by the ways in which racialised selves were constituted through fantasies about racial difference. Housing managers in Cape Town, for example, were excited by the different cultural practices of Malay tenants and found satisfaction in their ability to understand and accommodate these (fantasised) differences.

Embedded in all these processes of (mis)recognition is a moment which is crucial to the exercise of power: a moment in which the gaze fails adequately to apprehend or recognise the subject (although ironically this is exactly the moment when subjecthood *per se* is acknowledged).[14] It is at this moment that the subject (object of the gaze) can be seen to shape the surveillant relationship and determine the outcome of the interaction. If we return to the Panoptical model of disciplinary power, we find that this moment is already (if unproblematically) inscribed in the operation of surveillance: the subject 'internalising' a particular mode of discipline is necessarily an active, choosing subject. What needs to change in our analysis of 'noisy' surveillance (allowing for 'interference' and displacement of the gaze) is that the outcome cannot be predetermined. Indeed, if the exercise of power is to be successful and sustainable, subjects must choose for themselves to adopt (or adapt) certain forms of behaviour or belief. And it is in allowing for, and often requiring, personal choice – based on a specifically modern understanding of the humanity (freedom) of the subject – that surveillant practices are successful precisely at this moment of (mis)recognition and fragility. It means that practices of surveillance rest upon a recognition of the capacities and independence of the subject, which implies a moment of recognising the individuality of the subject: something which might sit uneasily with our understanding of racialised practices of domination, for example. This repeats my assessment of the Octavia Hill Women Housing Managers' strategies of management as involving a mutual inter-action and a recognition of the agency and humanity of the tenants: as, for example, in Octavia Hill's distinction between '*poor* people' and 'people'. An objectifying stereotype, although frequently present in these disciplinary contexts, is an in-sufficient basis for the successful operation of surveillant power, which depends instead upon a recognition of the individual agency of the subjects of power and their ability to choose.[15] The spatiality of surveillance, then, is characterised by disturbances, discontinuities, displacements, interference – it is 'noisy' – and is necessarily a mutual process, actively involving both the gazing subject and the subject of the gaze.

The historically specific and embodied nature of the inspection involved in housing management, then, is somewhat different from the initial stylised account of the apparently anonymous and ubiquitous workings of modern disciplinary power. Rather than the gaze being everywhere and nowhere, it is embodied in young, well-educated women from respectable backgrounds walking from house to house. The women housing managers brought to the surveillant gaze a particular form of looking, constituted as a result of their class background, their reasons for entering the profession (motivations) and the training, intentions and practices of their work itself. They also inherited a group of professionally and socially constituted interpretations of the tenants with whom they interacted. As people of particular classes and racial backgrounds, who could be divided into particular categories of tenant (unsatisfactory, good, apathetic, lazy, dirty, drunk, immoral), the possibilities for the managers' looking/seeing were at least partially channelled into a limited range of subject positions. But the central place of subjectivity in the process of transformation meant that the look had to see much more than this: it had to acknowledge subjectivity as well as the objectifications involved in profes-sional management (categorisation) and class and racial stereotyping. The project of transforming these tenants into responsible citizens rested upon their successful internalisation and choice of certain types of motivated behaviour – cleanliness, orderliness, punctuality, thrift, temperance, work discipline, careful child rearing, stable family relations.

In the context of South Africa, the Octavia Hill Women Housing Managers certainly understood their tenants through a racialised lens. As one woman manager noted, writing of the experience of management in South Africa for a British audience, the 'usual hazards connected with housing' were further complicated in South Africa by the existence of racial groups. Accepting the idea that Coloured people (the bulk of tenants on Cape Town) were a 'buffer group', economically and socially, between whites and Africans, she suggested that because of their particular positions 'a special type of housing has been provided for each group'. Nonetheless, she was mindful of the 'stupid and irrational' outcomes of this system of segregation, which left poor 'non-Europeans' far from the centre of the city and in poorly serviced new areas (Cheesman 1954: 8).[16]

Despite the racialised environment in which they operated (and partially produced in their generation of 'knowledge' and stereotypes of 'these people'), the familiar tactic which these housing managers adopted of negotiated relationships, gentle persuasion, tactfulness and sympathetic listening was an essential requirement for the sustainable exercise of power, if the desire was that the tenants themselves chose to change their behaviour and values. If the tenants were to become 'independent, useful and healthy citizens', it was to be 'through regular personal contact, and the confidence established thereby', through 'the "rent lady" becom(ing) the housewife's friend'.[17] Here it is the internal shift in values that is crucial, although it is the external markers of behaviour (clean floors, healthy and well-disciplined children, industrious husbands and good homemaking skills) which are the object of the look, and the criteria for assessment. At times, exasperation caused the women housing managers to demand or coerce certain actions on pain of punishment (which, they claim, sometimes managed to effect a more permanent change in attitude), but more generally the aim was the gentle and friendly rehabilitation of unsatisfactory tenants, bad mothers and inefficient housewives. Eviction, the ultimate sanction of the Octavia Hill Women Housing Managers, was seldom invoked and, in practice, quite difficult to effect.

Institutionally the Octavia Hill Women Housing Managers played out their sympathetic role through constant battles with the City Treasurer's department. Concerned to ensure the economic viability of the council's housing schemes, the City Treasurer tried to discredit the woman managers, who attempted to ensure that the financial and social needs of the tenants were given consideration. The precarious employment position of many tenants, as well as the complicated family circumstances of some, were high on the list of things that the housing manager felt important to take into account when setting rents or deciding on access to housing. As Constance Grant argued in one memorandum:

The tenants on Council housing estates are largely unskilled labourers whose household incomes are subject to wide fluctuations. It will never be possible to put into operation a foolproof system to ensure that every tenant is paying to the Council the maximum rental of which his income allows. Household incomes vary as wages increase; a man changes work, is sick, becomes unemployed, his wife works temporarily, his children find employment, fall out of work, marry and leave home. Time and again the forms . . . will be completed and checked only to be altered to cover the changes which have occurred.[18]

That this sympathetic gaze nonetheless failed in all sorts of ways is apparent from one case in which the women managers attempted to ascertain incomes in order to specify differential rents. Having successfully supported community demands to ensure that a small fishing community was rehoused in public housing near their original settlement, Octavia Hill Women Housing Managers in Cape Town attempted to determine the income of this poor community in order to assist in setting rents which would be affordable to families to be relocated into the new public housing development. The managers relied on their sympathetic knowledge of the community, employment conditions and particular family circumstances in their assessments of a tenant's ability to pay. In the end, though, they had to be quite open about the likelihood that the information they were being supplied with by the family was incomplete. Their position as negotiators between the local bureaucracy (primarily the Treasury department, keen to maximise returns on rentals) and the interests of the individual tenants was exemplified by the long battles that they had to protect poorer tenants from having rents raised to economic levels on the basis of the acknowledgement of this widespread duplicity.[19]

The housing manager may have symbolised the gaze, may have been the one looking, but her look was not entirely in her control, as the project of training tenants entailed an engagement with a subject who might have entered into the tenancy (perhaps as a result of a (forced) removal from a previous home) and even into the project of self- and home-rehabilitation with enthusiasm, but never without effect on its outcome and often with reasons for complying which were unfathomable to the manager. In one account of experiences with unsatisfactory tenants (in an inhospitable ex-army camp environment in post-war Britain), a manager reflects upon the strategies that she mobilised to encourage the tenants to be rehabilitated:

Practical help followed by friendship; constant nagging about rent payment and plain speaking about standards . . . and I could quote many cases where it has in time led to a friendship with tenants; to instil into them the desire to have a nice home for their children, particularly when they reach the courting age; by allocating improved accommodation occupants were sometimes roused to make improvements in their ways . . . I am tempted to go on yarning about some of the other families, but in these cases I cannot put my finger on the reason for the improvement . . . I never knew what changed these people, but change they did.

(D. C. K. 1956)

Aside from this dynamic of mis-recognition, it was also in a very real sense that the look of the housing managers was a 'noisy' look; it was usually associated with speaking and listening, with perhaps trivial conversation and sympathetic advice as well as with admonitions and instructions. The housing manager walked from house to house, frequently covering long distances, climbing staircases in flats, stopping to talk to groups of people gathered in the street, stopping for longer in some houses as problems were discussed. As located and embodied individuals, the manager and the tenant were involved in the mutual elaboration of a gaze born out of a complex mixture of the context in which they met, their personal relationship, class ambitions and practices, racialised constructions of self-expectations, professionalised standards around tenancy and housekeeping and socially derived racial stereotypes as to what was appropriate and expected for different race groups, such as house size, decorations, children's ambitions and proper rearing and education, husband's behaviour, economic and income expectations.[20]

We could speculate that the subject's desire to see herself being seen or recognised as a good housewife or as a good tenant was constituted through the interaction with the housing manager (with a suitable personality and an orientation to helping the poor) who personally represented the making visible of the family, home and habits of the tenant to a wider judgemental world. That none of the housing managers recall much recalcitrance, aggression, refusal or resistance to their initiatives suggests that for the vast majority of tenants, N. Rose's (1989: 10) account of the successes of expert knowledge and advice ring true: 'It achieves its effects not through the threat of violence or constraint, but by way of the persuasion inherent in its truths, the anxieties stimulated by its norms, and the attraction exercised by its images of life and self it offers to us.' So for reasons which cannot be reduced to a simple internalisation of the gaze, or to the punitive consequences of failing to do so, the subject of the gaze seeks (self-)recognition and inscribes her own concerns in her constitution as a certain type of subject: housewife, tenant, mother, housekeeper, homeowner – or perhaps in subject positions more easily recognisable as resistant, such as breadwinner or household head, rather than housewife, and beer brewer or shebeen owner, rather than obedient tenant. Certainly there were tenants who elected to refuse attempts to 'improve' their lives, whose response to efforts to assist with recovery from illnesses, addictions or conflictual family dynamics – again for reasons which cannot be known – did not meet with favourable reactions from the Octavia Hill Women Housing Managers.[21] But 'resistance', I would suggest, was not absent from the moments of apparent conformity. This does call into question a dichotomy between power and resistance – a question I return to in the final section.

It is my argument that in the spaces of mis-recognition created in the process of surveillant looking, the subject may (or may not) be changed: but not necessarily in predictable ways, and not in any sense that is fully determined by the initiator(s) of the gaze. Even further, a case can be made that if power is to be sustainable, it is precisely in the necessity of the subject's choosing, the subject's freedom, that this friendly surveillance both potentially fails and thereby succeeds.

Conclusion: power, freedom and subjectification

If there are relations of power throughout every social field it is because there is freedom everywhere.

(Foucault 1994: 12)

The friendship model of management developed by the Octavia Hill Women Housing Managers brings into view a model of power that emphasises friendship, consisting of mutuality, negotiation and a respectfulness for the subject, which seems extraordinary within a context of racial domination, and which is rather unusual in descriptions of power relations in general. However, there are a number of theoretical contexts where assessments of power relations have pointed to the significance of active subjects freely choosing to behave in particular ways and contributing to the elaboration of forms of power. One historically specific account of the place of freedom in the exercise of power is encapsulated in the principles of liberal forms of governance. I would suggest, though, that while this may constitute a particular historical and normative account of power relations in government, it also embodies a moment in the sustainable exercise of power which is much more general: the necessity of the subject of power choosing to conform to the requirements of power, and the consequent fragility of power relations at this moment of its greatest achievements.

The project of government within a liberal perspective is to find ways of regulating society without intruding upon the autonomy of individual subjects and of civil society in general. As Rose and Miller (1992) observe, in the case of liberalism:

Power is confronted, on the one hand, with subjects equipped with rights that *must not* be interdicted by government. On the other hand, government addresses a realm of processes that it *cannot* govern by the exercise of sovereign will because it lacks the requisite knowledge and capacities.

(N. Rose and Miller 1992: 179–180, original emphasis)

The task for liberal government, then, is to formulate strategies for governing 'at a distance'. This depends upon forging alliances with subjects, and soliciting their participation in the projects of government. Government thus depends upon linking private worlds and desires or interests with public ambitions. Key to this are strategies of calculation and the role of experts in negotiating the relations between public regulation and private domains (see also Miller and Rose 1990; Burchell *et al.* 1991). My concern here is not with the elaboration of forms of government able to manage this complex balance of rights and expertise, but with the initial normative premise of liberal government: that government should respect the freedom and rights of the individual. Rose and Miller (1992) develop Latour's (1986) idea that power needs to 'enrol' subjects if it is to be successful. But they pass very quickly from this point to Latour's more substantial contributions concerning the ways in which power can come to be stabilised in networks (such as bureaucracies), incorporating not only subjects but also strategies of calculation, communications, spatial distributions and texts which materialise the relationships and practices involved and ensure their 'normalisation'. I would like to consider more the implications of his initial points concerning the exercise of power.

Latour (1986) returns to what is the central problem of standard accounts of power: how do we explain why subject 'B' responds to the efforts of 'A' to exert power over her? Of course Latour (and Foucault) encourage us to consider power as far more than this dyadic interpersonal relationship: but power also involves social relationships, and it is the character of this interaction between subjects which the example of the Octavia Hill Women Housing Managers raises. For Latour, power is not something which can be possessed or held: it is only in its operation that it is realised. In the process of exercising power, it is the actions of others which determine its success or failure. As he writes, 'Power is not something you may possess and hoard. Either you have it in practice and *you* do not have it – others have – or you simply have it in theory and you do not have it' (Latour 1986: 265). He advocates a translation, rather than a diffusion, model of power: power is operationalised only in translation. Each person drawn into a relationship of power not only is responsible for its success or failure, but also is likely to shape it according to their own projects. Power in general, then, is successful only insofar as it manages to 'enrol' subjects to participate in ways that make sense to them, and to change where necessary in order to make this possible.

We could formulate this another way, which is to argue, following the quote from Foucault above, that power is premised upon freedom: without freedom, power cannot be exercised. Compared to a relationship of violence,[22] which acts upon bodies, or things, and destroys or closes off possibilities, a power relationship, Foucault (1982) argues, requires

> that 'the other' (the one over whom power is exercised) be thoroughly recognised and maintained to the very end as a person who acts; and that, faced with a relationship of power, a whole field of responses, reactions, results, and possible inventions may open up.
>
> (Foucault 1982: 220)

Power relationships, then, by definition, require an active subject upon which to act, and moreover, involve acts on actions, not bodies (persons), and in that relationship resides the necessity of freedom. The relationship of power involves an essential agonism between the exercise of power and the 'permanent provocation' implied by the necessary freedoms of the subject of power (Foucault 1982: 222). Following Latour, successfully enrolling subjects in the exercise of power means that very often power must already have incorporated the concerns of the subject, or accommodated their resistances. Power is, as I have suggested in the discussion of the Octavia Hill Women Housing Managers, already contaminated with resistance.

On the basis of his later studies of the technology of the self, Foucault pointed to the need for a stronger place for the agency of the subject of power. He suggested that we would have to take into account 'the points where the techniques of the self are integrated into structures of coercion or domination' (Foucault 1993: 203). More generally, we need to consider the ways in which stable and widespread forms of power which emerge are a result of that point of intersection: between the active subject of power and the agents of its exercise. From this viewpoint, resistance could be seen to be more widespread and more effective in shaping power relations – indeed, internal to them, rather than outside of the structures and techniques of domination (see O'Malley 1996; van Krieken 1996). Of course this will have wider implications for how we imagine a politics of transformation, for how we assess various institutionalisations of power relations, and for where we imagine resistance to be active or possible. Indeed, it suggests that a politics built upon the idea of resistance as always something other than power could have limitations.

Acknowledgements

I must first thank Sally Peberdy for her generosity in helping me in a last minute rush to collect information from the Pretoria Archives. Ronelle Rudman of the

City of Cape Town Records Department was immensely helpful in tracking down surviving documentation. Unfortunately late application of the Archives Act to local authorities, the trivial nature of tenancies in the opinion of archivists and the slowness of cataloguing vast numbers of files meant that her efforts did not yield much. We hope that there will be further research on the Octavia Hill Women Housing Managers when documentation permits. I must also thank Felix Driver for advice on some London reading matter, Steve Pile for helpful comments on an early draft of this chapter and comments from participants in seminars where I discussed the original paper (the Domination/Resistance conference in Glasgow, the UCL Geography Seminar series and the Institute of Comonwealth Studies). Diane Perrons, John Allen and the editors of this book also commented helpfully. Funding from the University of Natal and the London School of Economics ensured that I could be in the same location as the archival material for a while.

Notes and archival sources

1 I use the general term power here, mindful of the complexities of doing so, and concerned to engage critically the category even as I (ab)use it in this way. There are many different ways in which we could categorise different types of power: in terms of the origins, effects, objects, modalities, normative assessments, and so on (for a discussion of this see Allen 1999). None of these definitions seem to be adequately agreed upon in the literature. Two key questions are raised by this chapter. The relation between 'power' and 'resistance' which I discuss seems to presuppose that resistance is not power. Of course in a post-Foucauldian world, such a dichotomy is not possible: for if power is productive and not just bad, then mobilising resources, capacities and discourses in the interests of resistance is also a form of power (for a discussion of the relations between power and resistance see Pile 1997). Power, as I initially use it here, then, starts by implying a normative judgement of relative inequality in the relations of power. However, insofar as I argue that power itself incorporates those activities which oppose it (resistance) (and insofar as I am concerned with power *relations*), then the initial category is brought into question. It might also be considered important to distinguish the particular form of power relations involved and, insofar as this is necessary, I also base this on a normative judgement and am concerned with relations of domination: unequal power relations involving attempts to control people. Yet it will be clear that the argument I am presenting also undermines as much as it mobilises this categorisation of power.

2 'Coloured' refers here to an Apartheid-defined group of people, predominantly of mixed race.

3 Elsewhere I have pointed out how native administrators (effectively managers of African townships) were also involved in power relations which were far more ambivalent (and even sympathetic) than many political narratives about urban politics allow (Robinson 1996: ch. 7).

4 SABA (South African Archives Bureau, Pretoria) GES (Department of Health) 361 509/4x 'Octavia Hill Housing Management' submitted by Miss M. E. Hurst 3/3/1937.

5 SABA GES 361 509/4x 'Octavia Hill Housing Management: Training Schemes for South Africa', M. E. Hurst, no date (to Johannesburg Housing Utility Company); for the South African women to be trained in the system, overseas study was seen potentially to lead to them losing touch with South African conditions. Newcomers had to be prepared to familiarise themselves with these conditions.

6 SABA GES 361 905/4x Housing supervisor to the Town Clerk 1/7/1937.

7 SABA GES 361 905/4x Housing supervisor to the Town Clerk 1/7/1937.

8 CABA (Cape Archives Bureau, Cape Town) 3/PEZ (Port Elizabeth Town Clerk's Files) 4/2/1/1/356 Work of Octavia Hill Managers in Cape Town 12/8/1938.

9 SABA GES 361 509/4x 'Octavia Hill Housing Management' submitted by Miss M. E. Hurst 3/3/1937.

10 CABA 3/PEZ 4/2/1/1/356 Work of Octavia Hill Managers in Cape Town 12/8/1938.

11 CABA 3/CT (City of Cape Town, Town Clerk's Files) 1/4/9/1/1/18 Report from the Housing Estate Manager, Cape Town, 16/2/1935. An assistant was appointed in 1936 (approved, Housing and Estates Cttee Minutes 12/9/1935).

12 SABA GES 360 509/4e J. McClear, Sec, Society to Secretary for Health 21/4/1950.

13 As one housing manager wrote, 'Now that the average tenant can be regarded as a responsible citizen, the old technique must give way to a new, and while the need for good public relations is as great as ever, the close supervision exercised traditionally by the Octavia Hill trained manager is now only necessary in a minority of cases' (Clark 1957: 2–4). The growth of size of schemes and centralisation and bureaucratisation of management meant that relations with tenants changed. Mavenwe Penberty commented that in her experience inspections in houses went 'out the window' in the mid-1980s when they became too expensive and staff safety became a concern. The management as a result became more impersonal and their image changed to being the people who would put you out of your house (interview, August 1996).

14 This is a point which assessments of powerful looking (stereotyping, for example) have often pointed out as an attribute of domination. I want to suggest that this is not only a source of horror, but also a source of possibility.

15 In this light, Vaughan's (1991) discussion of disciplinary practices in an African context is both insightful and, in the light of my discussion here, in need of revision.

16 Other phrases which Cheesman (1954) uses to capture the cultural differences of Coloured people include: 'Coloured people are fond of gardens', 'they have a sense of humour', 'a cheerful people, and not argumentative'.

17 City of Cape Town Records Office, Staff Housing and Supervisor's Office H/14/3/2/9 Control of Housing Supervisor's Staff. Medical Officer of Health to Chairman and Members, Housing Committee 27/8/1945.

18 City of Cape Town Records Office, Staff Housing and Supervisor's Office H/14/3/2/9 Control of Housing Supervisor's Staff. C. Grant to Medical Officer of Health 18/9/1947.

19 Housing Committee Minutes, CABA 3/CT, attached correspondence, Medical Officer of Health to Town Clerk, 27/10/1945; Minutes of Housing and Slum Clearance Committee, 23/6/1938 (3/CT 1/4/9/2/1/1).

20 Stenson (1993) points out that the placing of many professional social worker–client interactions in the home, rather than in an office, where markers of professional authorisation abound (certificates of qualifications, the desk), was important in shaping the quality of the relationships created. The particualar spatiality of the interactions

between the housing managers and the tenants, I would suggest, was similarly important in making possible their claims to 'friendly power'.

21 As with one tenant, suffering from TB, who resisted efforts to heal his illness
and allegedly chose to sell his milk rations, was found by the TB Care Society to be
'evasive and full of excuses' – and the manager requested that the council evict him.
He was found by the OHWHMs not to 'conform to the usual standard required
for normal tenants' – City of Cape Town Records Office, H15/72 Application for
flats and houses under Council's Housing Schemes, 1939–1946; Medical Officer of
Health, to Housing Committee 6/11/1945.

22 In making an argument about power's operation as friendship, I have come up against
coercion, or the possibility of violence, a few times in the chapter (and in questions
about it). Clearly in this study the possibility of coercion existed alongside negotiation and friendship. One might find that the heterogeneity of power's operation also
extends to forms of power which rely almost entirely on violence, and that violence
may also run alongside more negotiated forms of power. I continue to make the argument I do here because the case of the OHWHMs was surprising enough to suggest
that, despite general accounts of the diversity of power relations (starting, perhaps,
with Weber), we have not taken the friendly or negotiated aspects of power, especially in relation to the activities of the state, seriously enough.

References

Allen, J. (1999) 'Spatial assemblages of power: from domination to empowerment', in
J. Allen and D. Massey (eds) *Human Geography Today*, Cambridge: Polity.

Brion, M. (1995) *Women in the Housing Service*, London: Routledge.

Burchell, G., Gordon, C. and Miller, P. (1991) *The Foucault Effect: Studies in
Governmentality*, London: Harvester Wheatsheaf.

Cheesman, J. (1954) Note in *Society of Housing Managers Quarterly Bulletin* 3(13): 8–10.

Clark, H. (1957) 'Problems of the housing manager today', *Society of Housing Managers
Quarterly Bulletin* 4(7): 2–4.

Crossley, N. (1993) 'The politics of the gaze: between Foucault and Merleau-Ponty',
Human Studies 16(4): 399–420.

Crush, J. (1993) 'Scripting the compund: power and space in the South African mining
industry', *Society and Space* 12: 301–324.

D. C. K. (1956) Note in *Society of Housing Managers Quarterly Bulletin* 4(1): 14–15.

Driver, F. (1990) 'Discipline without frontiers? Representations of Mettray Reformatory
Colony in Britain, 1840–1880', *Journal of Historical Sociology* 3(3): 272–293.

Foucault, M. (1979) *Discipline and Punish*, Harmondsworth: Penguin.

Foucault, M. (1980) 'The eye of power', in C. Gordon (ed.) *Power/Knowledge*, Brighton:
Harvester.

Foucault, M. (1982) 'The subject and power', in H. Dreyfus and P. Rabinow (eds) *Michel
Foucault: Beyond Structuralism and Hermeneutics*, Brighton: Harvester.

Foucault, M. (1993) 'About the beginning of the hermeneutics of the self: two lectures
at Dartmouth', *Political Theory* 21(2): 198–227.

Foucault, M. (1994) 'The ethic of care for the self as a practice of freedom: an interview', trans. by J. D. Gauthier, in J. Bernauer and D. Rasmussen (eds) *The Final Foucault*, Cambridge, MA: MIT Press.

Hill, O. (1877) *Our Common Land (and Other Short Essays)*, London: Macmillan.

Hill, O. ([1879] 1921) *House Property and its Management: Some Papers on the Methods of Management Introduced by Miss Octavia Hill and Adapted to Modern Conditions*, London: Allen and Unwin.

Kendall, T. and Crossley, N. (1996) 'Governing love: on the tactical control of counter-transference in the psychoanalytic community', *Economy and Society* 25(2): 178–194.

Kern, S. (1996) *Eyes of Love*, London: Reaktion.

Latour, B. (1986) 'The powers of association', in J. Law (ed.) *Power, Action and Belief*, London: Routledge and Kegan Paul.

Lewis, J. (1991) *Women and Social Action in Victorian and Edwardian England*, London: Edward Elgar.

Miller, P. and Rose, N. (1990) 'Governing economic life', *Economy and Society* 19(1): 1–31.

Mitchell, T. (1988) *Colonising Egypt*, Cambridge: Cambridge University Press.

Mulvey, L. (1989) *Visual and Other Pleasures*, London: Macmillan.

Nash, C. (1996) 'Reclaiming vision: looking at landscape and the body', *Gender, Place and Culture* 3(2): 149–169.

Ogborn, M. (1993) 'Ordering the city: surveillance, public space and the reform of urban policing in England, 1835–56', *Political Geography* 12(6): 505–521.

O'Malley, P. (1996) 'Indigenous governance', *Economy and Society* 25(3): 310–326.

Philo, C. (1989) '"Enough to drive one mad": the organisation of space in 19th century lunatic asylums', in J. Wolch and M. Dear (eds) *The Power of Geography*, London: Unwin Hyman.

Pile, S. (1997) 'Introduction', in S. Pile (ed.) *Geographies of Resistance*, London: Routledge.

Prochaska, F. (1980) *Women and Philanthropy in Nineteenth Century England*, Oxford: Clarendon.

Rendall, J. (1985) *The Origins of Modern Feminism: Women in Britain, France and the United States*, London: Macmillan.

Robinson, J. (1996) *The Power of Apartheid: State, Power and Space in South African Cities*, Oxford: Butterworth-Heinemann.

Robinson, J. (1998) 'Octavia Hill Women Housing Managers in South Africa: femininity and urban government', *Journal of Historical Geography* 24(4): 459–481.

Rose, G. (1995) 'Distance, surface, elsewhere: a feminist critique of the space of phallo-centric self/knowledge', *Society and Space* 13: 761–781.

Rose, N. (1989) *Governing the Soul*, London: Routledge.

Rose, N. and Miller, P. (1992) 'Political power beyond the State: problematics of government', *British Journal of Sociology* 43(2): 173–205.

South African Society of Women Housing Managers (SASWHM) (1959) *Housing Management in South Africa*, Cape Town: SASWHM.

Stedman-Jones, G. (1984) *Outcast London*, Harmondsworth: Penguin.

Stenson, K. (1993) 'Social work discourse and the social work interview', *Economy and Society* 22(1): 42–76.

Thomas, N. (1990) 'Sanitation and seeing: the creation of state power in early colonial Fiji', *Comparative Studies in Society and History* 32(1): 149–179.

Valverde, M. (1996) '"Despotism" and ethical liberal governance', *Economy and Society* 25(3): 357–372.

van Krieken, R. (1996) 'Proto-governmentalization and the historical form of organizational subjectivity', *Economy and Society* 25(2): 195–221.

Vaughan, M. (1991) *Curing their Ills: Colonial Power and African Illness*, Cambridge: Polity.

4

NOMADIC STRATEGIES AND COLONIAL GOVERNANCE

Domination and resistance in
Cyrenaica, 1923–1932

David Atkinson

In the modern period, questions of domination and resistance have arguably been cast in their starkest extent during European imperial interventions in the colonial world, and the social, political and cultural control exacted by the modern, totalitarian regimes of the twentieth century. This chapter considers one episode when a European imperial power incorporated a further slice of African territory as its colonial domain. Yet this instance of colonial expansionism was part of the totalitarian project of Fascist Italy, and the regime utilised all the modern technologies of warfare and social control to defeat the resistance of the indigenous population. The region concerned was Cyrenaica: the eastern coastal region of modern-day Libya, that stretches from the Egyptian border to the Gulf of Sirte and reaches into the Saharan interior to the south. My concern is with the efforts of Fascist colonial governance to subjugate and quell the resistance of the nomadic and semi-nomadic populations of Cyrenaica. Through this example, I suggest that the colonial conflict in Cyrenaica should not be regarded as a simple struggle between the colonisers and the colonised; but rather we should recognise a complicated and shifting matrix of relations of domination and resistance. Moreover the struggles were not only grounded in a series of spaces and territorialities, but also revolved around questions of mobility that were played out *across* the spaces of the North African desert. This chapter also constitutes an introduction to one of the least known, but perhaps one of the more lethal episodes in the history of modern European colonialism, for although the figures are vague and contested, estimates put the fatalities at somewhere between 30,000 and 70,000 (Santarelli *et al.* 1986)

Fascism, colonialism and domination

Since the early 1980s, our understandings of Fascist Italy have been augmented by an increasing number of studies that emphasise the various roles of culture, hege-mony and popular consensus in the regime's domestic governance through the late 1920s and 1930s (De Grazia 1981; Falasca-Zamponi 1992; Gentile 1996). From the mid-1920s, for example, the regime seldom resorted to the kinds of blatant public violence and physical oppression that characterised its early years. Whereas emphasis upon cultural persuasion was characteristic of metropolitan Italy, the regime's oppression of its colonial subjects in Africa was often acute and unforgiving. Once established, Italians represented their imperialism as a more benign and co-operative strain than that practised by other European powers (Bono 1989; Finaldi 1997). However, there is little doubt that during the actual conquest of colonial territories, the Italian record was scarred by instances of extreme ruthlessness and barbarity. As ever, European notions of authority and dominance were materialised in a far less subtle manner in Africa than in mainland Europe. In the Italian case, Fascism's use of mustard gas and machine guns against poorly armed Abyssinian troops in the Ethiopian campaign of 1935–1936 is only the most infamous example. Less well known is the campaign waged by the Italian colonial authorities in Cyrenaica, between 1923 and 1932.

The campaign was directed against the resistance of the indigenous Bedouin tribes and the Sanussi religious fraternity that provided political leadership for the Bedouin throughout the campaign. The Italians were concerned to establish their dominance over their new colonial territory, and to appropriate the land so that it might be distributed to Italian emigrant peasant families to be settled and farmed as Italy's 'Fourth Shore' (Del Boca 1991b; Segrè 1974, 1987). However, the resistance of the nomadic and semi-nomadic peoples of the coastal plateau and the interior stymied Italian plans for nine years. The Sanussi had previously constituted a loose form of governance in the region; after the Italian invasion, they led the Bedouin in a campaign of resistance that relied primarily upon their adaption of their traditional nomadic lifestyles. Above all, their mobility, their familiarity with the Cyrenaican environment and the support they received from the nominally 'conquered', semi-settled peoples of the coastal provinces rendered the Bedouin formidable opponents who frustrated the far superior numbers and equipment of the Italians. Eventually, the Italians applied some of the most modern and savage technologies of warfare and social control to defeat this resistance, ultimately incarcerating the entire population in a chain of concentration camps where *at least* 35,000 died in just two years (Abdullatif Ahmida 1994).

These extreme measures finally ended the resistance, but stand as one of the bleakest episodes in the history of European imperialism in Africa. The war had been fought over territory and was essentially a struggle over space; but it was also, crucially, a struggle contested *across* space, and fought out through issues of mobility. This chapter attempts to develop a more nuanced understanding of these events through a discussion of Bedouin mobility as a strategy of resistance to Italian colonial rule. At the same time, I consider the interwoven geographies of domination and resistance that characterised both sides of this struggle. Finally, I discuss ideas of the desert and mobility in Euro-American cultures and suggest that a tendency towards romanticising nomadism in contemporary social theory ought, perhaps, to be counterposed by an awareness of this war and other such attempts of the colonial period to conquer and sedentarise nomads by force.

Europe, the desert and nomadology

Europe and the desert

Since Classical times, the arid lands of the Sahara desert to the south of the Mediterranean Sea have constituted a region of enduring fascination for European societies. To peoples accustomed to more temperate climates, the apparent hostility of its environments combined with the vast scale of its area rendered the desert the antithesis of continental Europe. At the same time, from the Classical understanding of the Sahara as the edge of the known world (Mudimbe 1994; Romm 1992), through to more recent interpretations of an arid, impassable barrier that separated the Mahgreb from the territories of Sub-Saharan Africa, the Sahara has frequently been regarded as little more than a relentless stretch of hostile, barren, empty space. Throughout this cultural tradition, the only inhabitants of the region to make any sizeable impression upon European imaginations were Bedouin nomadic tribes and their herds, moving across the desert between waterholes and oases.

To many Europeans, these mobile peoples living amidst an elemental environment contrasted markedly to their own bounded, sedentary, industrial lifestyles. The worlds of nomads, it was thought, were not demarcated by the territorial boundaries or disciplined spaces of modern European societies. As such, throughout history they constituted one of Europe's Others and were frequently viewed with suspicion because of their difference. Yet simultaneously, they were also cast as romantic figures with apparently care-free, pre-modern lifestyles and an affinity with their natural environments. They provoked both fear and fascination for Europeans (Dawson 1994; Root n.d.). Obviously, there are far more complicated histories and flows of peoples, cultures, political influences, commodities and goods across the Sahara than these arbitrary European geographies allow. Yet at the same time, parts of the North African littoral, and the Saharan interior in particular, remained *terra incognitae* to western knowledges until well into the twentieth century (Atkinson 1996). The region was a casualty of generalised western imaginations whereby it was consistently represented and reproduced as a hostile, barren environment populated by a scattering of nomads. To Europeans of the imperial period, such lands were considered ripe for colonial occupation. As Root writes: 'Although there have always been people living in the Sahara, in colonial eyes the desert is landscape without culture, wild, uncultivated land that remains out of control' (Root n.d.: 29).

Cast as such, the desert served as little more than vacant space to be appropriated at will by the west. In nineteenth-century French art and literature, the Sahara was frequently represented as a wilderness that could be at once redemptive and restorative (Heffernan 1991). It was also imagined as an elemental, 'pure' environment, denuded of the layers and comforts of modern society. In the early twentieth century, for example, the desert served as a liminal space on the margins of western society. It attracted a host of 'travellers' who sought solitude, or the chance to fulfil their romantic fantasies of heroic, unhindered movement and escape from the more ordered societies whence they came. It became a stage for the embodiment and performance of nomadism by western subjects such as T. E. Lawrence or Wilfred Thesiger, who rode into the desert to escape European social boundaries (Dawson 1994). Equally, the Sahara provided a space for the transgressive cross-dressing and masculine, Arabic lifestyle chosen by the 'self-willed nomad', Isabelle Eberhardt (Abdel-Jaouad 1993). Crucially though, these fantasies were enabled by a long-standing European colonial discourse that constructed this land as both vacant and available for precisely this kind of physical and philosophical journeying. Equally, such travels were often enabled in material terms by the military force of European colonial powers. It is clear that historically, the desert has been imaginatively appropriated at will by westerners for their own purposes.

One such 'desert explorer', dubbed posthumously 'another potential T. E. Lawrence [with] the same dynamic qualities' (Driberg 1937: 7) was Knud Holmboe. An Arabic-speaking Danish Muslim, in 1930 this self-styled traveller resolved to drive across North Africa, from Morocco to Cairo, in search of the freedoms of desert travel (Holmboe 1937). Yet when he reached Italian Cyrenaica, rather than finding solitude, freedom and empty spaces populated by nothing more than a scattering of Bedouin communities, he found a territory under strict Italian martial law. All forms of movement were outlawed, the Bedouin tribes were herded into concentration camps, and any individuals caught outside the camps was liable to be shot on sight (Holmboe 1937). What Holmboe encountered in place of the wild empty spaces that he anticipated were some of the complex geographies of domination, resistance and mobility that I consider in this chapter.

Nomadology

Arguably, aspects of the European tradition of romanticising the desert and the figure of the nomad have never gone away. The success of the 1996 film *The English Patient* is only one more recent example. Even in the more rarefied realms of social theory there are traces of this enduring rhetoric. Academic interest in the ideas of nomadology has intensified markedly in recent years and the trope has emerged as a popular element of contemporary theory. As Cresswell (1997) points out, by contrast with earlier cultural theorists, contemporary writers celebrate mobility and movement endlessly, and the metaphor of the nomad enjoys significant currency at the heart of postmodernism. For many, it provides an idealised model of movement and displacement, and a metaphorical trope for non-fixity, anti-essentialism and mobility as resistance to the bounded spaces and orders of modern society. As such, the nomad provides a classic example of a 'deterritorialised' subject.

The popularisers of nomadism, Deleuze and Guattari (1986, 1987) ground their discussions of the term using the example of the resistance to fixed, feudal authorities of the stone-masons, carpenters and labourers who migrated around medieval Europe while building its Gothic cathedrals. Nevertheless, despite their chosen example, in other sections of their discussion, Deleuze and Guattari (1986) refer casually to two very different, non-European regions as the emblematic landscapes of the nomad. The western Asian Steppes, and particularly the arid, desert environments of North Africa, are enlisted as the archetypal nomadic environments. The nomads who illustrate their arguments move around between 'water points' (Deleuze and Guattari 1986: 50), and it is the nomad who 'clings to the smooth space left by the receding forest, where the steppe or the desert advance, and who invents nomadism as a response to this challenge' (Deleuze and Guattari 1986: 51). Furthermore, the nomad is said to make the desert as much as be made by it, and is compared to the shifting, rhizomatic vegetation that characterises arid lands (Deleuze and Guattari 1986: 53). This rhetoric clearly draws upon long-standing European traditions that associate these vast spaces and harsh environments on the margins of Europe, with nomads and nomadism – the exotic, mobile cultures that constitute 'others' to settled, European societies (Root n.d.). Although they call for historicised case studies, Deleuze and Guattari seem to reproduce the associations of nomads with steppes and deserts, and particularly, with the arid lands of North Africa (Cresswell 1997).

Other writers also incorporate this imagined, depopulated desert environment into their discussions of nomadology and movement. Kaplan (1996) critiques Baudrillard and Deleuze and Guattari in her extensive survey of the debate (see also Kaplan 1987). She claims that

> Mapping 'terra incognita' requires the open spaces and depopulated zones constructed by colonial discourse. While the 'dark continent'

signals Africa's imbrication in imperial modern culture's self-construction, the blinding white spaces of the desert present another opportunity for Euro-American inventions of the Self. From Isabelle Eberhardt to Jean Baudrillard, from T. E. Lawrence to David Lean, the philosophical/ literary trek across the desert leads to a celebration of the figure of the nomad – the one who can track a path through a seemingly illogical space without succumbing to nation-state and/or bourgeois organisation and mastery . . . the nomad . . . offers an idealized model of movement based on perpetual displacement.

(Kaplan 1996: 66)

In various western cultural traditions then, the 'blinding white spaces of the desert' are uncritically reproduced as the archetypal spaces of the nomad. Even today, these imagined places serve as a backdrop to modern theories of movement and resistance, with the desert a ready metaphor for contemporary theories of deterritorialised movement and flux. Quite aside from the dangers of generalisations and the over-simplistic counterposing of domination and resistance, in this tradition, the nomad is also often gendered as masculine: moving with impunity around the natural landscapes where 'he' lives. To borrow Deleuze and Guattari's terminology, the nomad's desert is smooth space: land available for unhindered movement, and unstriated by any extant social, political, economic or cultural geographies.

The problem with all of this, as Kaplan notes, is that 'Euro-American recourse to the metaphors of desert and nomad can never be innocent or separable from the dominant orientalist tropes in circulation throughout modernity' (Kaplan 1996: 66). There is a danger that, for privileged, western theorists,

The Third World functions simply as a metaphorical margin for European oppositional strategies, an imaginary space . . . This kind of 'othering' in theory repeats the anthropological gesture of erasing the subject position of the theorist and perpetuates a kind of colonial discourse in the name of progressive politics.

(Kaplan 1996: 88)

The risk is that nomadic peoples are seldom problematised, historicised or allowed any detailed social, political or cultural profiles in modern texts. They simply provide a convenient example of mobility and deterritorialisation. Cresswell argues that the figure of the nomad has been subjected to 'vast generalisations and misguided metaphorical play' (Cresswell 1997: 362). He points out: 'Such metaphorical reductions can serve only to negate the very real differences which exist between the mobile citizens of the postmodern world and the marginalised inhabitants of other times and places' (Cresswell 1997: 377). To avoid such casual appropriations of nomadic experiences and spaces, both writers call for situated, contextualised and provisional accounts of mobility – although Kaplan notes that previous attempts were partial and unsatisfactory (Kaplan 1996).

My intention in the remainder of this chapter is to provide a preliminary account of some of these marginalised inhabitants of another time and place, and particularly the ways that the nomadic Bedouin population of Cyrenaica developed 'nomadic strategies' to resist Italian occupation. I also relate the ways in which Bedouin nomadism and their deterritorialisation were constructed as 'problematic' by the Italian colonial authorities, and finally, the extreme methods that were employed to force these communities to abandon their traditional mobile lifestyle *precisely because* of the continuing resistance that their movement posed to Italian colonial governance. I suggest that our understandings of nomadism might benefit from a more contextualised and historicised awareness of some of the complex, entwined, contradictory, but nevertheless deadly geographies of domination and resistance that have marred Maghrebian histories.

Conquering Cyrenaica

The Italian conquest of Cyrenaica, Tripolitania and the Saharan regions to the South provides a complicated and regionally differentiated history of different degrees of domination being exacted over the indigenous peoples. As Thomas (1993) emphasised, colonialism was frequently negotiated amidst local and regional contingencies. Italian colonialism in Libya was no exception. In the early years, the Libyans were virtually untouched by European 'rule'; on later occasions, the Italians collaborated with groups within Libyan society to maintain order; and eventually, the Italians imposed direct, brutal, martial control upon the population. The story also differs between Libya's two main regions of Tripolitania and Cyrenaica, and the episode was also played out across the expanses of the Saharan interior. What follows is a brief account of the strategies that marked the first stages of Italian colonialism in Tripolitania and, especially, Cyrenaica.

Early Italian colonialism in Cyrenaica

Italian forces invaded the Ottoman territories of Cyrenaica and Tripolitania on the North African coast on 4 October 1911 (Malgeri 1970). Urged onwards by a growing colonialist lobby, the patriotism stirred by the fiftieth anniversary of the Italian state, and, some argue, to deflect attention from increasing domestic problems, Prime Minister Giolitti authorised the move with rhetoric about seizing Italy's fair share of colonial Africa (Del Boca 1986; Bosworth 1996). There was also talk of reclaiming the former granary of ancient Rome, and settling Italian emigrants under the Italian flag in Africa (Pistolese, 1932). The invasion was concluded in 1912, although not before the Italians had employed the most modern and technological machinery of warfare against the Turks and their Libyan supporters, including motor transport, radio and the world's first use of aerial bombing (Paris 1991; Wright 1989). The Ottoman empire sued for peace in October 1912. One fading imperial power was displaced by a more recent European imperialism as the Italians established themselves in the main cities and along the coastal littoral (Del Boca 1986, 1991a).

Italian involvement in the First World War from 1915 to 1918, and the social and political chaos that engulfed the country afterwards, meant that the Italian presence in Libya was never consolidated. Indeed, it became so precarious that Italian colonial agriculture at one stage totalled just eighty-nine allotments (Bosworth 1996). The situation changed in 1922 when Mussolini's Fascist movement emerged from the anarchy of Italian society with promises to restore stability and Italian prestige on the world stage. Clearly, their pretences to a more 'fitting' international profile for Italy depended, in part, upon Fascism's ability to 'pacify' their North African territories successfully. Consequently, Cyrenaica, Tripolitania and the Saharan regions would suffer greatly at the hands of yet another European intervention in Africa.

In Tripolitania, the western coastal region of Libya that adjoined Tunisia, Governor Giuseppe Volpi anticipated the regime's thirst for colonial territory, and, in response to some isolated, local resistance from tribal *shaikhs*, inaugurated the full-scale *reconquista* (reconquest) of inland Tripolitania in January 1922 (Segrè 1974; Volpi di Misurata 1926). Within three years, much of Tripolitania and the adjacent Saharan interior had been incorporated into the Italian colonial realm, although the region was not fully 'pacified' until 1928; even then, some 'rebels' retreated still deeper into the Sahara to the Fezzan. In one respect, any effective, co-ordinated, sustained resistance to the Italians had been critically compromised by the antagonisms and in-fighting between the Tripolitanian tribes, divisions which the Italians encouraged and developed (Abdullatif Ahmida 1994; Santarelli *et al.* 1986; Wright 1969). But whereas the Tripolitanian resistance had tried to hinder Italian mobility through attacks on the region's rail network (Maggi 1997), in a taste of the tactics that would later be visited upon Cyrenaica, the Italian military commander Rodolfo Graziani dispatched highly mechanised, mobile forces to attack the Arabs' *camps* rather than their military bases. Volpi reported back to Rome that Italian success was based upon 'using Arab tactics against Arabs' (Segrè 1974: 48), although the Italian exploitation of the rivalries and tensions between the Tripolitanian tribes was also significant.

In Cyrenaica the situation was different. In 1922, the Italian authorities controlled the ports and the coastal strip; although the inland plateau and the Saharan interior were composed of a mosaic of traditional tribal territories, these local units were transcended by the broader, regional layer of political governance provided by the Sanussi religious fraternity (whom I discuss shortly). This pattern of governance had been enshrined since 1917 when an Italian government, pre-occupied with the war in Europe, signed the Acroma accords with the Sanussi (Evans-Pritchard 1945a). The infrastructure of forces, camps and officials that underpinned the Sanussi authority was partially funded by the Italians and the two groups even shared some military camps along their mutual frontier (Santarelli *et al.* 1986). Clearly, as the Italians fought the Austro-Hungarian empire on their northern frontier, they were prepared to share power in Cyrenaica with the Sanussi order, and to collaborate with this single political entity rather then with the tribes and *Shaikhs* that were thus further subsumed beneath Sanussi authority. The Acroma treaty underpinned the position of the Sanussi, who in turn provided some measure of colonial stability for the Italians (Evans-Pritchard 1945a). Thus, as contingencies directed events, the Liberal governments of pre-Fascist Italy forged a mutually beneficial agreement that served Italian purposes and reinforced Sanussi authority over Cyrenaica and its peoples.

The collaboration disintegrated in the spring of 1923. On 6 March, the Italians seized a Sanussi camp and took control of the 'mixed-camps' they shared with the Sanussi without warning. Half the Sanussi regular troops were thus captured (Evans-Pritchard 1949). Subsequently, on 1 May 1923, the Italians unilaterally declared all Italo-Sanussi treaties to be void, and attacked their former partners (Evans-Pritchard 1945a). To a regime that was gradually silencing opposition at home, and one that was also slowly extending its control over the public sphere (Atkinson 1998), the notion of sharing power in the colonies with African 'subjects' fell far short of the ideal. Similarly, although the embryonic totalitarianism of Fascism would never be fully realised (Morgan 1998), one element of the regime's aspirations seems to have been unchallenged control over national and colonial territory. Yet their attempt to seize exclusive hegemony over Cyrenaica left the Italians contesting their presumed colonial realm with the ambitions and territoriality of the Sanussi.

The Sanussi fraternity

The Sanussi fraternity had been established as an Islamic Order in 1843 by al-Sayyid Muhammed bin 'Ali al-Sanusi, an Algerian cleric and intellectual who had studied and taught in Fez, Cairo and Mecca (Evans-Pritchard 1945a). The movement worked to encourage simple, austere Islamic observance throughout the territories of modern Libya, Chad, Eastern Sudan and the central Sahara (Peters 1990; Wright 1988). It was highly successful and over the next eighty years, the order gained followers rapidly and extended its influence from its historical centre in Cyrenaica across the Sahara to the Sudan, and along the Egyptian coast into the west of the Arabian peninsula (Evans-Pritchard 1945b). The Sanussi expanded by establishing lodges, called *zawiyas*, in populated regions, but especially at oases and other key nodes of trans-desert routes. The *zawiya* network served as centres of education and religious teaching, but also functioned as sites of poor-relief, banking, administration and commerce (Evans-Pritchard 1945b; Valenzi 1932). Thus, concomitant with their expansion, the Sanussi also became much more of a secular authority that, having won the support and loyalty of the majority of the semi-nomadic and nomadic tribes of North Africa, constituted a semblance of co-ordinated governance and political authority in the Saharan interior – at least, as viewed in European terms. In Evans-Pritchard's words, theirs was a 'Theocratic empire' (Evans-Pritchard 1945a).

Perhaps inevitably, the *zawiya* network began to exercise a degree of Sanussi control over the desert interior, and especially over trade. While the Ottoman empire had established control over coastal trade and some routes into the Sahara, the Sanussi lodges and the order's hierarchy were sustained by taxation upon trade in the interior (Rochat 1973) and the order struggled to preserve its hegemony over this relatively lucrative source of income. It is alleged that the Sanussi continued to control a trans-Saharan trade in slaves until as late as 1930 (Abdullatif Ahmida 1994; Wright 1989). Evidently, despite their support from many among the Bedouin tribes, the Sanussi were implicated in a clear geography of control and territoriality.

In addition, Sanussi territoriality was also partially constituted in response to earlier European interventions in North Africa. The fraternity had traditionally avoided contact with European desert-travellers, and earned itself much negative publicity as a result (Wright 1988). Abdullatif Ahmida (1994) claims that the Sanussi anticipated European interference in their affairs from the late nineteenth century, when European colonial designs upon Africa became unmistakably clear. When the French destroyed some of its lodges in Chad in the 1890s, it seems that the Order decided to resist any future European incursions into their space. Consequently, they were inextricably drawn into international political relations as they developed an anti-colonial, pan-Islamic ideology that aimed to unify the disparate tribes of North Africa. They also trained and prepared for any future European interventions (Abdullatif Ahmida 1994; Peters 1990). The influence of the Sanussi had its origins in their embedded history as Cyrenaica's leading religious order. However, they also developed into the major secular authority of the region, with a reach across the tribal divisions and territories of North Africa. Consequently, Cyrenaica and the adjacent Sahara were neither the empty, unpopulated landscapes of European imaginations, nor the vacant space that might provide the potential colonial territory that the Italians envisaged. Rather, these lands were already striated by various elements of taxation, control and authority, although these were not particularly evident to 'western' eyes.

When the Italians invaded in 1911, the Sanussi were inevitably concerned to maintain their domination over their desert territories. While some of the Bedouin tribes initially fought the Italians alongside Ottoman troops, upon the Italo-Ottoman peace of 1912, the Sanussi took over the leadership of the anti-colonial campaign themselves. They later also rallied to the *Jihad* (holy war) declared against the allies by the Islamic world in the First World War, and engaged the British in Egypt as well as the Italians. It is from this period that members of the fraternity began to talk in ever more literal terms of their *Hakuma al-Sanusiya* – their Sanussi government (Evans-Pritchard 1945a). The anti-colonialism and territoriality of the Sanussi therefore found expression from 1911. Although at one stage the order was content to maintain its hegemony over the Cyrenaican interior as established by the Acroma accords, it is hardly surprising that when the Italians renounced the treaty in 1923, the Sanussi were prepared to resist the seizure of 'their' territory.

The Italian leadership hoped for the relatively swift submission of inland Cyrenaica. In contrast to the situation in Tripolitania, however, the Sanussi provided a coherent degree of political leadership for the resistance, and many of the nomadic and semi-nomadic population rallied to their support. Although Peters (1990) revised Evans-Pritchard's (1949) analysis of the Sanussi to claim that the tribal *shaikhs* should be accorded some significance in any explanation of the practicalities of the struggle against Fascist Italy, the *symbolic* importance of Sanussi leadership seems to be broadly unchallenged. For example, Evans-Pritchard argued that: 'the Sanusiya comprised a symbol to which the Bedouin clung and which enabled them to withstand twenty years of privation, near-starvation and death, during the resistance to the Italians' (Peters 1990: 26).

If merely at a symbolic level, the Sanussi appear to have transcended most inter-tribal divisions. The Sanussi thus shifted away from their collaborative domination of the region in harness with the Italians, to provide the leadership for a sustained campaign of resistance that would frustrate Italian plans for nine years. It may have been the desire to preserve their territorial power that primarily motivated the Sanussi. Certainly, their moral leadership was compromised on occasions by the failure of many of the Sanussi hierarchy to become actively involved in the campaign, and more particularly by the 'taxes' and support that the Sanussi-led bands extracted from some of the semi-settled, *sottomessi* ('submitted', 'pacified'), tribes of the coastal provinces. Therefore, although the brutality of the Italians in Libya is perhaps the most salient aspect of this colonial struggle, it is important to remember that there were no simple or discrete categories of resistance and domination in this instance, but rather a series of complex and interwoven questions of oppression and resistance that were played out across the desert landscapes of Cyrenaica in a struggle for control over space.

Reconquering Cyrenaica

From 1923 until 1932, Italian forces waged an increasingly bitter and cruel colonial war against the resistance of the Cyrenaicans. Not all of Cyrenaica's population were actively involved in the conflict, however. The settled peoples of the coastal towns and cities were sympathetic to the cause and often resented Italian colonialism, but, perhaps due to the well-established Italian presence along the coast, they were seldom actively involved (Evans-Pritchard 1949). A little further inland, some of the semi-nomadic and sedentary Bedouin tribes likewise remained largely peaceful. These groups were the *sottomessi* who found themselves in a crucial, albeit unfortunate, position as the war developed. Further inland still were the more nomadic, desert Bedouin who had less entangled relations with Italian influences and less familiarity with European notions of space and territory. Under the leadership of the Sanussi, these groups provided the bulk of the resistance. After the initial battles, the Sanussi-Bedouin forces seldom totalled more than a thousand in number, and were divided into small, mobile groups. Yet the sustained and effective manner of their resistance was based upon their flexibility and their mobility in the Saharan desert. These 'nomadic strategies' and the responses they provoked from the Italians constitute the remainder of this chapter, although I connect this story to the geographies of domination, oppression and resistance that were consequent upon this struggle across space.

Nomad strategies: mobility as resistance?

In 1923, the Sanussi and their Bedouin supporters were far outnumbered by the Italians. Neither did they boast any of the conventional training, equipment and logistical support that the Italians shipped into the country. Sanussi-Bedouin forces comprised 2,000 untrained men. Their armaments were estimated at rifles for these troops with a further 3,000–4,000 among the tribes, and a few machine guns and pieces of artillery (Evans-Pritchard 1945a; Segrè 1974). By contrast, the Italians controlled all the ports and major towns, and mobilised some 20,000 soldiers – mainly colonial troops from Eritrea, who, as Christians, would become notorious for their harsh treatment of the Muslims (Evans-Pritchard 1949). Italian forces were also equipped with all the fearsome technologies of modern warfare, from mechanised columns, to aircraft fitted to deliver poison gas.

Unsurprisingly, in the first few months of the war when the Sanussi-Bedouin engaged the Italians in the standard military manner, they were heavily defeated (Abdullatif Ahmida 1994). It was clear that orthodox military campaigning was futile, and so, from mid-1923, the Sanussi-Bedouin began to resist the Italians on their *own* terms. The result was guerrilla warfare. The entire strategy was predicated upon the greater mobility of the Cyrenaicans, their familiarity with the landscapes and geographies of their region, and their ability to harass and attack the Italians at a number of different locations, but then to melt away into the supposedly pacified *sottomessi* in the coastal provinces (Rochat 1973). The Sanussi-Bedouin were led in these 'nomadic strategies' by a cleric called Omar al-Mukhtar. A highly capable and respected leader despite his advanced years, Omar al-Mukhtar commanded small, mobile groups called *muhafiziya* (or *dors*), although the Italians labelled them 'rebels' and 'brigands' (Evans-Pritchard 1949). However, the main theme of the resistance from late 1923 to the end of the struggle in 1932 was the ceaseless movement and evasion of the *muhafiziya*, and their constant harrying of the Italian colonial forces.

Both Patton (1988) and Muecke (1984) adapted Deleuze and Guattari's theories of nomadology to argue that nomadic groups have employed movement and mobility – what they call 'nomadic strategies' – in political struggles. I suggest that this very quickly became the case in Cyrenaica. In addition, I argue that it was precisely because the armed resistance was fought over vast spaces, by small, flexible, mobile groups that were familiar with movement around these environments, that the anti-colonial struggle lasted for so many years. Although Evans-Pritchard tended to romanticise the Sanussi-Bedouin resistance, he was well aware of the significance of the adoption of 'nomadic strategies'. He celebrated the success of this new Bedouin strategy:

> the Sanussi were fighting in their own country and the Italians had to adapt themselves to the kind of fighting which seldom fails to upset the orthodox military mind. Ordinary tactics are useless against an enemy who wanders at will over country with which he is familiar, among a population all friendly to him, and whose tactics are little more than the three guerrilla imperatives, strike suddenly, strike hard, get out quick.
>
> (Evans-Pritchard 1945a: 71–72)

A strategy based upon sustained mobility completely sidestepped the tactics and strategies of the Italians. European warfare was predicated upon the accumulation of territory through the capture of fixed points, and the advancement of a 'front' into enemy lands. The *muhafiziya* subverted these notions by refusing to engage the Italians along a front and defend a recognised territory, but by attacking their enemies wherever they could. The *muhafiziya* groups were small and highly mobile, partially to ease the logistics of supplies, but also so that they could cause maximum nuisance to the Italians in precisely this manner (Evans-Pritchard 1949). Evans-Pritchard continued:

> The smallness of the Sanussi units and their mobility confused the slow and unwieldy Italian columns. If they split up they were liable to be surrounded and annihilated; if they kept together they lost the advantages

of surprise and mobility . . . The Italians found that the blows they struck
at the enemy often struck at the air [for the] Bedouin retired to less
accessible regions or circulated gaily between the Italian garrisons.

(Evans-Pritchard 1945a: 72)

Rather than engage the Italians in the European manner that the Italians
anticipated, the Cyrenaicans exploited their traditional mobility and familiarity
with the desert landscapes to maximise their resistance. They ensured that there
was no easily identifiable enemy force that could be located and engaged. In so
doing, the Sanussi-Bedouin operated as if moving across smooth space, without the
limitations and constraints of European notions of territorial warfare. At the same
time, they undermined Italian notions of 'pacified' territory by operating in areas
that the Italians thought they had already 'conquered'. Even in the last year of the
war, when the Sanussi-Bedouin numbered fewer than 700 and attacks were only
a fraction of earlier periods, there were still 53 recorded 'engagements' and 210
'skirmishes' (Evans-Pritchard 1949).

The unorthodox strategies of the Sanussi-Bedouin clearly perplexed and
annoyed the Italian military officers and colonial officials. Written sources express
their frustration and indignation. Attilio Teruzzi, the Italian Governor of Cyrenaica
from 1927 to 1929, complained bitterly that his forces were not fighting against
a 'traditional enemy' which might be defeated in one orthodox military engagement, but one that had no identifiable form or bases. In Teruzzi's words:

the rebels are not tied down to anything, are not bound to any
impediment, have nothing to defend or protect, and can show themselves
today in one place, tomorrow 50km away, and the following day 100km,
to reappear a week later, to vanish for a month.

(Evans-Pritchard 1949: 172)

Or, Corrado Zoli, an official at the Cyrenaican colonial ministry and later
Governor of Eritrea, wrote of the Bedouin that:

[This was] an elusive enemy who kept the Italian forces in constant
movement and alarms by endless surprises, incursions, raids, and
ambushes; making use of his great mobility, powers of dispersal, tactical
independence, and perfect knowledge of the insidious terrain, to avoid
decisive encounters.

(Evans-Pritchard 1949: 173)

In response to the nomads' strategies, the Italians were compelled to increase their
own mobility. They also revealed a foretaste of the terror tactics that would later

be inflicted upon the civilian population in their uncompromising response to the guerrilla strategies and amorphous nature of their enemies. Given their inability to locate and target the mobile *muhafiziya*, in late 1923, they launched surprise mechanised attacks on the less mobile camps of the Bedouin, indiscriminately killing those found there and also destroying their herds and food stores. When the rainy season waylaid overland transport, the camps were bombed and straffed from aircraft (Evans-Pritchard 1949; Wright 1969). In addition, the Italians also dropped poison gas on some camps (Del Boca 1996). The Italians claimed 800 Bedouin fatalities, 230 captured and 1,000 wounded between March and September 1923 alone. In addition, they killed or confiscated 700 camels and 22,000 sheep, the livestock upon which the whole economy and well-being of the tribes depended (Evans-Pritchard 1949). If these early months of the war had revealed the horrendous casualty rates that the Cyrenaicans would later suffer, they also demonstrated that the Europeans were willing to employ all possible modern technologies against the Bedouin. Similarly, given their failure to engage their elusive enemy, the Italians were prepared to visit warfare upon the more static camps of the Bedouin, and to target the families of the *muhafiziya* and their herds. This proved a first taste of the reprisals that would be prompted by the 'nomadic strategies'.

Questions of mobility continued to dominate the war into 1924. In March of that year, the Italians adapted their strategy again and attempted to counter the mobility of the Sanussi-Bedouin groups by increasing their own flexibility and movement. They established a series of bases for highly mobile, mechanised patrols that criss-crossed the country (Evans-Pritchard 1949). The intention was to harry the Bedouin constantly and to prevent them from settling or regrouping at any time. The patrols also destroyed all the Bedouin herds and crops which they encountered (Abdullatif Ahmida 1994; Rochat 1973). Although this tactic proved relatively successful in the short term, ultimately, it was the Italians who were drained by the constant movement (Segrè 1974). The *muhafiziya* continued to attack smaller patrols, but would spend more time hidden among the *sottomessi* nearer the coast – making Italian reprisals difficult without provoking further resistance from the supposedly pacified groups of the coastal regions (Evans-Pritchard 1949).

The sottomessi

Situated between the Italian bases along the coast and the desert interior with its roaming Sanussi-Bedouin bands, the nominally pacified *sottomessi* played a central role in the war. They also found themselves in a uniquely vulnerable position. The literature is agreed that the majority of the indigenous population of Cyrenaica resented Italian colonialism and supported the resistance (Del Boca

1988; Evans-Pritchard 1949; Santarelli *et al.* 1986). However, as the buffer between the 'passive' populations of the coastal towns and the militant tribes of the interior, the nomadic and semi-nomadic *sottomessi* of the coastal hinterland were the focus of particular attention from both the Italians and the *muhafiziya* as each side sought, and often compelled, their assistance in the struggle (Santarelli *et al.* 1986).

Italian policies towards the *sottomessi* differed across the region, but all were designed to ensure that these groups would not support the resistance (Evans-Pritchard 1949). In some areas, surveillance and control of the *sottomessi* were strict and unrelenting. They were forced to camp in designated areas near Italian bases, so that they could be observed and inspected at any time. Their movements were restricted and their horses were confiscated to hinder any rapid movements. Punitive measures, including executions, accompanied any evidence of complicity with the resistance (Evans-Pritchard 1949). In other districts, the control was less severe, but the Italians attempted to turn the *sottomessi* against the *muhafiziya* through propaganda and subsidies (Del Boca 1988). In addition, they frequently armed militia groups from among the *sottomessi* to encourage resistance to the 'brigandage' of the Sanussi-Bedouin 'rebels'. This was in response to the support demanded from the *sottomessi* by the *muhafiziya*, who regularly raised 'taxes' and requisitioned supplies from the 'pacified' population. In this way, the Italians hoped to engineer conflicts and subsequent blood feuds among the population, and therefore, to divide and conquer as their Roman ancestors had done (Evans-Pritchard 1949).

However, the Bedouin appear to have resented the Italian presence far more than they nurtured inter-tribal rivalries, and the *sottomessi*'s support for the resistance seems to have been sustained throughout the conflict. Men from these communities often replaced *muhafiziya* members who were killed, and the pacified population continually channelled funds and intelligence, as well as supplies of food, horses, guns and ammunition to the fighting groups. Thanks to the *sottomessi*, the Italians themselves were the source of much of the armoury of the *muhafiziya* (Wright 1969). In addition, the Sanussi organised a taxation system by which a tithe was levied upon all Cyrenaicans. This funded the resistance effort and the caravans that were brought into Cyrenaica across the Egyptian border (Santarelli *et al.* 1986). However, although the Sanussi would even provide written receipts, the resistance fighters were not always supplied willingly. There were continual examples of domination *within* the broader Bedouin resistance as the *muhafiziya* routinely extracted taxes or 'religious dues' from the *sottomessi*. The record is vague on this issue, but it seems that the Sanussi-led forces would seize supplies and funds by force if necessary, or exact reprisals upon communities that resisted them (Evans-Pritchard 1949). Again then, any notion that the resistance was an entirely homogeneous movement is fractured on this question, and it was these disputes that the Italians tried to exploit by arming some elements of the *sottomessi*.

Nevertheless, despite these tensions and the unfortunate position of the *sottomessi* as subject to both Italians and *muhafiziya* demands, in general the settled and semi-nomadic *sottomessi* provided crucial support for the resistance. Their camps were places of rest and refuge for active resistance fighters. They also provided some degree of safety and relative anonymity from Italian surveillance – for the *muhafiziya* could pass as peaceful herders among their families in camp (Rochat 1973; Santarelli *et al*. 1986). Certainly, the Italians struggled to tell them apart and to separate 'rebels' from supposedly 'pacified' Bedouin (Graziani 1932, 1937; Pace 1932). This situation was unfortunate for the *sottomessi* in other ways. One consequence was the increasing targeting of the camps by the Italians. Here, it was women, the aged and the young who suffered these attacks more frequently than elements of the male population who were away with the *muhafiziya*. Here again we find fractures within the resistance movement, this time along lines of gender, age and vulnerability.

Despite these cleavages, in general, the continuing support of the *sottomessi* for the resistance forced the Italians to recognise that *all* of the population were involved in the struggle in some respect. In 1932, Graziani, the man who would eventually defeat the Sanussi resistance, admitted that:

> In essence, the Cyrenaican rebellion, was an expression of hostility to our rule that has been developed and consolidated in the peoples' spirit by the Sanussi . . . all of the population of Cyrenaica participated in the rebellion – on the one hand, the potential [rebels]: the so-called submissive population, on the other hand, those that are openly in the field: those that are armed. All of Cyrenaica, in a word, was rebellious.
>
> (Graziani 1932: 56–57)

It was the recognition of the implicated nature of all Cyrenaican society that prompted the strategies through which Graziani eventually crushed the resistance.

Theorising nomadism: Italian attitudes

The war dragged on throughout the 1920s. Gradually, the Italians managed to pacify much of the coastal hinterland and to push their influence further into the wooded valleys and ravines of the Cyrenaican steppe, and towards the edge of the Sahara proper to the south (Santarelli *et al*. 1986). However, despite continual heavy losses, the *muhafiziya* remained an enduring problem for Italian colonial ambitions. In Italy, the Fascist regime had cemented its social control by 1925, and the regime became increasingly angered by the resistance in Libya. A string of officials were appointed to try to resolve the brutal colonial war in the desert; throughout this period, a particular Italian rhetoric developed in response to the resistance of the Sanussi-led nomads.

The nature of nomadic society and the category of nomadism itself became problematised as deviant and dangerous, and Italian incomprehension at their lifestyles and non-fixed territoriality led to nomads being portrayed as uncivilised and backward. As Evans-Pritchards writes:

> The Italians detested the Bedouin. Long years of campaigning against guerrilla bands, under Omar al-Mukhtar, who refused to submit in spite of heavy losses, had irritated them more than governments are usually irritated by Bedouin and made them increasingly flamboyant and brutal. In the whole Italian literature on Cyrenaica I have not read a sentence of understanding of the Romany way of life and its values. Because they lived in tents without most of the goods of the peasant, and even more the townsman, regard as a sign of civilisation, the Italians spoke of them as barbarians, little better than beasts, and treated them accordingly.
>
> (Evans-Pritchard 1946: 12)

The idea that the nomads were rooted in their natural desert landscapes and lived anti-modern lives also took hold among some Fascists. One colonial official, Biagio Pace, hinted at mysterious, deep-seated psychological reasons behind the resistance, and speculated upon the seemingly transcendental control of Omar al-Mukhtar over the population (Pace 1932). Such sentiments also informed the new military commander of Cyrenaica. A fresh, hard-line approach had been signalled in 1928 when Rome appointed Field Marshal Pietro Badoglio to the joint governorship of both Tripolitania and Cyrenaica (Segrè 1974). After a failed attempt to negotiate a settlement with Omar al-Mukhtar, Badoglio resolved that a military solution was his only viable option. To this end he appointed Graziani in March 1930, with the remit to finally defeat the Bedouin by whatever means. As mentioned earlier, Graziani had successfully quelled Tripolitania in the early 1920s by employing mobility and attacks upon Bedouin camps, so using 'Arab tactics against Arabs'. But in addition to his reputation for desert warfare and his experience of anti-tribal campaigning, Graziani would also propose his own theories about the nomads (Graziani 1932, 1937, 1948).

Graziani revealed a conventional European, conservative distaste for the Islamic Bedouin: 'The problems of nomads were not new in the history of colonisation', he wrote, 'all ancient and modern nations have had to exercise their authority and dominion over them' (Graziani 1932: 275). He reflected broader Fascist ideologies by rationalising nomadism as a pathological, 'primitive' condition of these people. He described a 'typical' Bedouin:

This is the nomad, anarchist, the lover of the most complete liberty and independence, intolerant of any restraint, a headstrong, ignorant, unconquerable, bluffing and boastful hero, it is enough for him to possess a rifle and a horse; [and] under the pretence of the necessity of moving his tent, he will disguise the desire to withdraw himself from every governmental contact and control.

(Graziani 1932: 189)

Furthermore, the Bedouin were

Rebellious against every tie of discipline, used to wandering in immense, desert territories, bold in mobility and ease of movement, and pervaded by a fascination with independence, they are always ready for war and raiding, the nomads have always resisted every governmental restraint.

(Graziani 1932: 191)

In conclusion, although Graziani seems to reveal a reluctant fascination with the Bedouin lifestyle and their unfettered movement (as he perceived it) across the desert, he also labelled this group a threat to the 'security and peace' of the colony (Graziani 1932). Their 'natural' anarchism and resistance to any European order or governance meant that the Bedouin simply could not be governed in any recognisable European manner. As such, the Bedouin situation was represented as being virtually intractable: how could Italy hope to settle its emigrant population in the colony while the incorrigibly rebellious Bedouin roamed around the region at will?

Furthermore, Graziani's theories reinforced other writing in Fascist Italy that justified Italian imperialism by reference to the 'primitive', 'ungoverned' peoples who lived in Italian colonies. Such conclusions were published as part of Italy's 'scientific' colonial surveys of Libya in 1937 (Atkinson 1996; Lando 1993). While his colleagues adapted eugenic sciences and the study of cephalic indexes to analyse southern Libyans (Cipriani 1937; Gini 1937), the geographer Emilio Scarin studied the 'population characteristics' of the Fezzan in the Saharan interior. He too found a people unsuited to the structures of modern society and government. He wrote: 'given the particular constitutions of families of the Arab type, it is an impossibility for the Fezzanesi to live independently, because of their indolence, [and] their poverty' (Scarin 1937: 609; see also Scarin 1934). The populations abutting southern Cyrenaica were thus dismissed as uncivilised and 'backwards'. These are only some examples of the kind of rhetoric that served to justify the brutal and uncompromising measures that Graziani would enact in Cyrenaica. They also provide some insight into Italian frustrations with the Bedouin conceptions of space and territory that jarred so fundamentally with

European conventions of fixed, bounded space. I suggest in this final section that the mobility employed by the anti-colonial resistance combined with this Italian distrust of nomads to tragic ends.

Confining nomadism

As Graziani had realised, the entire population was involved in the rebellion in some respect or other. He consequently talked disparagingly of the whole of Cyrenaica as

> a poisoned organism, but which has a festering bubo at one point on the body. In this case, the bubo is the fighting-band of Omar al-Mukhtar, [but this] results from the infection of the entire body. To heal this sick body, one must destroy the origins of the illness that cause the bubo.
>
> (Graziani 1932: 64)

His contempt for the nomads, their apparently anarchic lifestyles and their 'infection' of the entire country was compounded by his irritation at their ability to forestall Italian plans for Cyrenaica. Armed with his instructions to finally defeat the Sanussi-Bedouin at whatever cost, Graziani initiated what the Italians now called a 'War without quarter' (Graziani 1948: 63). The Sanussi-Bedouin resistance bands were harried still more: Graziani increased the mobility of his forces yet again, while desert wells were poisoned to restrict the movement of the *muhafiziya* around the interior (Wright 1969). The Sanussi lodges in these regions were also closed, and their *Shaikhs* were exiled to the Italian prison-island of Ustica. However, it was the 'pacified' *sottomessi* who now became the prime focus of Italian efforts to crush the rebellion.

Although reprisals against individuals had been harsh, and the surveillance and control of certain *sottomessi* communities had been stringent, it had proved impossible for the Italians to stem the flow of recruits, supplies and funds to the *muhafiziya*. In consultation with his officials and superiors in Rome, Graziani determined that if *all* the population was implicated in the rebellion, then the entire nomadic and semi-nomadic tribal population of Cyrenaica had to be brought under total Italian control before the *reconquista* could be completed (Graziani, 1937; Santarelli et al. 1986). To this end, he authorised fierce new measures against the *sottomessi*. To halt the supply of arms to the *muhafiziya*, the tribes were disarmed and a military court was flown around the country, dispensing summary 'justice' to any inhabitants found in possession of arms, or suspected of assisting the rebels (Rochat 1973). However, the most lethal measure was the creation of a series of concentration camps that would eventually contain over 100,000 people and 600,000 livestock (Bosworth 1996; Del Boca 1988).

Figure 4.1 The concentration camp at el-Abiar
Source: Graziani 1937: 272–273

The first camps were established in January 1930, and more were constructed in
the following months. Throughout this period, the population of Cyrenaica was
systematically rounded up and marched into confinement (Rochat 1973; Salerno
1979). Even the *sottomessi* communities who had provided the least overt resistance
to the Italians were nevertheless herded into the camps (Evans-Pritchard, 1949).
By the summer, the entire nomadic and semi-nomadic population were crowded
into barbed-wire encampments. The image of the concentration camp at el-Abiar
(Figure 4.1) demonstrates the nature of their confinement; el-Abiar was about 50
kilometres inland from Benghazi. It was one of the smaller camps and conditions
there were among the best in the system: it boasted two teachers, a first-aid station
and a medical tent (Graziani 1948; Ottolenghi 1997; Santarelli *et al*. 1986). At the
lowest estimate, it held 3,100 people, who were forced to pitch their tents in a
kilometre-square enclosure, upon an ordered, rectilinear pattern with broad
'corridors' that were designed to aid the surveillance of the Bedouin (Santarelli
et al. 1986). By contrast, Ottolenghi (1997) claims a total of over 1,200 tents,
with Bedouin family groups in each totalling some 8,000 inmates. Whatever the
figures, the camp and its barbed-wire fences materialised European notions of a
bounded territoriality; they finally forced the Bedouin to live within a disciplined,
controlled, fixed space – in contrast to their traditional conceptions of group
encampments and unfettered movement across territory.

Just as Bedouin senses of territory and mobility were crushed by the camps, so too the needs of their livestock were disregarded. The Bedouin herds were permitted to graze only within a given distance of the camps or they were confiscated. Inevitably, the stocks were decimated when the allocated grazing land was exhausted (Moore 1940). The herds had constituted the foundation of the Bedouin pastoral economy, which was subsequently ruined. Guerri (1998) considers this to be a deliberate policy, claiming that

> this was a true genocide that was perpetrated not only militarily, but also through a systematic extermination of the herds that would decimate their livestock resources, the only source of survival for the pastoral herders.
>
> (Guerri 1998: 299)

A further corollary was a change in the traditional Bedouin diet within the camps. From a diet of regular meat and plentiful milk, the Bedouin were forced to survive upon the desultory rations of tinned food that the Italians supplied to the prisoners (Rochat 1973). The Italians boasted of providing the first taste of sanitation and 'western' healthcare for the nomads (Pisenti *et al.* 1956), although even the most basic hygeine provision was often lacking and typhus was endemic in the camps (Santarelli *et al.* 1986). The change in diets, the insufficient rations and the cramped conditions in the camps meant that many Bedouin also succumbed to disease, ill-health and malnutrition. The mortality rates were appallingly high and the *lowest* estimates of fatalities in the camps start at 35,000 (Abdullatif Ahmida 1994). Quite aside from the deliberate destruction of the Bedouin livestock, the camps themselves were particularly lethal places. This was no accident. Graziani wrote to Badoglio in August 1930 claiming that

> the government is calmly determined to reduce the people to most miserable starvation if they do not fully obey orders. The same severity will be meted out to all those outside who act on their behalf.
>
> (Santarelli *et al.* 1986: 78)

It seems clear that the Italians were prepared to persist with this policy for as long as it took for the Sanussi-Bedouin resistance to be contained and crushed.

Graziani's primary intention had been to cut the supply lines to the rebels; with the incarceration of virtually all of the *sottomessi* population he succeeded. The Sanussi-Bedouin bands had been fighting for over seven hard years, and when the Italians managed to choke still further their support from the 'pacified' population, their situation became even worse. Moreover, Graziani also planned to restrict the *muhafiziya*'s final source of support. All remaining supplies were delivered via caravans that brought arms, food and other goods from across the Egyptian frontier, where sympathetic groups, Sanussi lodges and the covert support of the Egyptian authorities combined to provide assistance to the rebels (Santarelli *et al.* 1986). The Italians also solved this problem with barbed wire. Graziani ordered the construction of a 282 km long fence, to run from the Mediterranean coast southwards along the Egyptian frontier (Figure 4.2) (Graziani 1937; Pace 1932). The fence was constructed in six months and finished by September 1931, measuring 30 feet wide and 5 feet high. It was patrolled by mechanised units operating from a series of forts and by aircraft from a series of airfields. A telephone system ensured the coherence of the surveillance along the length of the fence (Wright 1969; Zoli 1949). Although incongruous in the midst of the Saharan landscape – particularly given the use of modern military and communications technologies – here again, Italian conceptions of fixed, impassable boundaries were eventually materialised, in this instance, by territorialising the desert interior along Italian lines.

Although some supplies still reached the diminishing *muhafiziya*, assistance from Egyptian sources were now also largely denied to the Sanussi-Bedouin. Commentators are agreed that these measures signalled the end of the rebellion. When Omar al-Mukhtar was captured, tried and executed in September 1931 in

Figure 4.2 The fence along the Libyan–Egyptian frontier
Source: Graziani 1937: 320–321

the concentration camp at Soluch, before an audience of 20,000 Bedouin who had been forced to attend, the campaign was virtually over. Without their charismatic leader, most remaining rebels were caught or fled to Egypt. Badoglio declared the rebellion vanquished on 24 January 1932 (Santarelli *et al.* 1986).

After the war

Throughout the 1930s, the Italian regime would invest remarkable sums of capital into the creation of Italy's 'Fourth Shore' – a colonial realm to the south of the Mediterranean stocked by settler families who farmed and cultivated the coastal regions (Del Boca 1988; Fowler 1972; Fuller 1992; Ipsen 1996; Segrè 1974, 1987; von Henneberg 1994, 1996). Yet beyond 1932, the Italian literatures mention little of the Sanussi and the Bedouin: decimated and crushed, they were allotted no significant roles in the making of this Fascist utopia. However, their dangerously amorphous sense of territory and their lack of respect for European-style boundaries ensured that after nine years of conflict, the colonial authorities were unwilling to allow the population simply to disperse across the areas of Cyrenaica now earmarked for Fascist settlers. The Italians debated maintaining the camps as permanent settlement sites for the Bedouin, but realised that the conditions would inevitably wipe out the remaining Cyrenaicans (Santarelli *et al.* 1986). Instead, most of the surviving population were released from the concentration camps in 1932, although their oppression continued. They encountered restrictions upon their movements and the spaces that they might occupy. Movements of tents and peoples were observed and recorded. Even everyday tasks and journeys were policed by Italian sentries who permitted movement only with an appropriate travel-permit (Evans-Pritchard 1946). Unregulated mobility still carried the penalty of imprisonment.

In later years, the Italians would portray themselves as unique among the European colonial powers in their sensitivity towards Islam and the cultures of their Libyan subjects (Bono 1989; Evans-Pritchard 1946). However, from 1932 onwards, a whole series of regulations and restrictions constrained the Bedouin spatially, and effectively denied them their traditional lifestyles. They were forced into the structures of European bounded spaces and subjected to Italian terri-torialities. Once established at such great length, Italian control over its colonial domain, and domination over the defeated Bedouin resistance, was sustained through the continual policing and disciplining of space.

Conclusion

In 1924, writing in an early edition of *Foreign Affairs*, an Italian called Carlo Schanzer reassured the American readership that

Under the friendly guidance of the Italian Government the patriarchal simplicity of tribal life in Cyrenaica has been gradually [improved, yet] . . . where the benefits of civic organization are refused by recalcitrant natives who want to continue under the tyrannical and arbitrary rule of their petty feudal lords, Italy takes such measures to re-establish order as seems advisable in the given case.

(Schanzer 1924: 456)

Tribal life in Cyrenaica may well have been patriarchal, and the various rivalries and oppressive practices of petty feudal lords, or the Sanussi fraternity, undoubtedly complicated the lives of individuals and groups within Bedouin society. However, while these extant instances of domination are significant and argue against a simple narrative of colonising Fascists oppressing an heroic, nomadic resistance, the gradual imposition of Italian colonial domination over Cyrenaica seems to me to have only compounded the entangled matrix of oppression and resistance in the region. In particular, as the 1920s progressed, the 'measures to re-establish order' that the Italian colonial officials found 'advisable' in response to the Sanussi-Bedouin resistance became more and more severe each year. By some estimates, between half and two-thirds of the Cyrenaican population died in the Italo-Sanussi wars between 1911 and 1923 (Evans-Pritchard 1949). The majority of these deaths occurred as a result of the incarceration of the entire nomadic population in the concentration camp system of 1930–1932. Even official Italian figures admit that the population of Cyrenaica declined from 225,000 in 1928 to 142,000 in 1931 (Segrè 1982). By any standards, this is an appallingly high casualty rate, and Del Boca is surely justified in reminding us that: 'In no other Italian colony did the repression assume, as in Cyrenaica, the character and the dimensions of an authentic genocide' (Del Boca 1988: 183).

These horrific consequences developed from a conflict over territory that was fought out through questions of mobility across space. Underpinning this struggle were the differences between Italian and Bedouin conceptions of space and territory. European notions held that Cyrenaica and the Sahara were empty spaces and that all nomadic peoples could be classified in the same manner. In fact, there were a series of hierarchies and tensions within the Bedouin resistance, and the Bedouin population had an extant series of territories and striations that they defended against the Italians.

By contrast, the Italians desired their colonial subjects to be fixed, controlled and submissive – and the movements of the Bedouin and the nature of nomadism were constructed as threats to colonial order. The Bedouin, however, eventually resisted the Italians by *accentuating* their mobility as a conscious strategy: they fought the war with no regard for the boundaries that the Italians were trying to define. Although they perplexed and enraged their enemy as a consequence, they also resisted superior Italian forces for nine years. The conflict was concluded only after the Italians had invested huge amounts of effort, funds and time into defeating a much smaller enemy; ultimately, the colonial authorities had resorted to total institutions to quell and control the Cyrenaican population.

This history of the Cyrenaican war is little known, but it does seem to compromise contemporary inter-war imaginations that romanticised the desert as a liberating space of free movement and liberty, and which celebrated the Bedouin as enjoying simple, pre-modern lifestyles. Likewise, when modern theory reproduces the casual metaphor of the desert nomad as an example of a de-territorialised subject, it too runs the risk of eliding some of the very brutal histories of sedentarisation that have marked North Africa.

Acknowledgements

My thanks to the editors for their comments and their patience and thanks also to Philip Morgan. Unless cited directly, I have used the English versions as followed in Santarelli *et al.* (1986) for all Libyan terms and proper nouns. Unless indicated otherwise, all the translations are mine; unfortunately, all the mistakes are my responsibility too.

References

Abdel-Jaouad, H. (1993) 'Isabelle Eberhardt: portrait of the artist as a young nomad', *Yale French Studies* 83(2): 93–117.

Abdullatif Ahmida, A. (1994) *The Making of Modern Libya: State Formation, Colonization, and Resistance, 1830–1932*, Albany, NY: SUNY Press.

Atkinson, D. (1996) 'The politics of geography and the Italian occupation of Libya', *Libyan Studies* 27: 71–84.

Atkinson, D. (1998) 'Totalitarianism and the street in Fascist Rome', in N. Fyfe (ed.) *Images of the Street: Planning, Identity and Control in Public Space*, London: Routledge.

Bono, S. (1989) 'Islam et politique coloniale en Libye', *Maghreb Review* 13(1–2): 70–76.

Bosworth, R. J. B. (1996) *Italy and the Wider World, 1860–1960*, London: Routledge.

Braidotti, R. (1994) *Nomadic Subjects*, New York: Columbia University Press.

Cipriani, L. (1937) 'Abitanti: Caratteri Antropologici' in Reale Società Geografica Italiana, *Il Sahara Italiano: Parte Prima: Fezzàn e Oasi di Gat*, Rome: Reale Società Geografica Italiana.

Cresswell, T. (1997) 'Imagining the nomad: mobility and the postmodern primitive', in G. Benko and U. Strohmayer (eds) *Space and Social Theory: Interpreting Modernity and Postmodernity*, Oxford: Blackwell.

Dawson, G. (1994) *Soldier Heroes: British Adventure, Empire and the Imagining of Masculinities*, London: Routledge.

de Grazia, V. (1981) *The Culture of Consent: Mass Organization of Leisure in Fascist Italy*, Cambridge: Cambridge University Press.

Del Boca, A. (1986) *Gli Italiani in Libia: Tripoli bel suol d'amore 1860–1922*, Milan: Mondadori.

Del Boca, A. (1988) *Gli Italiani in Libia: Dal fascismo a Gheddafi*, Milan: Mondadori.

Del Boca, A. (1991a) *Le Guerre Coloniali del Fascismo*, Rome: Laterza.

Del Boca, A. (1991b) *Guerra Italiane in Libia e in Etiopia: Studi Militari 1921–1939*, Treviso: Pagus.

Del Boca, A. (1996) *I Gas di Mussolini: Il Fascismo e la Guerra d'Etiopia*, Rome: Riuniti.

Deleuze, G. and Guattari, F. (1986) *Nomadology: The War Machine*, trans. B. Massumi, New York: Semiotext(e).

Deleuze, G. and Guattari, F. (1987) *A Thousand Plateaus*, Minneapolis: University of Minnesota Press.

Driberg, J. H. (1937) 'Introduction', in K. Holmboe, *Desert Encounter: An Adventurous Journey through Italian Africa*, trans. H. Holbek, London: Harrap.

Evans-Pritchard, E. E. (1945a) 'The Sanusi of Cyrenaica', *Africa* 15(2): 61–79.

Evans-Pritchard, E. E. (1945b) 'The distribution of Sanusi lodges', *Africa* 15(4): 183–187.

Evans-Pritchard, E. E. (1946) 'Italy and the Bedouin in Cyrenaica', *African Affairs* 45: 12–21.

Evans-Pritchard, E. E. (1949) *The Sanussi of Cyrenaica*, Oxford: Clarendon.

Falasca-Zamponi, S. (1992) 'The aesthetics of politics: symbol, power and narrative in Mussolini's Fascist Italy', *Theory, Culture and Society* 9: 75–91.

Finaldi, G. (1997) 'Adowa and the historiography of Italian colonialism', *Modern Italy* 1(3): 90–98.

Fowler, G. L. (1972) 'Italian colonisation of Tripolitania', *Annals of the Association of American Geographers* 62(4): 627–640.

Fuller, M. (1992) 'Building power: Italian architecture and urbanism in Libya and Ethiopia', in N. Alsayyad (ed.) *Forms of Dominance: On the Architecture and Urbanism of the Colonial Enterprise*, Aldershot: Avebury.

Gentile, E. (1996) *The Sacrilization of Politics in Fascist Italy*, trans. K. Botsford, Cambridge, MA: Harvard University Press.

Gini, C. (1937) 'Condizioni demografiche', in Reale Società Geografica Italiana, *Il Sahara Italiano: Parte Prima: Fezzàn e Oasi di Gat*, Rome: Reale Società Geografica Italiana.

Graziani, R. (1932) *Cyrenaica Pacificata*, Milan: Mondadori.

Graziani, R. (1937) *Pace Romana in Libia*, Milan: Mondadori.

Graziani, R. (1948) *Libia Redenta: Storia di Trent'anni di Passione Italiana in Africa*, Naples: Torola.

Guerri, G. B. (1998) *Italo Balbo*, Milan: Mondadori.

Heffernan, M. J. (1991) 'The desert in French Orientalist painting during the nineteenth century', *Landscape Research* 16(2): 37–42.

Holmboe, K. (1937) *Desert Encounter: An Adventurous Journey through Italian Africa*, trans. H. Holbek, London: Harrap.

Ipsen, C. (1996) *Dictating Demography: The Problem of Population in Fascist Italy*, Cambridge: Cambridge University Press.

Kaplan, C. (1987) 'Deterritorializations: the rewriting of home and exile in western feminist discourse', *Cultural Critique* 6: 187–198.

Kaplan, C. (1996) *Questions of Travel: Postmodern Discourses of Displacement*, Durham, NC: Duke University Press.

Lando, F. (1993) 'Geografia di casa altrui: l'Africa negli studi geografici italiani durante il ventennio fascista', *Terra d'Africa* 73–124.

Maggi, S. (1997) 'The railways of Italian Africa: economy, society and strategic features', *Journal of Transport History* 18: 54–71.

Malgeri, F. (1970) *La Guerra Libica*, Rome: Edizioni di Storia e Letteratura.

Moore, M. (1940) *Fourth Shore: Italy's Mass Colonization of Libya*, London: Routledge.

Morgan, P. J. (1998) 'The prefects and party–state relations in Fascist Italy', *Journal of Modern Italian Studies*, 3(3): 241–272.

Mudimbe, V. Y. (1994) *The Idea of Africa*, Bloomington, IN: Indiana University Press.

Muecke, S. (1984) 'The discourse of nomadology: phylums in flux', *Art and Text* 14: 24–40.

Ottolenghi, G. (1997) *Gli Italiani e il Colonialismo: I Campi di Detenzione Italiani in Africa*, Milan: Sugarco.

Pace, B. (1932) Il Fascismo e la Riconquista della Libia, *La Rassegna Italiana* 15(3): 58–72.

Paris, M. (1991) 'The first air wars: North Africa and the Balkans, 1911–13', *Journal of Contemporary History* 26: 97–109.

Patton, P. (1988) 'Marxism and beyond: strategies of reterritorialization', in C. Nelson and L. Grossberg (eds) *Marxism and the Interpretation of Culture*, Urbana: University of Illinois Press.

Peters, E. L. (1990) *The Bedouin of Cyrenaica: Studies in Personal and Corporate Power*, Cambridge: Cambridge University Press.

Pisenti, P., Sardi, A., Canevari, E., Bocca, M., Belfiori, M., Villari, L., Pascal, P. and Teodorani, V. (1956) *Graziani*, Rome: Revista Romana.

Pistolese, G. E. (1932) 'La Libia nella politica Mediterranea Italiana', *La Rassegna Italiana* 15(3): 142–151.

Rochat, G. (1973) 'La repressione della resistenza araba in Cirenaica nel 1930–31, nei documenti dell'archivio Graziani', *Il Movimento di Liberazione in Italia* 110: 3–39.

Romm, J. (1993) *The Edges of the Earth in Ancient Thought: Geography, Exploration and Fiction*, Princeton, NJ: University of Princeton Press.

Root, D. (n.d.) 'Sacred landscapes/colonial dreams: the desert as escape', *Lusitania* 1(4): 25–32.

Salerno, E. (1979) *Genocidio in Libia*, Milan: Sugarco.

Santarelli, E., Rochat, G., Rainero, R. and Goglia L. (1986) *Omar Al-Mukhtar: The Italian Reconquest of Libya*, trans. J. Gilbert, London: Darf.

Scarin, E. (1934) *Le oasi del Fezzàn: Richerche ed osservazione di geografia umana*, Bologna: Zanichelli.

Scarin, E. (1937) 'Insediamenti umanie tipi di dimore nel Fezzàn e Oasi di Gat', in Reale Società Geografica Italiana, *Il Sahara Italiano: Parte Prima: Fezzàn e Oasi di Gat*, Rome: Reale Società Geografica Italiana.

Schanzer, C. (1924) 'Italian colonial policy in Northern Africa', *Foreign Affairs* 2(3): 446–456.

Segrè, C. G. (1974) *Fourth Shore: The Italian Colonization of Libya*, Chicago: University of Chicago Press.

Segrè, C. G. (1982) 'Libya (Cyrenaica and Tripolitania)', in P. V. Cannistraro (ed.) *Historical Dictionary of Fascist Italy*, London: Greenwood.

Segrè, C. G. (1987) *Italo Balbo: A Fascist Life*, Berkeley: University of California Press.

Thomas, N. (1993) *Colonialism's Culture: Anthropology, Travel and Government*, Cambridge: Polity.

Valenzi, F. (1932) 'La Senussia in Cirenaica ed il suo Patrimonio', *Rivista delle Colonie Italiane* 4: 432–3

Volpi di Misurata, G. (1926) *La Rinascita della Tripolitania*, Milan: Mondadori.

von Henneberg, K. (1994) 'Piazza Castello and the making of a Fascist colonial capital', in Z. Çelik, D. Favro and R. Ingersoll (eds) *Streets, Critical Perspectives on Public Space*, Berkeley: University of California Press.

von Henneberg, K. (1996) 'Imperial uncertainties: architectural syncretism and improvisation in fascist colonial Libya', *Journal of Contemporary History* 31: 373–395.

Wright, J. (1969) *Libya*, London: Ernest Benn.

Wright, J. (1988) 'Outside perceptions of the Sanussi', *Maghreb Review* 13(1–2): 63–69.

Wright, J. (1989) *Libya, Chad and the Central Sahara*, London: Hurst.

Zoli, C. (1949) *Espansione Coloniale Italiana, 1922–1937*, Rome: L'Avvia.

5

THE NEIGHBOURHOOD AS SITE FOR CONTESTING GERMAN REUNIFICATION

Fiona M. Smith

Wir wollen raus!
Wir bleiben hier!
Wir sind das Volk!
Wir sind ein Volk!

We want out!
We are staying here!
We are the People!
We are one People!

(Key slogans of the GDR
'revolution' of 1989–1990)

The events of autumn 1989 in the German Democratic Republic (GDR), and across eastern Europe, marked an explicit redefinition of political subjectivity. This was manifested most clearly in the mass demonstrations of people chanting 'We are the People' (*wir sind das Volk*). This redefinition of the political was achieved partially through a mobilisation of spatial tactics which simultaneously redefined the spaces of the political, claimed the right to determine these spaces and thereby redefined the political itself. The transgression of state borders by many (we want out; *wir wollen raus*), the refusal of others to leave and their demands that the population not be expelled (we are staying here; *wir bleiben hier*) and the use of public streets for protest – all shaped specific local geographies of protest. People experienced the possibility of challenging structures of domination and control: 'we learned we were in the position to bring down a government . . . It was beautiful and very easy' (leaders of Neues Forum civic movement in Leipzig, December 1989, quoted in Neues Forum 1989: 26–27). As the character of mass demonstrations subsequently altered and those marching demanded unification with western Germany, chanting 'we are one People', '*wir sind ein Volk*', a further redefinition of political space and political subjectivity (quite literally) took place.[1]

The political trajectories subsumed even under these four short slogans mark German (re)unification as a multiply contested process involving, in common with other post-communist transformations, the 'reworking of the political subject' (A. Smith and Pickles 1998: 12).[2] In line with a range of writers, I argue in this chapter for a consideration of the dimensions of contest shaping post-communism and for a serious engagement with the geographies of such processes: 'social location and cultural process cannot be left out of our attempts to understand the contours of post-communist transition' (Staddon 1998: 351; see also A. Smith and Pickles 1998; A. Smith and Swain 1997; Pred and Watts 1992; Bauman 1992). Specifically here I ground discussion of post-communist transition in empirical material from fieldwork on neighbourhood activism in the eastern German city of Leipzig in the early 1990s.

In contrast to this consideration of diversities and differences, the approach which has been a dominant interpretation of post-communist 'transition' arises from 'a reworked modernisation theory' and assumes the end of the Cold War reduced the choices available for development to one only, namely western capitalism (A. Smith and Pickles 1998: 4). These forms of 'neo-evangelistic rhetoric' (Derrida 1994 quoted in A. Smith and Pickles 1998: 7) or transition 'fantasies' (Sidaway and Power 1998: 408) are underpinned by a moral economy which depicts a particular model of modernisation, marketisation and democratisation as not only correct but also inevitable. Contrary to these normative assertions, this chapter contributes to the growing range of critical studies which explore the non-linear nature of transitions (A. Smith and Pickles 1998: 2), whether in post-communist states or elsewhere (Radcliffe and Westwood 1996). Indeed, as Hebdige (1993: 270) argues, 'After 9 November 1989 one thing, if nothing else, is certain: nothing, absolutely nothing . . . is *ever bound to happen*' (original emphasis). Both the neo-liberal theses of the right and certain theories of the left on transition between phases of capitalism (A. Smith and Swain 1997) give little consideration to the variety of transition and transformation processes, to their highly contested nature (other than in debates between those accepting and those failing to recognise the inevitability of the 'correct' form of development), or to the fundamental significance of the geographies of these processes. Studies which stress the existence of multiple 'pathways' of post-communist development address the absences of such work by exploring the institutional bases for particular trajectories of change. They note the limitations and possibilities offered by structural legacies of all forms (economic, political, social organisation, and so on) and examine how such factors come together in distinctive regional 'assemblages' to produce a variety of 'transitions' (Harloe 1996; Láng-Pickvance *et al.* 1997; A. Smith and Swain 1997; Grabher and Stark 1998).

However, I argue here that the working out of transformation processes does not simply coalesce in specific locations. Rather, struggles over the 'highly divisive issue of what normatively constitutes "transition"' (Staddon 1998: 350) operate in and through geographies which shape these processes and which are themselves contested, (re)defined and transformed. The geographies of domination and resistance which constitute post-communist transformation are multiply contested and intimately intertwined. This examination of urban renewal in the neighbour-hoods of Leipzig suggests the need to recognise how people and institutions actively seek to redefine and relocate the processes of reunification in ways which are not adequately defined either by convergence with 'standard' western processes (Mayer 1995), or by reactions against the communist past in the GDR or by rejections of western development. The Treaty of Unification and its consti-tutional emphasis on the social market economy (article 1, para. 3) apparently confirmed a singularly rapid and definite extension of the pre-existing western German system to the territory and population of the (former) GDR, leaving only questions of the 'proper' administration of transitional arrangements. However, far-from-smooth subsequent developments indicate that the act of reunification 'did not end history but displace it in a new narrative plot involving a renewed struggle over legitimation of narrative standpoint' (Borneman 1992: 312).

Drawing on fieldwork carried out in Leipzig between 1991 and 1993 and on secondary sources dealing with the period before that, this chapter illustrates how the meanings and practices of the 'political' in this particular post-communist transition are contested and that their contestation is a complex and multi-dimensional process which defies the drawing of any simple lines of 'domination' versus 'resistance'. The relations of domination/resistance which constitute reunification converge in specific sites, such as the neighbourhood. Diverse relations intersect in this site with a plurality of identities and are mediated through a variety of institutional contexts to shape the neighbourhood and the social relations surrounding it. However, individual neighbourhoods are also identifiable places, the materialities of which may disrupt or reconfigure these relations of domination/resistance. Thus, the complex and often contradictory intersections of these multiple factors create spaces in which the processes surrounding the neighbourhoods are themselves contested in 'struggles for interpretative power' (Radcliffe and Westwood 1996: 46). These somewhat abstract notions are explored here through an example of the intersection of the debates and processes around local activism in Leipzig in a specific neighbourhood example and then through the interconnected geographies of this activism in relation to the contested spatialities of 'the neighbourhood', 'the city' and 'East' and 'West'.

'Democratic participation by the citizens is indispensable for a functioning city': [3] debating neighbourhood change in Leipzig

At a public meeting in April 1993 in an inner city neighbourhood in Leipzig, the local 'citizen initiative', the city council's Urban Renewal Office (URO) neighbourhood team, a representative of the city's publicly owned social housing company (Leipziger Wohnungsbaugesellschaft: LWB) and a range of elected councillors, residents and other interested parties came together to debate the dilemmas raised by plans to upgrade the area. This was part of a widespread programme to replace the area-clearance planning of the GDR with more environmentally and socially sustainable forms of development. The debates discussed here illustrate key arguments over the possible way forward and lead me on to consider understandings of 'the neighbourhood', 'the residents' and 'the citizens' and of the nature of transition involved in local struggles over the future of the neighbourhoods. It is possible to identify a range of standpoints taken at the meeting but although they indicate different interpretations of these processes, individuals and organisations do not (necessarily) consistently adopt single positions. Furthermore, as the later part of the chapter argues, the effects of particular actions often escape their intentions.

This neighbourhood, home to about 4,500 people in 1989, and constituted largely of badly degraded nineteenth-century working-class tenements, was one of around twenty areas in Leipzig designated for urban renewal status by the local council democratically elected in May 1990. Under existing and expanded federal programmes, it was eligible for a range of public funding. To access this funding, planning had to demonstrate local consultation. In this neighbourhood, although not in all designated areas, a private 'renewal agent', in this case a planning consultancy from western Germany, was appointed to draw up proposals. These appointments were deemed necessary because of the lack of personnel in the city administration experienced in the newly introduced federal legislation (Osterland 1994; Wollmann 1991) and because of the extent of problems facing the city. There had been serious proposals in the late 1980s to demolish 53,000 of the 104,000 pre-1919 buildings in Leipzig in a widespread programme of inner city clearance and reconstruction.[4] These plans were shelved after 1989 and moves were made to develop new programmes. However, the processes of policy formulation and implementation were problematic and, for example, by spring 1993, the contract with the renewal agent for this particular neighbourhood had been terminated because of disagreements with the city planning department. At the time of the meeting, then, a neighbourhood team from the city council's own URO was in charge.

Having spent the years since 1989 campaigning for long-delayed renewal, neighbourhood residents were increasingly frustrated with the problems of the

Figure 5.1 Postcard produced by a neighbourhood initiative: collapsed residential
buildings, September 1992
Source: Bürgerinitiative Neustädter Markt e. V

Figure 5.2 Postcard produced by a neighbourhood initiative: the ambiguous beginnings of
local renewal, September 1992
Source: Bürgerinitiative Neustäder Markt e. V

proposed plans which would see considerable rent increases and which, they felt, left tenants largely at the mercy of property owners. In addition, because of the complexity and delays of the system of post-GDR property restitution, ownership of many buildings was often as yet unclear, causing further delays in any start to repairs (Hannemann 1993; Reimann 1997; F. M. Smith 1996b). A series of postcards produced by the local residents' initiative highlighted the continued degradation of the building stock and frustration at the lack of progress. One postcard shows a residential building in the area which collapsed in September 1992 while still inhabited (Figure 5.1). Another (Figure 5.2) shows what is an apparent sign of progress, a renovated building which houses the URO neighbour-hood office on the ground floor. However, the empty upper floors tell a story of lack of progress since they were to be used as temporary housing for families relocated during renovation work elsewhere. The effects of delays and disappointments mounted, as these extracts from later discussions with residents show.

[In] the past so much was promised . . . I know my building was supposed to be renewed in 1980 and it has been delayed until now and there is not much done apart from a bit of paint, a few patches on the roof. None of it is proper. And that, that makes the people so depressed, because they have always heard just promises and the promises continue.

(woman, group discussion, May 1993)

Everyone had notice to move out for 1990 and then [after 1989] every-one who lived here said, 'Well okay, you can hold on for two years.' Everything was always a bit delayed in the plan economy, you know! [Laughs] But some time at the beginning of the 1990s [something would happen] . . . and then nothing happened and the disappointment was really huge.

(woman, group discussion, May 1993)

The pressure of expectations was so great in 1989 . . . and now three years have passed, four years, and not much has been achieved. A few facades have been painted, a few houses fell down, many have moved away because nothing happened . . . The people are continually knocked down and have to get up again.

(woman, group discussion, May 1993)

At the public meeting in April 1993 the limitations and frustrations of the current situation were expressed by the URO team who then sought to explain the 'realities' of the financial and legal situation for the city administration and for tenants and owners. In their view this made impractical some suggestions from the citizen initative and some proposals from the former renewal agent for alternative and innovative forms of renewal.

> [Where building owners make use of federal subsidies and therefore agree to rent controls] in the private buildings, if there is just normal renovation, no luxury, you get to rents of 12–15 Marks [per square metre], just completely normal Federal German average standards . . . Buildings which are funded entirely from private capital [are] the ones where, let's say, the tenants are caught out [with much higher rents] and where there is no way of avoiding them to any great extent.
>
> (URO team)

> The city has 11 renewal areas, or will even have 21 in the end . . . you can see it is the same situation everywhere. In GDR times many moved out to new buildings, it's mostly the elderly, single people who stayed. Then I don't know how the City should ever be able to carry out the renewal if there aren't important contributions from the rents.
>
> (URO team)

Although they were candid about the need for wider political decisions to change the legal and financial frameworks for development, the local URO team stressed that there was no alternative but for current action to operate within existing financial and legal limitations. This would involve 'competent advice sessions for tenants' and good planning in those cases where owners made use of federal subsidies. Both could 'cushion' residents from the impacts of change. Generally, however, they argued that residents would have to realise the extent of change which was in prospect: 'I think it is high time that the population is really clear that the rent levels [between properties] will diverge and will diverge considerably.' This view was echoed by a local liberal (free market) councillor, who argued succinctly that 'the time of equal rents is past'.

Some proposing these arguments saw good administration of the existing legislation as the only possible response until such time as the legislative framework was altered to deal with the extremely poor urban environment and the lack of public finance – both circumstances for which the federal legislation was not designed.

> I analysed that there is not a single flat where the tenants have applied to buy their houses. There is a reason for this. The buildings are kaput. You cannot renovate them yourselves. That is an illusion . . . We cannot *force* any tenant or force the LWB to give us the flats for self-help projects. This initiative which comes from the population . . . with respect – it, it, it is just not possible any more.
>
> (URO team, original emphasis)

Others saw present difficulties more in terms of a temporary problem of fitting communist anomalies into a well-functioning western system.

> We simply have to recognise that in 1990 we took on the legal system of the Federal Republic, that we took on the employment markets, that we took on their conditions . . . That means that now we simply have to accept that rents to cover costs will lie around 20–30 Marks. Some tax deductions could get it down to around 15 Marks. After that there is only social housing.
>
> (Liberal councillor)

While these are in some senses statements of a situation, they are also normative and evaluative statements which prioritise particular courses of action and draw on quite distinctive understandings of 'transition'. They have in common an appeal to clear divisions between 'then' (the communist past) and 'now' (the western, social-market economy). This defines each as a distinct system and relegates practices and expectations which draw in any way on GDR experiences to the realm of that which is no longer valid.

On another occasion a URO planner explained to me what she saw as the lack of understanding among the population of the changes in the framework of urban renewal towards a system which can offer to facilitate actions by owners but cannot prescribe action or carry it out itself.

> Renewal is the responsibility of the owner. We can only *support* the owners. [This new relationship] is something which the people don't get. In the past the URO simply started . . . at one end and worked like a conveyor belt along the roof, or whatever, regardless of whether it was private or not . . . And the people must now slowly realise that is not possible today any more. And . . . that is difficult. When they come now we have to say, 'we are sorry but it is the owner's responsibility' . . . Now it's getting better. At the beginning . . . it was much more difficult. They came and said, 'My roof is leaking – I want it repaired.' Then you just had to say, 'we can't do anything' . . . Many still don't understand. [They say] 'he's getting a new roof, so when is it my turn?'
>
> (URO planner, interview, April 1993, original emphasis)

In contrast to these views, the local residents' initiative argued at the public meeting that accepting market forces and the limitations of the present system was not an option, since both would have what they defined as unacceptable outcomes.

> One can see two possibilities for development . . . and we hope that *neither* happens. One is the tendency [with further decay] towards a

129

sinking number of flats and residents . . . and the better-off and higher-qualified will leave the area. That means the homogenisation of the neighbourhood will continue and the population structure will be even more one-sided [i.e. poor] than it is now. . . . The second tendency is the . . . absolute improvement of the value of the area [through private development] so that the current residents won't be able to afford it.

(Citizen initiative, original emphasis)

Along with the now-sacked renewal agent, a councillor representing the civic movement, Bündnis 90/Die Grünen, and a representative of a Berlin-based organisation promoting self-help housing strategies, the local residents' initiative argued that the URO was not flexible enough in its approach to local renewal and should move away from practices which relied on funding renewal only through private owners or privatising state-owned housing.

If there is no money we still think [we could achieve more] with newer forms of renewal, that is, with renewal more aimed at the tenants, renewal together with the tenants, with closeness to the citizens to create certain conditions such as self-help programmes.

(Citizen initiative)

Somehow one must also try to give new ways of thinking about it. And what we want to suggest . . . is that we strengthen tenant initatives here. They did much in the GDR times too, many people . . . We are used to it, it is nothing new for us . . . We have to escape from these thought structures that we go to the bank, get money, renew the house, then sit back and cash in.

(Civic movement councillor)

In fact, one man in the public at this meeting explicitly rejected any form of market solution to the housing problems. He drew on socialist ideals of housing removed from market relations and suggested that given the difficulties and insecurities of the new market-based system, many local people could not understand its supposed superiority to the previous system.

I ask myself whether we can accept such conditions. What is to happen to the people? Where are they to go? . . . The question of social housing, of security in life . . . is completely wrongly directed. When housing . . . is regarded in the first instance as a commodity then I think many people will fall by the wayside . . . The people here will probably mostly never understand. You can come with rights, with explanations, with laws.

They work from their experiences. It possibly wasn't a good experience [in the GDR], you know, lots of decay, but it had a logic of security.

(Local man)

While his observation reflected some serious concerns, its implications for how local planning could proceed were limited, given the macro-political context. However, others sought not to contrast the new situation with the possibilities of the past, but to suggest a more hybrid form of development which saw the possibility of incorporating some experiences from the GDR (such as involvement in housing self-help strategies) with an understanding of western forms of development as something which was already highly contested. The local residents' group and others, including the western German planning consultancy, stressed that it was possible to find a variety of approaches in western Germany already.

At the moment unfortunately there is no concept here. A new one will take two years . . . but please, there are already experiences elsewhere. Why don't they get used? Why do you just take the laws and say, 'right now we have to create cost-covering rents'? . . . We should perhaps be arranging it in a way which is just to the people . . . In the Federal Republic, or old federal states, there are projects where they have tried out renewal 'close to the citizens', in Frankfurt, for example . . . where the tenants themselves said how the renewal was to happen.

(Citizen initiative)

This is far from a simplistic 'anti-western' stance. It draws on an awareness of diversity within and across the boundaries of 'East' and 'West'. These arguments fundamentally disrupt any notion of the process of reunification as a move to a predetermined, fixed western model in which all the answers had been established (for good or ill) and they posit both the 'goal' and the process of transition as things open to negotiation and interpretation. Such arguments draw on an alliance of local experience, political ideas and planning information on a range of experiences from eastern and western Germany. They move the debate away from how best to implement locally an already existing western system to fundamental questions about what form planning and urban renewal could and should take, asking whose interpretation was valid and about who should have the right to decide. In the debates of the public meeting, we find what are quite literally the 'struggles for interpretative power' mentioned earlier (Radcliffe and Westwood 1996: 46), struggles which, furthermore, affect the material, social and political conditions of life for local residents.

In contrast to what might be seen as more administrative and technocratic approaches from the URO team, local residents and their 'allies' at this meeting

131

saw urban renewal as a 'technique of transition' operating through particular planning and legal 'instruments' (A. Smith and Pickles 1998: 2–4), in and through which power relations were constituted but which were also open to negotiation and even contestation. At a later group discussion local initiative members argued that, strictly in terms of the planning regulations, consultation had taken place, but that problems lay in the inability actually to move forward material processes of renewal in a way which would not result in the displacement of the local population.

> Citizen participation is set by law – that the citizens should be involved. As long as the renewal agent was involved . . . they talked to the citizens . . . *But*. Nothing has actually happened in the renewal until now.
>
> (man, group discussion, May 1993, original emphasis)

Western planning structures and the instruments of privatisation and marketisation were, they argued, failing to deal with the situation presented precisely in the areas where they promised most – namely in the improvement of the material conditions of life and in the implementation of a democratised system of government. A quotation from a longer printed discussion of the problems of what 'consultation' might mean illustrates that rather than the universally positive development described by cheerful discourses of modernisation and liberalisation, there was increasing awareness of the limitations of the promises of western democracy and of disparities in implementing rights of involvement.

> 'Those affected [by planning] should be encouraged to participate in the renovation (para. 137, Federal Planning Act)' . . . because it's really partly their business. But it was no doubt also clear to the legislators that those affected are . . . also obstacles who disrupt the nice plans of the town planners . . . That's why the law is so vague. [In practice] firstly citizens were encouraged to participate so that [the planners can say they] follow the law. Secondly, the town planners did what was necessary so that the obstacles which come up in their planning don't get too big. You know – the main obstacles are the people, those affected, the tenants, or the like.
>
> (anonymous, printed in neighbourhood newsletter, December 1992)

Although URO planners were also frustrated at the slow progress of renewal, their statements at the public meeting that certain approaches were 'no longer possible' or were 'inappropriate' functioned to limit debate and to assert their own analyses as correct. The URO team did consult widely with the local population (attending this and other meetings, staffing the local neighbourhood office, offering advice and consultation for tenants and property owners). However, it was on the fundamental parameters of what such consultation should mean and what the 'point' of renewal was that disagreement arose.

> We need . . . renewal from the point of view of the population, of the population who live here and those who want to move here . . . But that would mean, eh, the renewal process happened with the citizens. And closeness to the citizens means that one speaks with the citizens. That means too that the responsible departments speak directly with the citizens and not just [make pronouncements] at large events.
>
> (Citizen initiative)

These questions were addressed in terms of the relative positioning of the citizens/ residents/tenants (all three terms were used) *vis-à-vis* professional planners and building owners in particular. The city housing company (LWB) representative and the liberal councillor, for example, emphasised the need for the vast majority of residents who were tenants to know their legal rights and to use them. In fact this was a significant issue across virtually all organisations working on housing issues in Leipzig at this time and formed the basis for considerable co-operation between official agencies and local activist groups (F. M. Smith 1996b). However, the extent to which this was seen as the only possible strategy, or one among many, varied. Those adopting an approach to reunification which assumed it to be an adaptation to a largely unchanged western system tended to emphasise the actions of individual citizens with respect to their legal rights as the *only way forward*. Others, including some officials in the city council's housing allocation and administration department (Wohnungsamt) suggested that this view was overly narrow. They argued for attention to be given to collective rights − such as the right for the neighbourhood population as a whole to be heard and for its overall social composition to be a legitimate cause for concern − *as well as* to personal rights associated with tenancies and ownership. In this they challenged what they saw as the limitations of the existing praxis of 'consultation' in some neighbourhoods and by some sections of the city administration.

> The problem is that at the moment particular repair measures are being carried out which the people aren't even being *informed* about. That's the problem . . . That means you are not even working with the people here . . . You have to work together with the people. You can't just wait for the initiative to come from them because they don't own their flats.
>
> (Citizen initiative, original emphasis)

These arguments interweave with those about the significance of the neigh-bourhood as a focus for renewal. If reunification is regarded as 'transition' to a by-and-large predetermined destination, local areas need to be administered in a way consistent with that goal. Individual neighbourhoods are only part of a 'larger picture' and planners or the city council use their ability to see this to make strategic decisions 'for the good of the city' (interviews with Social Democratic Party (SPD) and Christian Democratic Union (CDU) councillors, April 1992). In contrast, local activists and others argue that neighbourhoods are significant to the immediate concerns of their residents, that they they constitute meaningful social space and not only physical space. At a public meeting on planning in September 1992 in another neighbourhood, a member of the public argued the area was 'not just houses but people'. Demands that the views of the local population be heard, indeed that they be placed as the central focus of local renewal, thus disrupt claims by a range of professionals to know best. Such a manoeuvre postions reunification as a contestable set of processes. Before moving on to place this example in its wider context, however, I want to stress that this contestation does not operate on simple divides between professionals and tenants, or between eastern and western views. In particular, the notion of shaping urban development in Leipzig 'for the good of the city' was a goal which was supported by virtually all of those involved in these debates. What is at work here is instead a complex set of (often temporary and place-specific) alliances and tensions which emerge between differing *interpretations* of how this aim can be realised, and how 'the good of the city' might be defined. These relate, as we have seen, to debates about the form and purpose of urban governance and how transition is to be understood: 'what is at stake [is] the very reconceptualisation of "the political" *as such*; the political imaginary of post-communist "governmentality"' (Staddon 1998: 351).

(Dis)entangling the politics of neighbourhood change

This section of the chapter moves on to position the debates in this one neighbourhood in the context of the fractures and alliances involved in shaping urban renewal and its politics in Leipzig more generally. In particular, it explores the shifting discourses around urban renewal and the fluid relations of these discourses to a variety of institutions, organisations and individuals. While the interpretation of power implicated in this analysis is one which views power diffusing throughout social relations, it is also appropriate to note the limits which different people seek to draw around the extent of 'the political' and the different relations of power implicated in these processes. The approach here does not seek to identify single lines of 'pure' resistance or domination, where power lies with one or the other, although clearly some relations will tend more towards relations of domination than others. Rather this approach argues relations of domination and resistance are 'never in a position of exteriority in relation to power' (Foucault

1978: 95) and that, in relation to each other, 'neither domination nor resistance is autonomous' (Haynes and Prakash 1991: 3). Instead they are 'always entangled in some configuration' (Routledge 1997: 70). This interpretation of power as diffuse and the relations of domination/resistance as intertwined in a multiplicity of sites and social processes matches notions of the 'decentred self' – where human subjectivities are understood as multiple, fractured and shifting – and of the 'fractured and decentred social' – where no single structure of domination defines all lines of conflict (Radcliffe and Westwood 1996: 2–3). Stressing the interplay of forms of power and the spatialities of their contestation, this section recognises the existence of 'many liberties' and a 'plurality of antagonisms' (Radcliffe and Westwood 1996: 44).

The problems in this neighbourhood were not found uniformly across Leipzig. In each neighbourhood the nature of political contests and local problems varied with differing combinations of factors such as housing stock condition, socio-economic profile of the population, speed and dimensions of property restitution, time-frames for planning and local activism, the effects of capital (speculation, continued disinvestment, pressures for land-use change) and past and present forms of local activism (F. M. Smith 1996a; cf. Láng-Pickvance *et al.* 1997). Debates in this one neighbourhood nevertheless constitute part of wider and ongoing debates about urban change and the politics of the right to influence that change. In autumn and winter 1989/90, alongside demands for political reform, the mass demonstrations and newly public debates included demands for reform in urban policies. Leipzig's severely decayed and polluted urban environment was discussed in public events and in the local and national media. Well before the election of a city council in May 1990, newly independent architects and planners as well as activists (many were both) were developing agendas for local change which centred on discourses of 'careful renewal' and 'citizen involvement' (Initiativgruppe 1. Leipziger Volksbaukonferenz 1990). In reaction against the shortcomings and centralised control of GDR urban policies (Marcuse and Schumann 1992; Reuther *et al.* 1990a, 1990b; Rostock 1991) these agendas stressed the importance of existing neighbourhoods (rather than city-centre prestige projects and massive area clearance), the 'expertise' found in local citizens' knowledge and the need to avoid political dogma in favour of rational planning led by independent professionals (see F. M. Smith 1999). These concerns intersected with the resignation of the city council and the emergence of a range of civic action groups across Leipzig, both in neighbourhoods and more broadly thematically based (Stührmann 1991; Verein für ökologishes Bauen 1992).

As we have seen, tensions emerged in the post-reunification period in neighbourhood development and city planning as the discourses of the 'revolution' were carried forwards in the changing context of the (imperfectly implemented) normative and regulatory framework of (western) German legal, political and administrative systems, and of the (re)introduction of capitalist market relations to the city and region, producing socially and spatially highly uneven patterns of investment and disinvestment (Rat der Stadt Leipzig 1991; Schmidt 1991, 1994). Leipzig's 'pathways' of transition were already strongly influenced by the agendas and practices of this in-between period when bodies such as the Leipzig 'Round Table', the city's interim government, established a model of collaborative work 'for the good of the city', an aim in fact common to many post-GDR local administrations (Osterland 1994). The Round Table sought to elaborate forms of local self-governance and to avoid narrow party interests in policy formation and implementation. Its achievements in the short period of its existence were widely applauded.

> The city must deal with its own problems itself. What looked like a democracy game at the beginning became serious work *for the good of the city* . . . The idea of the Round Table does not belong [in a museum]. Rather the future councillors, some of whom possibly sat at this table, should maintain this approach: unselfish argument for the sake of the issue, distance from party tactical considerations, producing decisions which help our plundered city back to its feet.
>
> (Editorial, *Leipziger Volkszeitung* [local daily paper] 3 May 1990, original emphasis)

The Round Table allowed, indeed required, the involvement of members of the new civic movements and the quickly reforming party structures to introduce key policies and allowed these forms of political organisation to establish their legitimacy (Rink 1995). Innovations and new personnel were introduced to the city administration in environmental policy, education, urban renewal issues, equal opportunities policies, and so on. After local elections in May 1990 a power-sharing administration was formed, led by the newly re-established SPD with the reformed CDU and Bündnis 90, but excluding, in particular, the reformed communist party, the PDS. This period marked a time of great openness to new ideas and increased opportunities to implement them, particularly with the end of the GDR cadre system, but it also brought potential threats. New personnel in key positions in the city, often fresh from the new civic movements sought to implement their new styles of politics. Existing administrative personnel were allocated to new responsibilities and many were eager to exercise their new-found independence from party dogma. Subsequent pressures on public expenditure (the city faced bankruptcy at one point), threatened staffing cuts and political screening all meant city employees shared the general experience of uncertainty and restructuring. In this context city officials were pleased to have opportunities to exhibit their democratic credentials by consulting the population. The complexity and enormity of the task, which would have stretched even a well-oiled and experienced administration, left considerable scope for other channels of influence and early agenda-setting. As SPD Mayor Lehmann-Grube, one of the limited number of senior '*Westimporte*' (western imports) and formerly chief executive of Hanover City Council said, 'the old ways do not function any longer and the new ones do not function yet' (*Leipziger Volkszeitung* [local daily paper] 8/9 December 1990).

As well as changes in the city administration, several city-wide ecological and urban development projects and initiatives also developed over this period. They generally established strong linkages with the city planning administration and related departments and the overlap of personal careers between civic movement, administration and new initiatives was in some cases quite considerable. Several of the key figures in these groups had been influential in the earlier discussions of the 'careful renewal' agenda in the city and were able to use a series of professional contacts to influence the city's agendas. These groups were closely involved in the wide series of workshops and public debates on which the planning departments embarked after 1990, holding key policy-setting events together with the city (Elsässer 1991; Stadt Leipzig 1992). Representatives of these groups stressed the 'great chance' (interview, November 1991) for new forms of development to replace the mistakes of the 'East' but avoid the 'wrong' type of development the 'West' had experienced (interview, October 1991). They also chided some in the administration for what they saw as a failure to learn from more up-to-date practices in the west: 'some people have kept very . . . eh, backward looking ideas of planning' (interview, November 1991).

However, it would be a mistake only to see pressure for openness to civic involvement and new ideas as coming from outside the administration. For example, the head of the city's Education Department contributed to the programme for one of the major events mentioned above where he argued,

> The future belongs to this form of politics . . . I myself come from the civic movement . . . Citizen movements are for me an institutional instrument for citizens to take direct influence on politics and to concern themselves with alternatives. Citizen movements are not just reactive but active.
>
> (Wolfgang Tiefensee in Verein für ökologisches Bauen 1992: 18)

Similarly, the head of the Culture Department argued that citizen initiatives provided 'spaces for democratic experience' (Georg Girardet in Verein für ökologisches Bauen 1992: 18).

At the neighbourhood level, a series of the type of residents' initiative discussed in the first part of this chapter developed. Thirty-five were mentioned in the local paper from January 1990 to May 1993. They generally advocated similar agendas to those of the city-wide thematic groups, but their overall strategies were less focused on professional issues of planning praxis and more rooted in demands to 'save' their particular areas and to be active collectively and individually where this had not been possible before. They argued that neighbourhoods previously scheduled for renewal were in danger of being 'forgotten' in new plans or that the plans being drawn up failed to improve the generally poor housing and infra-structural conditions while protecting residents from displacement pressures following property privatisation.

> [The area] had been discredited with the administration [but now] the city has received from us some idea of the area, it is prepared to stand up for its ugly corners.
>
> (group leader, interview, November 1991)

> It is not my nature just to wait and be put out of my flat. One has to do something about it even if it is just something small . . . and I'm certainly in favour of the citizen initiative trying to do something to let the people who have lived here for a long time stay here.
>
> (woman, group discussion, May 1993)

This emergent neighbourhood activism not only predates *and* postdates the post-reunification planning structures, but also reacts against perceived problems and proactively seeks to shape the frameworks in which development proceeds. Far from being seen as a 'problem' to be overcome (Osterland 1994), the presence of such activist groups was regarded by many in the city administration as proof of Leipzig's civic and democratic development, with mention of them even being made in the economic development literature (Rat der Stadt Leipzig 1991). In some neighbourhoods the establishment of a citizen initiative was encouraged (in some cases facilitated) by the local branch of Bündnis 90, as part of their development of civic politics, or even by community arts workers employed by the city, in an effort to develop forums for expressing local opinions and developing neighbourhood projects.

Neighbourhood activist groups worked to create local public spheres, holding events which, in using a range of local public spaces, such as halls, squares, parks and schools, displayed to the local population and to a wider public the active construction of local agendas. They also explicitly set out to act as channels through which consultation on local development could take place, creating spaces where plans could be displayed and debated and where information could be exchanged between official bodies and the local population (Figure 5.3). In some cases the working relationship was good enough to allow the citizen group and the

Figure 5.3 Public consultation on planning proposals at a neighbourhood festival organised by a neighbourhood citizen association, Leipzig, May 1992
Source: Fiona M. Smith

planning officers to share an office. Several neighbourhood groups also took on direct service provision (such as after-school childcare), stepping in where facilities were lost in the transition from GDR to Federal Republic or exploring new forms of social action.

Clearly, then, neighbourhood action groups were part of a broader intersection of people, ideas and institutions, many of whom shared common goals. While at times the lines of conflict between, say, a neighbourhood and the city council might harden, over this period as a whole relations were much more fluid, with shifting alliances developing between a range of institutions, organisations, parties, individuals and sectors of the city administration. Some lines of tension also emerged within and between civic groups themselves. Notably these included debates around the interpretation and implementation of what appeared to be shared agendas. The agenda of 'careful renewal' had been established almost as a totem for urban planning. While it was widely supported, some civic groups were more able to define its characteristics than others, largely because of the professional backgrounds and greater cultural capital of their members. Very few people questioned its importance but in one discussion with members of the neighbourhood initiative from the first part of the chapter doubts were raised about whether it was an agenda appropriate for the social groups and physical conditions of their particular neighbourhood.

WOMAN 1: It doesn't make sense for many people. For me personally, for example, it is better to move away and not just to be active locally, otherwise you'll still be waiting ten to fifteen years. That's just too long . . . I don't want to stay living here. I can imagine very well how it might be . . . a neighbourhood done on ecological lines, where there are nice play areas and you have your bio-dustbin. I can imagine it all very well . . .

FMS: In retrospect, do you think the people in the area would have prefered the demolition and new construction to have happened after 1989 as had been planned?

WOMAN 2: I think so, . . . because they would finally have had something visible or tangible to show for it, and would finally have the pleasure of living in a decent flat . . .

WOMAN 3: It isn't so important for the people how the building looks. The main thing is I have a flat, which is dry and warm and I have warm water. And that it is affordable.

(group discussion, May 1993)

Given the poor conditions in this and other neighbourhoods, where even the basic conditions described by the last speaker were not guaranteed, the ideal of careful renewal seemed to some like a fanciful waste of energy.

There was debate within and between groups not only about the idea and praxis of careful renewal, but also about the desirability of a type of civic politics which some saw as always aiming at consensus and failing to take a stance on particular issues. At a meeting of leaders from a range of groups, one man whose group worked with young people in a neighbourhood rife with conflict and indeed violence between left-wing and right-wing factions argued that, in this situation, not being ready to argue strongly for one position or another was not a sufficient response (meeting of group leaders, October 1992). In fact, over time neighbourhood groups increasingly became frustrated in their dealings with the city planning processes about what they saw as a lack of willingness to engage in fundamental discussions. As the earlier discussion showed, neighbourhood groups argued that the administration's claims to be able to administer change in a non-conflictual way failed to take into account the need to prioritise between a variety of needs among the population and across different parts of the city. While most of the groups set out deliberately to generate inclusive strategies for their own local populations (conducting extensive social surveys of the composition and needs of the neighbourhoods' residents and small businesses, for example) the groups themselves had to work out how to deal with the idea that 'the citizenry' was not a uniform body but that different social groupings and different interests would produce a 'plurality of antagonisms' (Radcliffe and Westwood 1996: 46). However, most groups identified a fundamental problem in attracting social groups with problems, such as the elderly, those on lower incomes or teenagers, to be active since these groups were also least willing or able to engage in local activism.

Theoretical literatures disagree fundamentally on the 'value' of local action. Harvey (1989) famously argued that while social movements are able to control local place, capital increasingly controls space. Pickles (1993) suggests that locally focused environmental activism in Bulgaria might be a retreat or forced displacement from involvement in wider structural changes. In contrast, Marcuse (1990: 520) argued that in the then-still GDR bases of resistance included the strength of local community 'in the sociological sense because of the lack of socio-economic segregation in the city; and the tradition of local interest representation (both from the broad Left tendency and from the new social movements)'. In fact this very ambivalence characterises much of the debate among local groups themselves about the function and effectiveness of their action. Local action was for many explicitly a form which drew on the 'citizen politics' of the GDR revolution, or was a forum for individual and collective action which was possible now in a democratic system and had not been possible or worth it before.

> I am someone who was never active in the past, because I knew that I would always run up against brick walls, so I just saved myself the trouble. Only now, after 1989, when I think that we should use the chances of a democracy, it's now that I found the courage to involve myself too and to protest.
>
> (woman, group discussion, May 1993, original emphasis)

> We have to give the people encouragement, . . . strengthen people and say 'You have certain rights. Don't . . . just give up'.
>
> (younger woman, group discussion, May 1993)

> I am one of the people who welcomed the change under the motto of 'We are the people' – that means we want to be *involved* and not be ruled by others. Then when the other [western German] laws were forced on us it wasn't *my* change-over any more. But then one is even more responsible for attempting to influence things through the citizen association.
>
> (older man, group discussion, May 1993, original emphasis)

These extracts suggest that local activism allowed people to 'try out' the new forms of democracy, to work out how to 'do' democracy and to contest the ways in which it operated. This involved asserting the rights of the citizens to be taken seriously and was based on a conscious redefinition of what 'politics' should be and how local areas and their populations should participate.

My aim was really that I should be part of the development of the neighbourhood . . . and the second thing is really to participate in creating 'closeness to the citizens', so that not only the administration has its say, but that . . . we too are allowed to contribute.

(woman, group discussion, May 1993)

They are shooting out of the ground because everyone thinks 'a citizen association – that isn't something political, is it?' Anyone can go there. It's not a particular direction, it's not PDS, it's not SPD, not even Bündnis 90. Here everyone is welcome . . . and here I can do something for the area where I live directly.

(woman, group discussion, May 1993)

Those involved in these groups set out to reshape 'the political' in a variety of ways. Principally they intended this type of political action to be different from that in the past but also to escape from the party political constraints of both the GDR period and western party management and, as suggested by the example in the neighbourhood (pp. 125–134), to be distinct from an individualistic conceptualisation of rights of citizenship. In that sense they expanded the political from narrower, more limited definitions. 'Failure' was laid at the door of official agencies which at times were accused of acting in the same old ways as people knew from the GDR, the ultimate stamp of disapproval. This form of politics did, nevertheless, include formal political rights, as groups advocated that people vote in local elections for the 'right' kind of democratic inclusion (public meeting, September 1992). However, as we have seen, tensions arose periodically within and between groups, and at city-wide events involving a range of the groups, about how this politics could and should be implemented, and in particular whether it was meaningful in the context of the material and social conditions of those who lived in these neighbourhoods.

Several local groups, despite all that they had done and achieved, were rather pessimistic about what could be done, 'whatever you do, it is just a different front to what there was before. You have the right to say something but in the end it only works for some things' (woman, group discussion, May 1993). Personal pressures of unemployment and housing uncertainty and a rejection of all forms of political organisation were cited as key reasons: 'many just don't want to be organised any more. They had enough in the past' (man, group discussion, May 1993). For some people, then, the pressures of coping with the disruptions of the restructuring and the desire to escape from the over-politicisation of life in the GDR meant they sought to limit the extent of the 'political'. Others interpreted being involved in their neighbourhood as a positive difference from the politicisation of life in the GDR.

The debates in the neighbourhood discussed in the first part of the chapter show that attempts to limit the extent to which neighbourhoods were regarded as 'political' sites were also found in the discourses and practices of those who viewed expert decision-making in the best interests of the city as a preferable approach. However, different positions of relative power constitute these attempts to define the neighbourhood as an apolitical site, while those who retreat from political action out of frustration or because of the intensely practical difficulties of their lives do so simply in order to be able to cope. City officials, for example, who seek to avoid troublesome debate do so from a postion reinforced by structures of professionalisation and the systems of planning regulation. Given the insecurities of public sector employment they face, though, these planners by no means only operate in relations of domination. Furthermore, most would be dismayed to find that their actions could be interpreted as working against the 'interests of the citizens' or 'the good of the city'.

Conclusion

In these localised yet fundamental debates and processes, we find the very contours of post-communist and post-reunification governmentality shaped, negotiated and contested. According analytical primacy to a version of transition which offers narrow western liberalisation and modernisation as the only possible outcome falls into the trap of producing an undifferentiated subject and object position in reunification. It elides difference within eastern Germany and alliances across eastern and western Germany and beyond, and it denies the spaces of conflict, resistance and acquiescence. This chapter moves away from tendencies to describe eastern Germany either in forms of 'glorious resistance' or as a landscape of despair and insurmountable problems (Hörschelmann 1997; Pratt 1992). Furthermore it shows that debates about the nature of transition are definitely not limited to academic spheres. They form instead the very basis for post-communist transition, the processes of contestation at once similar to those over policy formation in western states and at the same time drawing on very different experiences: of the GDR, of its removal, and of the effects and expectations of sudden and rapid transformation.

People work in a variety of connected, sometimes contradictory, often complex ways with a range of definitions of how this transition is to be implemented. In this the geographies which people construct, perform and contest have powerful outcomes. Hebdige's argument (1993: 270–1) that in 1989 'all those ideological investments, all those ideological and geopolitical maps inscribed as if in stone around [the Berlin Wall], [all] melted into air' is overstated. The notions have not 'melted into air' but are transformed, reworked and reinscribed in other powerful ways, whether at the local or the geopolitical scale. In this processes of the 'neighbourhood' may actually set limits to, 'even constitute' ways in which transition works (A. Smith and Pickles 1998: 17). In going beyond 'accepting the dominant discourse of [post-communist transition] on its own terms' (Mani 1992: 394), we see a range of people and institutions, exploring a range of responses to transition: active adoption of liberal western models, acquiescence, resignation, outright rejection. As has been seen here also, the possibilities arise for what might be called 'a productive hybrid "betweenness", relocation and reinscription', not simply 'revising or inverting the dualities, but revaluing the ideological bases of division and difference' (Bhabha 1992: 58–60).

None of these 'maps' unproblematically on to space but, in general, notions of absolute space – of the pure spatiality of the 'West' which admits no possibility for a hybrid system (either in principle or in frustrated resignation), particularly one mixing with the 'East', its long-time 'other' – connect in the local sphere to policies and practices which, while genuinely attempting to work for the good of the city as a whole, nevertheless prioritise professional expertise and subordinate local demands to a wider goal. In contrast, local activist groups and others re-defined local involvement, critiquing officials when they failed to honour promises to work closely with the citizens, repositioning themselves as active participants in transition, both locally and in a broader sense. In so doing they were part of wider agendas which sought to contest the nature of change in Leipzig. This suggests that to regard activism in Leipzig's neighbourhoods as 'only' about the local is to fail to understand the complexities in and through which relations of domination and resistance are produced, and to neglect the significance and highly contested nature of the geographies of these relations. The intertwining of relations of domination/resistance operates in and through unstable and shifting processes of alliance and opposition and around varying shades of interpretation about the possibilities of material, social and political change in post-communist transition and reunification. This chapter has therefore sought not only to draw out (to dis-entangle) the politics of change as they converge on, and are contested in and through the sites of the neighbourhoods of Leipzig, but also to insist on the necessity of understanding these processes as fundamentally entangled in relations of power.

Acknowledgements

I would like to thank the editors for their encouragement in writing this chapter, and particularly Paul Routledge and Ronan Paddison for their comments on an earlier draft. This research was generously funded by the Carnegie Trust for the Universities of Scotland and benefited greatly from Ronan Paddison's supervision.

Notes

1 The following provide a range of first-hand and academic accounts of the events of 1989–1990 in Leipzig in particular: Förster and Roski (1990); Grabner *et al.* (1990); Hänisch *et al.* (1990); Neues Forum (1989); Schneider (1990).
2 See F. M. Smith (1994) for a discussion of debates around the terminology of (re)unification in Germany. 'Reunification' is predominantly used here because of its explicit reference to a previously existing united German state.
3 Taken from a poster advertising a meeting to establish the Citizen Initiative Waldstrassenviertel in February 1991.
4 The figures were discussed by a planning expert, Professor Topfstedt, at Leipzig's first major public discussion on reorienting urban policy, transcripts of which are published in Initiativgruppe 1. Leipziger Volksbaukonferenz (1990) and were also reported in the local press (*Leipziger Volkszeitung*, 25/26 November 1989).

References

Bauman, Z. (1992) *Intimations of Postmodernity*, New York: Routledge.
Bhabha, H. K. (1992) 'Postcolonial authority and postcolonial guilt', in L. Grossberg, C. Neilson and P. Treichler (eds) *Cultural Studies*, New York: Routledge.
Borneman, J. (1992) *Belonging in the Two Berlins: Kin, State, Nation*, Cambridge: Cambridge University Press.
Derrida, J. (1994) *Specters of Marx: The State of the Debt, the Work of Mourning and the New International*, London: Routledge.
Elsässer, R. (ed.) (1991) *Lebensraum – Ökologische Stadtentwicklung – eine Chance für Leipzig*, Leipzig: Hinkel und Junghans.
Förster, P. and Roski, G. (1990) *DDR zwischen Wende und Wahl*, Berlin: LinksDruck.
Foucault, M. (1978) *The History of Sexuality: An Introduction, Vol. 1*, Harmondsworth: Penguin.
Grabher, G. and Stark, D. (eds) (1996) *Restructuring Networks in Post-Socialism: Legacies, Linkages and Localities*, Oxford: Oxford University Press.
Grabner, W., Heinze, C. and Pollack, D. (eds) (1990) *Leipzig im Oktober*, Berlin: Wichern-Verlag.
Hänisch, G., Hänisch, G., Magirius, F. and Richter, J. (eds) (1990) *Dona nobis pacem: Fürbitten und Friedensgebete Herbst '89 in Leipzig*, Berlin: Evangelische Verlagsanstalt Berlin.
Harinemann, C. (1993) 'Stadterneuerung = Bevölkerungserneuerung? Anmerkungen zu sozial-räumlichen Konsequenzen der Berliner Stadterneuerung', in Arbeitskreis Stadterneuerung (ed.) *Jahrbuch Stadterneuerung 1993*, Berlin: Technische Universität Berlin.
Harloe, M. (1996) 'Cities in the transition', in G. Andrusz, M. Harloe and I. Szelenyi (eds) *Cities after Socialism*, Oxford: Blackwell.
Harvey, D. (1989) *The Condition of Postmodernity*, Oxford: Blackwell.
Haynes, D. and Prakash, G. (1991) 'Introduction: the entanglement of power and resistance', in D. Haynes and G. Prakash (eds) *Contesting Power: Resistance and Everyday Social Relations in South Asia*, Delhi: Oxford University Press.

Hebdige, D. (1993) 'Training some thoughts on the future', in J. Bird, B. Curtis, T. Putnam, G. Robertson and L. Tickner (eds) *Mapping the Futures: Local Cultures, Global Change*, London: Routledge.

Hörschelmann, K. (1997) 'Watching the East: constructions of "otherness" in TV representations of East Germany', *Applied Geography* 17: 385–396.

Initiativgruppe 1. Leipziger Volksbaukonferenz (1990) *Tagungsergebnisse der 1. Leipziger Volksbaukonferenz*, Merseburg: MebuDruck-H. Rößler.

Láng-Pickvance, K., Manning, N. and Pickvance, C. (eds) (1997) *Environmental and Housing Movements: Grassroots Experience in Hungary, Russia and Estonia*, Aldershot: Avebury.

Mani, L. (1992) 'Cultural theory, colonial texts', in L. Grossberg, C. Neilson and P. Treichler (eds) *Cultural Studies*, New York: Routledge.

Marcuse, P. (1990) 'Social, political and urban change in the GDR: scarcely existing socialism', *International Journal of Urban and Regional Research* 14: 515–523.

Marcuse, P. and Schumann, W. (1992) 'Housing in the colours of the GDR', in B. Turner, J. Hegedüs and I. Tosics (eds) *The Reform of Housing in Eastern Europe and the Soviet Union*, London: Routledge.

Mayer, M. (1995) 'Urban restructuring, new forms of exclusion, and the role of social movements', paper presented at European Research Conference on Future of European Cities.

Neues Forum (1989) *Jetzt oder nie – Demokratie! Leipziger Herbst '89*, Leipzig: Forum Verlag.

Osterland, M. (1994) 'Coping with democracy: the re-institution of local self-government in Eastern Germany', *European Urban and Regional Research* 1: 5–18.

Pickles, J. (1993) 'Environmental politics, democracy and economic restructuring in Bulgaria', in J. O'Loughlin and H. van der Wusten (eds) *The New Political Geography of Eastern Europe*, London: Belhaven.

Pratt, M. L. (1992) *Imperial Eyes*, London: Routledge.

Pred, A. and Watts, M. (1992) *Reworking Modernity: Capitalisms and Symbolic Discontent*, New Brunswick, NJ: Rutgers University Press.

Radcliffe, S. and Westwood, S. (1996) *Remaking the Nation: Place, Identity and Politics in Latin America*, London: Routledge.

Rat der Stadt Leipzig (1991) *Wirtschaftsstandort Leipzig: Business Location Leipzig*, Leipzig: Dezernent für Wirtschaft, Amt für Wirtschaftsförderung.

Reimann, B. (1997) 'The transition from people's property to private property: consequences of the restitution principle for urban development and urban renewal in East Berlin's inner-city residential areas', *Applied Geography* 17(4): 301–314.

Reuther, I., Doehler, M. and Zerche, I. (1990a) *Ansätze für einen Bereichsentwicklungsplan*, Leipzig: Wissenschaftliche Planungsgruppe 'Stadtentwicklung Leipzig'.

Reuther, I., Doehler, M. and Zerche, I. (1990b) *Wie behutsam kann die Stadt erneuert werden?*, Leipzig: Wissenschaftliche Planungsgruppe 'Stadtentwicklung Leipzig'.

Rink, D. (1995) 'Bürgerbewegungen und Kommunalpolitik in Ostdeutschland', in S. Benzler, U. Bullmann and D. Eissel (eds) *Deutschland-Ost vor Ort: Anfänge der lokalen Politik in den neuen Bundesländern*, Opladen: Leske und Budrich.

Rostock, J. (1991) 'Zum Wohnungs- und Städtebau in den ostdeutschen Ländern', *Aus Politik und Zeitgeschichte*, B29/1991: 41–50.

Routledge, P. (1997) 'A spatiality of resistances: theory and practice in Nepal's revolution of 1990', in S. Pile and M. Keith (eds) *Geographies of Resistance*, London: Routledge.

Schmidt, H. (1991) *Die metropolitane Region Leipzig*, Vienna: Institut für Stadt- und Regionalforschung, Österreichische Akademie der Wissenschaften.

Schmidt, H. (1994) 'Leipzig zwischen Tradition und Neuorientierung', *Geographischer Rundschau* 46: 500–507.

Schneider, W. (ed.) (1990) *Leipziger Demontagebuch*, Leipzig: Kiepenheuer.

Sidaway, J. D. and Power, M. (1998) '"Sex and violence on the wild frontiers": the aftermath of state socialism in the periphery', in J. Pickles and A. Smith (eds) *Theorising Transition: The Political Economy of Post-Communist Transformations*, London: Routledge.

Smith, A. and Pickles, J. (1998) 'Introduction: theorising transition and the political economy of transformation', in J. Pickles and A. Smith (eds) *Theorising Transition: The Political Economy of Post-Communist Transformations*, London: Routledge.

Smith, A. and Swain, A. (1997) 'Geographies of transformation: approaching regional economic restructuring in Central and Eastern Europe', in D. Turnock (ed.) *Frameworks for Understanding Post-Socialist Processes*, Occasional Papers 36, Department of Geography, Leicester University.

Smith, F. M. (1994) 'Politics, place and German reunification: a realignment approach', *Political Geography* 13: 228–244.

Smith, F. M. (1996a) 'Contested geographies of reunification: neighbourhood activism in Leipzig, 1989–1993', unpublished PhD thesis, University of Glasgow.

Smith, F. M. (1996b) 'Housing tenures in transformation: questioning geographies of ownership in eastern Germany', *Scottish Geographical Magazine* 112: 3–10.

Smith, F. M. (1999) 'Discourses of citizenship in transition: scale, politics and urban renewal', *Urban Studies* 36: 167–187.

Staddon, C. (1998) 'Democratisation and the politics of water in Bulgaria: local protest and the 1994–5 Sofia water crisis', in J. Pickles and A. Smith (eds) *Theorising Transition: The Political Economy of Post-Communist Transformations*, London: Routledge.

Stadt Leipzig (1992) 'Workshop Leipzig Stötteritz', *Beiträge zur Stadtentwicklung*, 4.

Stührmann, H. (1991) 'Bürgerinitiativen und Stadtteilentwicklungsplanung in Leipzig', unpublished manuscript, Göttingen University.

Verein für ökologisches Bauen (1992) *Bürger gestalten ihre Stadt: Vereine und Bürgerinitiativen stellen sich vor*, Leipzig: Verein für ökologisches Bauen Leipzig e. V.

Wollmann, H. (1991) 'Kommunalpolitik und -verwaltung in Ostdentschland: Institutionen und Handlungsmuster im "paradigmatischen" Umbruch. Eine empirische Skizze', in B. Blanke (ed.) *Staat und Stadt*, Opladen: Westdeutscher Verlag.

6

SPORT AS POWER

Running as resistance?

John Bale

In seeking to emphasise the global in addition to the local as a context for a broader social and cultural geography, Cosgrove and Rogers (1991) drew attention to the potential of such themes as 'westernisation' and 'global cultural experiences and events' as possible foci for such a geography. Among themes related to a locus on territoriality and nationalism they suggested that 'myths of nations' and 'national passion' form potentially fruitful lines of inquiry. Underlying such 'passions' and 'experiences', however, lie the realities of tangled power relations; I would suggest that representational sport is a site where entanglements of domination and resistance are found. It is these forms of power which form the subject of this chapter. In it I examine one dimension of globalised sport as a potential and actual site of resistance. I do this by examining the nature of representational sport and the way in which sports theorists have viewed it as a site where power relations are 'played' out. I suggest that modern, serious sports pose certain problems when interpreted as sites of resistance and argue that transgression is a more appropriate term to use in relation to 'resistance-like' activities on the part of athletes and other actors in the sports business. I briefly outline the character of representational sport in the context of colonial Kenya and illustrate it as a source of power and domination. I highlight the problematic nature of dealing with the concept of resistance in such a highly rule-bound and conventional form of body culture as sport and then use three short case studies involving Kenyan athletics and consider the extent to which they can be read as sites of cultural resistance in the context of various national and global pressures. In this way, the chapter exemplifies transgressions and resistance, not in riots and violence but in more mundane – perhaps even banal – forms such as disobedience in the lives of sports workers of various kinds. Nevertheless, banal forms of disobedience, like banal forms of oppression, are felt and often do matter (Berg 1998).

Sport as power

In contrast to recreational sport, representational sport can be defined as one where athletes represent their town, state or country, and where spectators feel that the athletes represent them against other groups (Brownell 1995). Today representational sport is often highly commodified and is essentially part of the globalised entertainment industry. It is characterised by the kinds of spectacular global cultural experiences alluded to by Cosgrove and Rogers (1991), including the Olympic Games and the World Cup (in soccer). The kind of sport I have in mind is almost the polar opposite of disport – the word from which 'sport' derives and which was formerly used as a synonym for frolicsome or play-like activities. Alternatively, it could be suggested that the kind of dis-play described in this chapter is the antithesis of play (Stone 1970). Those who participate in achievement sport are sports workers. Their financial rewards may vary but they are motivated by an ideology of achievement and 'production' – improving performance, winning and record breaking (Eichberg 1997). It is a serious business. A paradigm for this configuration of sport is the 100 metre sprint. Those who compete in such a contest, which is recorded to one-thousandth of a second, have no time to make mistakes, no room for laughter and no scope for play. According to Olsson (1980: 198), such people 'ignore what cannot be counted. They have left their hearts at home, for otherwise they cannot be objective. They feed on a diet of certainty and they get upset by ambiguity'. This is the distinction between racing and running – perhaps between life and career. Thrift's (1997) writings on dance suggest that it can be seen as an example of play. When dance becomes sportised, however, it can be read as an example of work; in becoming sportised 'play-dance' has been subverted.

Serious sport can be seen as a form of power and dominance. The roots of representational sport in many modern nation-states lie in imperialism and colonialism. Sports such as football, cricket and athletics were introduced to the colonies of the European realm as a means of social control at the local level and of developing sports workers for the emerging European and global systems of sports. A recurring metaphor in colonialist texts is that the African body was something to be 'processed' by the colonial machine. In Lord Cranworth's (1912) *A Colony in the Making*, a photograph of a prospective member of the King's African Rifles, semi-naked in native clothing, is captioned as the 'raw material'. It was juxtaposed to 'the made article' – an immaculately uniformed soldier (Bale and Sang 1996: 97). Similarly, from the 'noble savage' could be prepared the global athlete. In the corporeality of the African, the European saw the Olympian of the future. A cartoon in a 1923 issue of the French sports magazine *L'Auto* revealed that the semi-naked native who was (in 1922) carrying a spear and shield and surrounded by a landscape of palm trees would (by 1925) be transformed into a uniformed and sports-shod athlete bedecked with a victor's garland, carrying a cup and standing in the unmistakable sportscape of the athletic arena (Deville-Danthu 1997: 24).

During the first quarter of the twentieth century the social control function of western sports was graphically illustrated in many corners of the colonised world, including Kenya Colony. In 1907, for example, Dr John Arthur, who had taken charge of physical education and games at the Church of Scotland's Mission in Thogoto, Kenya, observed that

> the game of football played in the afternoon, was played for moral benefit as much as recreational relief, . . . to stiffen the backbone of these boys by teaching them manliness, good temper and unselfishness – qualities amongst others which have done so much to make a Britisher.
>
> (quoted in Bale and Sang 1996: 77)

Elsewhere it was noted by the Colonial Secretary, Henry Moore, that

> the government attaches great importance to African athletics and to the encouragement of a spirit of local emulation, if for no other reason than because such pursuits provide not only a substitute for political intrigue but also a legitimate channel into which might be diverted inherent instincts which otherwise lead to raiding and stock-thieving.
>
> (quoted in Bale and Sang 1996: 79–80)

What is impressive about these statements is the explicitness of the power relations between colonial policy and Kenyan people. Control was clearly the *intention* of those in power.

The iconography of the resulting cultural landscape symbolised the power of western sports. The schools where sports were practised as part of a European curriculum, for example, had their sports fields and geometrically inscribed running tracks and football fields. The running track, with its uniform plane and imperial measurements was, in its own small way, an imperial strategy involving the mastery of space. Whereas traditional body cultures had taken place within the homestead or places utilised for day-to-day activities, the sports field required a special space, separated from the daily activities of the people and their social life. The 100 yards straight was an analogue of the railway line from Mombasa to Nairobi; it was a colonial 'reaction to the winding African footpath'; each was a landscape of power, a 'straight line, a man-made construct . . . indicative of order and environmental control' (Cairns 1965: 78). Events were scheduled for specified times; they were quantified and recorded; they took place in segmented spaces.

The record(ing) symbolically replaced the oral traditions of an earlier Kenyan society and as Africans' spatiality was erased and place became more of an athletic space, so pre-modern Africans became an athletic 'other', their movement culture being defined in terms of the imperialist.

By the end of the 1930s bureaucratically organised athletics events were relatively well developed in Kenya and inter-territorial competitions were being held. At the same time, indigenous body-cultures were in decline. In the Kenyan case it was not until the mid-1960s, however, that the dreams of the colonial rhetoricians could be said to have become sporting realities. It was during the Mexico City Olympics of 1968 that Kenyan athletes indicated that they could soon assume dominance in the athletics arena in middle and long distance running events. Since then, runners from Kenya have served to re-present Kenya to a global audience as a virile, indeed, a successful nation. The successes of Kenyan athletes have, it is sometimes argued, both boosted and bonded the nation. Kenyan athletes have projected their nation on to the world – rather than the African – stage by winning many Olympic championships and breaking numerous world records. At the same time, Kenyan running success has often been 'explained' by occidental observers in stereotypical terms, the successes of African runners being devalued by recourse to environmental deterministic forms of explanation which view them as 'natural athletes' (Bale and Sang 1996: 142–147).

Achievement sport is regarded by many critical observers as an ideologically conservative phenomenon. It can be argued that whereas forms of representation such as literature, drama and dance each provide a site of resistance for colonised peoples, the protocols of achievement sport prevented analogous forms of resistance from taking place. Meaningful participation in athletics requires adherence to very precise rules and regulations, and little or no scope is found for improvisation or alternative interpretations. The dimensions and rules of a 100 metres race in London must be the same – exactly the same – as one in Lagos; deviations from the themes of competition, achievement and the record are simply not permitted. It is hardly surprising, therefore, that achievement sport has been rendered as a Foucauldian 'prison of measured time' (Brohm 1978). In literature and other cultural genres, hybrid forms can be regarded as examples of post-colonial cultural and political resistance. The fact that the adjective 'post-colonial' can readily be applied to literature and dance but not so readily to sport, reveals the difficulty of conceptualising 'resistance' in sport. Once body cultures become hybrid (as in Trobriand cricket where indigenous people have appropriated cricket and modified it to fit their own culture) they eliminate themselves from the world of sport – though not disport.

In the global systems of such conservative – indeed right-wing (Hoberman 1992) – organisations as the International Olympic Committee (IOC) and the International Amateur [sic] Athletics Federation (IAAF), how might resistance be manifested? After all, the IAAF is quite explicit about its neo-colonial ambitions, its aim being to 'help remove cultural and traditional barriers to participation in athletics' (Abmayr 1983). Hence, African body culture, for example, is seen as something to be 'removed'. The imperialising power of this organisation is reflected in the rhetoric of its 'mission statement'. It alludes to the propagation of an 'athletics culture' – 'the environment in which athletics *must grow* and develop. Each member federation *must* see that athletics in its country is broadly based and strong as possible'.[1] Only in rare instances were native corporealities adopted by the coloniser (e.g. lacrosse in North America) and in such cases they were subsequently 'sportised' by applying standard rules, regulations and governing bureaucracies. At a local level it is true that some early examples of resistance were registered in colonial writing as in cases where attempts to organise sports events degenerated into fights between the winner and the rest (Bale and Sang 1996: 175). In Kenya ethnic groups such as the Maasai were much more resistant than others in adopting any European forms of culture. As a result, few of Kenya's major athletes today are Maasai. Where western sports were successfully implanted, however, subsequent resistance seems to have been non-existent, even during the early 1950s when Mau Mau rhetoric advocated a 'return to ancient customs' (Leakey 1954: 27). During the twentieth century a distinction has emerged which has separated 'sport' from 'disport'. Increasingly, sport – and certainly achievement sport – has become the antithesis of disport, rather then being its synonym. It has become more difficult to 'play' in sport, though such a form of transgression is not yet totally impossible in all sports as the behaviour of aberrational 'clown'-like sports workers testifies.

Sport and resistance

The dominance by which, first, agencies of the colonial power implanted western sport in the colonial world, and second, how the international governing bureaucracy exerted global control, are each often thought of as having another side. This is the ability of colonised people to appropriate aspects of the colonial culture in articulating their own culture. I am not talking here about the hybridisation of a western form of culture. Rather, it is through achievement sport that not only may the 'masters' be defeated at their own game but also, in so doing, national identity and place-pride may be forged. Indeed, representational sport is often justified by political leaders as providing a focal point for national identity and, as Ehn (1989) has noted, at major sports events national symbolism becomes 'over-explicit'. These are examples of the appropriation of western culture, rather than resistance to it, although (as I shall show later) such appropriation can have transgressive qualities. At a more local level, the notion of *resistance* to such attempts as those of the IAAF to erode regional cultures in the Kenyan context can be related to the writing of Ngugi wa Thiong'o (1981), whose work has focused almost entirely on drama, literature and dance as forms of cultural resistance against Europeanised forms of such cultures. His strategy is basically to return to African languages and indigenous modes of drama and writing as a means of countering the hegemony of the colonial legacy. When he mentions sports as such hegemonic legacies he refers only to high status activities such as golf and polo: introduced by colonialists, 'black pupils now do the same, only with greater zeal: golf and horses have become "national institutions"' (Thiong'o 1981: 64–65). Why does he ignore the most internationally visible form of Kenyan movement culture, that of middle and long distance running? Is it because drama, literature and dance permit the script to change – to include a radical voice? Kenyan athletes, it is true, frequently defeat white athletes from the former rulers of empire. But they are defeating them at an activity which is undertaken essentially on neo-colonialist terms. The colonial body can be said to be *licensed* in the double sense of being allowed to compete against the (neo)colonial power but remaining under that power's control. To overthrow the ideology of achievement sport would mean the rejection of the ideology of the society in which it flourishes. As Leys (1982) points out, it would therefore be inconceivable to use the dramatic and theatrical potential of the Olympic Games to subvert any system seen as oppressive. In other words, it is the deeper, structural shortcomings of sport which deny the possibility of it being a site for resistance. Some cultural forms (such as literature, theatre and dance) may possess critical elements and provide a forum for resistance but achievement sport is not one of them. This is because in sport the script never changes; there is little or no room for improvisation. As Brohm (1978: 178) puts it, 'sport is a positivist system and as such always plays an integrating and never an oppositional role'. (On sport and hegemony see Morgan 1994.)

Yet a number of instances of so-called resistance have been cited in a sports context. Several observers have seen the 'Black Power' salutes of black American athletes at the Mexico City Olympics in 1968 and the 'Black September' massacre of Israeli athletes at the Munich Olympic Village in 1972 as classic forms of resistance through sport. However, each of these, while undoubtedly intentional, occurred at the margins of a sporting occasion. While intended as a political statement of resistance to the condition of blacks in the USA, the black athletes had already won their Olympic medals before making their gestures of defiance. The murder of the Israeli athletes was not actually (to the best of my knowledge) undertaken by sports participants at all. It was sport which provided a highly visible site for such an attack.

Arguably the most widely cited scholar on the subject of sport as resistance is the late Trinidadian-Marxist, C. L. R. James (1969), who viewed 1950s West Indian cricket victories over their English 'masters' as forms of resistance. James's view has been introduced to a geographical audience by Hall (1996):

James records with pride how cricket became a site of resistance. He records with pleasure the symbolic victories scored in post-war cricket by the West Indies, as the 'underdogs' turned the tables and defeated their colonial masters at Lords, the 'home' of English cricket (for the first time and at the same moment – the 1950s – as the tide of Caribbean decolonisation and the independence movement was gathering pace).

(Hall 1996: 178)

A similar view is presented by the anthropologist, Klein (1994), in his analysis of baseball in the Dominican Republic. While dominated by the USA it can, in various ways, be seen as a form of resistance.

In Dominican baseball resistance is manifested among other ways in the symbolic preferences of the fans, in the way baseball is covered in the press, and in various actions by Dominicans.

(Klein 1994: 152)

Klein also notes how American baseball interests have, in fact, gained overall power, with their 'farm clubs' and baseball academies. The imbrication of dominance with resistance is clearly illustrated in such an example.

A further example of resistance in sport comes from the work of Eichberg (1997), who views the emergence of non-sportised forms of body culture as forms of resistance against status quo views of achievement sport. A claimed resurgence of national body cultural practices (e.g. revived folk-games), open-air activities such as jogging, new expressive activities and new architectural sporting structures such as water worlds rather than the geometric swimming pool, are used to support his views.

However, using definitions proposed by Cresswell (1996), the examples provided by James, Klein and Eichberg might be better defined as transgressions. Unlike resistance, transgression 'does not, by definition, rest on the intentions of actors but on the *results* – on the "being noticed" of a particular action' (Cresswell 1996: 22–23, original emphasis). I take resistance to mean the *intentional* 'action against some disliked entity with the intention of changing or lessening its effect' (Cresswell 1996: 22). This is not to say that unintentional acts cannot contribute to change, as Cresswell (1996: 176) points out, the 'limits to transgression lie in the fact that it is not enough to constantly deconstruct and destabilize'. Among the 'disliked entities' which sports workers might theoretically seek to resist (in, for example, post-colonial Kenya) are the colonial or neo-colonial state(s) which imposed modern sport on them, the global or national bureaucracies (e.g. IOC, IAAF) within which sports workers ply their labour, national or international injustices (e.g. racism) and (theoretically) the nature and functions of achievement sport itself.

James's approach is seen today by other Caribbean writers as an inappropriate reading of a form of cricket which has become a well-organised, global, travelling circus which renders problematic

> any *uncomplicated* notions of cricket being a vehicle for the expression of impulses and experiences whose character is determined by the contingencies of place and time, but also . . . by the 'stage' of historical development [which] a particular nation happens to be in at the time.
>
> (Surin 1995: 325, original emphasis)

Even in the 1950s, however, it seems hardly likely that West Indian cricketers can be interpreted as 'resisting' neo-colonialism or one of its prime symbols, sports, when they played the quintessentially English game themselves. West Indian cricketers, in achieving their early victories in particular, may have 'crossed some line that was not meant to have been crossed' (Cresswell 1996: 23) but their prime objective was to win a game whose rules and regulations are determined by a neo-colonial global bureaucracy. West Indian cricket victories could be read as being analogous to some of Ingrid Pollard's photographs in which, as a black British photographer, she views herself as out of place in particular 'English' contexts (Cresswell 1996: 167–169). Each trespasses on unexpected terrain; in each the black skin of the other is placed in the context of European-ness. But they represent a form of 'resistance that appears to occur only within socio-spatial limits predefined by people in positions of relative power' (Cresswell 1996: 23). Their actions could not take place without the pre-existing spaces which they transgress. It is not simply the defeating of a (former?) colonial power that can be seen as a form of transgression. Simply *by appearing* in sporting competitions assumed to be the preserve of another group can be seen as an act which involves moving into the 'space' of a more dominant group. The African Americans Arthur Ashe and Althea Gibson and the Australian Evonne Goolagong, in the case of tennis, could be given as examples. Klein's approach suffers from similar criticism while Eichberg's view might be countered by noting that the intention to encourage, say, jogging cannot be shown to be aimed deliberately against achievement sport. Indeed, such play-like 'leisure' activities all too easily become sportised, commodified and over-serious.

What might be viewed as a paradigmatic example of sporting resistance is found in the fictional *The Loneliness of the Long Distance Runner*, a novel by Allan Sillitoe (1961). Resistance in this case is graphically illustrated by the case of Colin Smith, the borstal boy who intentionally loses a long distance race while well in the lead and which his captors (the establishment) expected and encouraged him to win. In order to resist, however, Smith had to remove himself from the protocols of sport. His act is not marginal to the event in which he is participating but in order to resist he had to stop racing. In such a context, merely running would become a form of resistance. Victory/winning would represent compliance; losing (intentionally) would demonstrate another kind of victory, that of resistance. In such a context, the entangled nature of domination and resistance is again graphically exemplified. Leys (1982) regards the kind of agency which is fictionally depicted in Sillitoe's book as being virtually impossible in the real world, being incompatible with the ideology which drives achievement-oriented athletes to compete at the highest level. The number of such cases similar to that of Colin Smith from the world beyond fiction are restricted to a small number of athletes who have for short periods withdrawn themselves from sports events for religious or political reasons. By withdrawing themselves from sport, however, they cannot then be said to be *using* its spectacle and visibility as a site for resistance. Hence, it can be argued that any sport-based resistance can be carried out only if the theoretical identification of 'sport' with 'work' is relaxed.

Kenyan running: being noticed

I now want to move back to Kenyan running and explore the ways in which the actions of Kenyan athletes and others in Kenya can be viewed as transgression or resistance. Kenyan body-culture has been seen as a medium within which trans-gression and resistance may take place; the 'meaning' of the Kenyan body can be involved in the construction of geographical knowledge – notably, knowledge of the 'athletic other'. Modern sport was introduced to Kenya as a form of social control but by the early 1950s Kenyan athletes had reached a standard worthy of competition at the global scale. Kenyan middle and long distance runners are now widely regarded as being among the best in the world. I will take three examples of transgression and/or resistance drawn from the geographical history of Kenyan running since the 1950s.

In the early days of their participation at the international level, Kenyan runners were seen by western observers as being 'out of place'. Since before the days of Jesse Owens, black athletes were traditionally viewed by the white athletocracy as 'born sprinters' – 'raw talent', 'quick off the mark', 'great speed but little stamina' (Hoberman 1997). To be even taking part in long distance races was, for black athletes, to be transgressing the norms of 'athletic space' – just as the West Indian cricketers had done with their presence on English cricket fields in the 1950s. But it cannot be seriously argued that the intention of the Kenyan runners was to resist the hegemonic structures of either neo-colonialism or global sport. It was in 1954 that the first Kenyan athletes competed in Europe. The event was the British championships at the White City Stadium in London. In the press previews of the event, no mention was made of any Kenyan long distance runners. At the start of the six mile championship, the spectators were surprised to see a barefoot

Kenyan runner, Lazaro Chepkwony, lining up among the UK elite. They were even more surprised when he dashed away from the rest of the field and opened up a large gap, only to slow down and let the British runners catch him. He would make another fast burst and repeat this form of running for about three miles of the race before dropping out with a leg injury. From the viewpoint of the British spectators, this confirmed that the black runner was out of place in a long distance race. The next day, another Kenyan, Nyandika Maiyoro, lined up for the three mile race. He adopted similar tactics, building up a 40 metre lead after a mile but he was eventually caught and he finished in third place. The response of the English press to these performances reveals the Africans' transgressive nature. The rhetorical mode of negation was adopted by the British press (see also Spurr 1994). The *Manchester Guardian* noted that the six mile event was 'bedevilled' by Chepkwony's sudden bursts; the three mile race was 'made confusing' by Maiyoro's 'ludicrously fast pace'. *The Times* reported that it was 'inevitable' that Maiyoro would be overtaken by the English runners (Bale and Sang 1996: 4–6). Clearly, the two Africans were viewed as athletic 'others'. They also reflected the problem of how the western writer could coherently represent the apparently strange reality of the movement culture of the African runner. By labelling the Africans as 'deviant' they were accommodated, albeit being recognised as failing to subscribe to the protocols of running as racing.

Two decades later, various changes in the global sports system had resulted in the Kenyan athletes beating the Europeans at their own game, just as the West Indians had done at cricket. Kenyans first achieved particular success at the Mexico City Olympics in 1968 when they won several gold medals. The transgression of the Kenyans into the white athletes' 'sports space' was again incorporated by labelling it as deviant. One journalist wrote of the Mexico Games as the 'unfair Olympics' because the Africans, who lived at high altitude, had an unfair advantage over athletes from the west. This is a theme which has subsequently been applied to many of Kenya's athletic successes. By winning, therefore, the Kenyans may appear to resist the traditional dominance of both western athletes and the governing bureaucacies of the sports business but they do so within the confines of those very organisations. Furthermore, no allusion was made to the many economic and cultural advantages long possessed by the European and American athletes. In these examples the 'out-of-placeness' and the 'being noticed' of such sports workers seem clearly to exemplify Cresswell's (1996) interpretation of transgression.

Rootedness and mobility

Modern sport is governed by bureaucracies at the national and global levels. Such bureaucracies have been able to exert considerable power over when and where sports workers ply their athletic trade. Athletes face considerable controls on their freedom of mobility through the bureaucracies, national and global, which govern them. However, Kenyan athletes may now be in the vanguard of radical change in the relationship between themselves and their country.

Some Kenyan athletes live most of their lives in Kenya. Others roam the world as migrant-workers, moving from competition to competition, from workplace to workplace. Some of these athletes seem to lead a 'decentred' existence, under-mining the norms of order and authority which have been traditionally taken to characterise the modern sports system. Representing Kenya may be rejected in favour of (to an extent) representing the self. This may happen when the national governing body attempts to dictate when and where an athlete may compete. Since the early 1990s the Kenyan governing body of athletics has been anxious about athletes choosing to compete overseas rather than run in their own national championships. In 1994 attempts were made by the Kenyan authorities to require written requests for permission to compete abroad. Indeed, this is the IAAF rule. This attempt to restrict the mobility of sports labourers has proved unworkable and as a result the Kenyan authorities were seen to be losing control of 'their' athletes. In 1994 the Kenyan AAA ruled that athletes would not be considered for the Commonwealth Games if they did not first compete in the national championships. Many athletes preferred to make money on the European Grand Prix circuit than represent their country. The link between athlete and nation was effectively broken. I noted earlier that in representational sport the athlete acts as an intermediary between the people and the state. In athletic representation, the group partakes vicariously in the success of one of its parts, the athlete. This assumes that the part represents the whole and can do so because the athlete shares a fundamental likeness with the group. In the Kenyan case the foundations of this assumption are collapsing. What does 'Kenya' mean to athletes who work for several thousand dollars rather than compete for their country? What does 'representation' mean when athletes choose to represent countries other than their own?

Clearly, such athletes are not reflecting the norms of traditional sports behaviour. They are transgressing these norms, crossing boundaries of affiliation and representation. They reflect a lack of conformity, independence and a less uniform and less conventional attitude to the world of sports, but they continue to display images from a dominant culture. When they race they have changed their affiliation not only from Kenya to, say, Denmark, but also from Kenya to Reebok or Nike. After the Kenyan-Dane Wilson Kipketer won the world 800 metres championship in the colours of Denmark, an article appeared in a Danish magazine with the following headline: 'Kenyan or Danish? The king of runners is quite certain: "I am Wilson Kipketer"' (quoted in Bale and Sang 1996: 184). Traditionally athletes have been restricted by the 'container' of the state.

Here we see examples of international resistance against state-imposed limitations on labour mobility. In Taylor's (1994) words, we are now witnessing a 'leaking container' and a possible reduction in the significance of the territorial state in achievement sport.

Resistance in the press

In this section I present a form of more intentional resistance by moving outside the athletic arena – where I have suggested that resistance is better rendered as transgression – and present some Kenyan views of occidental attitudes. Irony and humour can be potent modes of resistance. Cartoons are a particularly well-known medium for critique and I include readings of two examples from the sports pages of Kenyan newspapers to illustrate a form of deliberate fun-poking at western attitudes. The cartoons reflect a kind of occidentalism and were included in the Kenyan press at the time of the World Athletics Championships in Tokyo in 1991. They amount to a mocking (resistant?) East African view of western perceptions of Kenyan running.

A long tradition of racist and environmentally deterministic rhetoric surrounds attempts to 'explain' the apparent athletic successes of black athletes. In the Kenyan case the dominance of the nation's athletes in events requiring aerobic capacity is often explained by the fact that they were born and raised at high altitude. However, many high altitude countries produce no world-class athletes. These 'explanations', therefore, ignore historical and geographically specific factors and cultural traditions. The first of these alternative perspectives depicts a situation in the 1990s but pokes fun at a long tradition. As Hoberman (1992) points out:

in the colonial context, assessments by European observers of their 'primitive' subjects reflected the one sided nature of this comparative project: the men from Paris, London and Berlin were the scientists, while the indigenous populations of Africa . . . provided an endless stream of interesting subjects.

(Hoberman 1992: 36)

Have things changed very much? The cartoon in Figure 6.1 examines the popular and widespread notion that Africans are in some way biologically or genetically 'different' from people in Europe or North America. Note that the two scientists – the 'experts' – are white, authoritative, rational and part of an international system of sports-medical research. The modern, clinical environment of the laboratory also serves symbolically to mock western science – which cannot solve the perceived problem. The cartoon reflects the existence of several western

159

Figure 6.1 'I've checked everything, Sir . . . his bones, lungs, heart . . . there's nothing
extra to make him run faster'
Source: *The Sunday Nation* (Nairobi) August 1991

Figure 6.2 'I hear in Kericho they used to run 20,000 metres to and from school'
Source: *The Daily Nation* (Nairobi) September 1991

'sports-medical' projects which have been seeking to 'explain' the success of
Kenya's long distance runners. At the same time there has been a re-emergence of
an interest in 'racial science'. The Kenyan response, shown in Figure 6.1, is a clear
reaction to such scientific endeavours.

The second cartoon (Figure 6.2) mocks the European impulse to romanticise
the virility of the idealised Kenyan runners. It shows two Kenyan athletes (carrying

the national flag) close to lapping the rest of the field in the world 10,000 metres championship held in Tokyo in which Kenyan runners occupied the first two places. The national icon, the flag, is placed next to the global icons of a number of multinational corporations. 'Kenya' has already passed Sanyo and will soon overtake National and Seiko. Kenya, at the economic/cultural periphery, is seen defeating the representatives of the multinational global sports economy. The two discussants in the infield are clearly identified as white and are trying to explain the success of the Kenyan runners. This time they fall back on the romantic myth of the simple, eternal, essential Africa – of young children running long distances across the plains to school.

Conclusion

These three examples form a sort of continuum between transgression and resistance. In the first – the black athletes at the White City in the 1950s – it seems to be a clear example of transgression; the Kenyan athletes were 'out of place'. Intentionality was present in their desire to compete (and to seek victory) but not, I suggest, in any desire to overthrow the status quo. Yet unintentionally, this is what they did. They signalled that black athletes could be long distance runners and could deny the stereotypical explosive-sprinter image. But while winning in sport might appear to be a form of resistance it is entangled in the domination of neo-colonial global bureaucracies. Transgression is dependent on the dominance through which it is achieved. The second example is less clear. The Kenyan athlete who became a Dane could be read as deliberately changing his nationality in order to deny the dominance of the Kenyan athletics bureaucracy over him. His act was one of agency over the structures of national and global sports control. In both these cases, however, the acts of transgression/resistance took place within the neo-colonial structures of the IAAF. Given its hegemonic power, the changes resulting from the kinds of transgressions noted above could be said to be cosmetic. The third example of the cartoonist's view of western sport is more obviously one of resistance but from outside the arena of sports itself.

In the light of this, and previous comments, it may be possible to argue that in the case of modern achievement-oriented sport resistance takes place at its margins whereas transgression is the most that can be expected during the sporting performance itself. This is not so far from the view of Leys (1982: 310), who has suggested that as a general principle the most that could be hoped for might be de-sportisation of play rather than the liberation of sport.

Note

1 Taken from http://www. iaaf. org/iaaf/dev. html website.

References

Abmayr, W. (1983) 'Analysis and perspectives of the project in Kenya', in *Women's Track and Field in Africa*, report of the first IAAF congress on women's athletics, Darmstadt, Deutscher Leichthletik Verband, 532–538.

Bale, J. and Sang, J. (1996) *Kenyan Running: Movement Culture, Geography and Global Change*, London: Frank Cass.

Berg, L. (1998) 'Banal oppressions', paper presented at the Association of American Geographers' annual meeting, Boston, MA.

Brohm, J-M. (1978) *Sport: A Prison of Measured Time*, London: Ink Links.

Brownell, S. (1995) *Training the Body for China*, Chicago: University of Chicago Press.

Cairns, A. (1965) *Prelude to Imperialism*, London: Routledge.

Cosgrove, D. and Rogers, A. (1991) 'Territory, locality and place', in C. Philo (ed.) *New Words, New Worlds: Reconceptualising Social and Cultural Geography*, Lampeter: Social and Cultural Geography Study Group.

Cranworth, Baron (1912) *A Colony in the Making: Sport and Profit in British East Africa*, London: Macmillan.

Cresswell, T. (1996) *In Place/Out of Place: Geography, Ideology and Transgression*, Minneapolis: University of Minnesota Press.

Deville-Danthu, D. (1997) *Le Sport en noir et blanc*, Paris: L'Harmattan.

Ehn, B. (1989) 'National feeling in sport; the case of Sweden', *Ethnologia Europea* 19(1): 57–66.

Eichberg, H. (1997) *Body Cultures: Essays on Sport, Space and Identity*, edited by J. Bale and C. Philo, London: Routledge.

Hall, S. (1996) 'New cultures for old', in D. Massey and P. Jess (eds) *A Place in the World*, Milton Keynes: Open University Press.

Hoberman, J. (1992) *Mortal Engines*, New York: Free Press.

Hoberman, J. (1997) *Darwin's Athletes*, Boston, MA: Houghton Mifflin.

James, C. L. R. (1969) *Beyond a Boundary*, London: Stanley Paul.

Klein, A. (1994) *Sugarball: The American Game, the Dominican Dream*, New Haven, CT: Yale University Press.

Leakey, L. (1954) *Defeating the Mau Mau*, London: Methuen.

Leys, C. (1982) 'Sport, the state and dependency theory', in H. Cantelon and R. Gruneau (eds) *Sport, Culture and the Modern State*, Toronto: University of Toronto Press.

Morgan, W. (1994) *Leftist Theories of Sport: A Critique and Reconstruction*, Urbana: University of Illinois Press.

Olsson, G. (1980) *Birds in Egg/Eggs in Bird*, London: Pion.

Sillitoe, A. (1961) *The Loneliness of the Long Distance Runner*, London: Pan.

Spurr, D. (1994) *The Rhetoric of Empire*, Durham, NC: Duke University Press.

Stone, G. (1970) 'American Sports: play and display', in E. Dunning (ed.) *The Sociology of Sport*, London: Frank Cass.

Surin, K. (1995) 'C. L. R. James' material aesthetic of cricket', in H. Beckles and B. Stoddart (eds) *Liberation Cricket: West Indies Cricket Culture*, Manchester: Manchester University Press.

Taylor, P. (1994) 'The state as a container; territoriality in the modern world system', *Progress in Human Geography* 18(2): 151–162.

Thiong'o, N. (1981) *Detained: A Writer's Story*, London: Heinemann.

Thrift, N. (1997) 'The still point: resistance, expressive embodiment and dance', in S. Pile and N. Thrift (eds) *Geographies of Resistance*, London: Routledge.

7

ENTANGLING RESISTANCE, ETHNICITY, GENDER AND NATION IN ECUADOR

Sarah A. Radcliffe

One notable feature of recent years in Ecuador has been the emergence of a social class of people dressing as 'indigenous'. Large numbers of Andean rural and low-income urban dwellers continue to wear the ponchos, straight skirts and shawls that are visible markers for 'indianness': my interest lies not directly with them, but with the increasing number of professionals, urbanites and others whose change in clothing style is an element of identity renegotiation. Luis Macas, leader of the Confederation of Indigenous Nationalities of Ecuador (CONAIE), a Spanish-speaking lawyer recently elected to Congress, wears poncho, plait and white trousers. Likewise, Nina Pacari, another lawyer and CONAIE leader, wears the female equivalent (see Figure 7.1). While clothing does not comprise an explicit confederation policy, CONAIE's political programme calls for a rejection of assimilationist state policies that deny the 'pluricultural' dynamism of indigenous nationalities. Articulations between state actions and discourses, and popular senses of affiliation, raise questions about the entanglements of power, representation and resistance around such new national and cultural identities.

With a focus on the political culture of contemporary Ecuador, this chapter addresses itself to the meanings and complexities of power behind the readoption of indigenous clothing. Others who are not indigenous leaders are involved in this trend, as a case study outlines. More than a formal political issue, the adoption of 'indian' clothing appears at first glance to represent an act of political-cultural resistance, arising as part of a strengthening indigenous social movement. Certainly, indigenous confederations resist indian populations' marginalisation and partial citizenship within nationhood, as 'indianness' for much of the republican period has been constituted as 'non-national'. Drawing on recent geographical debates around the interconnected and mutually influential facets of domination and resistance (see Sharp *et al.*, Chapter 1 in this volume; Radcliffe 1999), this

Figure 7.1 CONAIE leadership
Source: Sarah A. Radcliffe

chapter examines Ecuadorian political culture around dress not only as an outcome of resistance, but also as a complex interplay of power relations.[1] Viewing the entanglements of gender, nation, ethnicity and identity as a result of processes of hybridity has particular relevance in this discussion of visual referents to identity (such as clothing), as it prompts us to unpack the dominating/resisting histories and geographies behind appearance. In this sense, indigenous clothing reasserts a representational economy, by reclaiming the space of the nation for a previously marginalised group, but in a scenario where domination and resist-ance are not easily – or meaningfully – separable. Central to these hybridities are gender relations – particularly the power of tradition associated with the figure of the indigenous woman – and geography, in this case complex national, rural and transatlantic interchanges.

Indigenous clothing in contemporary Ecuador is one facet of the complex 'cultural implications of the geography of modernity' (Lomnitz 1996: 68). Ecuadorian geography of modernity is primarily instantiated through processes of hybridity. Hybridity is utilised here as a term to refer to the nature of cultural mixing, the bringing together of tradition/modernity, indigenous/*mestizo*, national/non-national in power-invested processes. Context-specific histories and geographies lie behind new cultural formations, and hence the 'need to think through, rather than around, hybridity' (Hall 1996: 251; cf. Ahmad 1992), while adding a sensitivity to gender relations (Moore-Gilbert 1997: 20). Rather than represent pristine 'local' indigenous cultures, the readoption of indigenous clothing arises out of entangled geographies, national politics, gender relations and notions of tradition. In these entangled processes, the representational elements that appear so secure and enduring are, in practice, the outcome of spiralling relations of domination/resistance. Present day indigenous, national and gender identities thus illustrate a wider sphere of meaning-production, breaking down taken-for-granted boundaries around spaces, bodies or nations, and the powers associated with them.

The Ecuadorian nation-state appears to have interpellated citizens succesfully around particular meanings of 'nation' (Breuilly 1982), by means of the self-representations available in museums, popular mass culture and the habits of commemoration (Gillis 1994; Radcliffe and Westwood 1996). In multi-layered national identities, it is Amazonian promise and new commoditised social relations that offer the most accessible elements (Radcliffe and Westwood 1996). Other, non-state arenas for identity constitution, such as cuisine,[2] language and clothing, provide a means by which indigenous populations articulate their complex positioning vis-à-vis the nation-state, the market and social relations of modernity. With demands for land-title, bilingual education, constitutional changes, recognition of a pluri-national society and new development agendas, the indigenous movement has defined a clear position in Ecuadorian political life. Such transformations rest upon a network of national confederations and widespread mobilising capacity (resulting in the 1990 Indigenous Uprising, and uprisings in 1994 and 1995) (Zamosc 1994; Albán Gomez et al. 1993), as well as local and regional organisation.

This chapter is structured around a discussion of the entangled relations around Ecuadorian indigenous cultural politics, and the tracing of a geography and history to that cultural politics, specifically the adoption of indigenous clothing by Andean social actors. First, the indigenous–state relations are introduced, before a discussion of the spatial and social dimensions of hybridity is outlined. The next section examines power relations in the context of Ecuador's relationship to the indigenous clothed body, and its intertwined history with a number of powers and corresponding resistance. The gendering of indigenous identity is traced by means of a history of a transatlantic, national and rural visual economy, followed by some concluding points.

Indigenous–state relations

One of the forces reinventing the nature of indigenous societies and identities in Latin America has been the power of nation-building (Urban and Sherzer 1991). In this context, indigenous identities are 'unintelligible apart from their struggle with the state' (Abercrombie 1991: 111), as the nation-state attempts to shape 'indigenous-ness' within its remit of forging the national community, while indigenous groups infuse representations and discourses with elements of their own making. However, as the state project of dealing with indigenous groups has changed over time, depending on the prevailing political and economic agendas, so too has the nature of 'indigenous-ness' changed.

The period of modern developmentalism, starting in the post-war period with the Andean Mission of the United Nations (1960), carried with it an agenda of *mestizaje*, namely the idea of assimilating indians within modern, class-based and consumer culture in which indians would lose their traits as they became 'whiter'/*mestizo* (Stutzman 1981; Silva 1991). Under this discourse and associated practices, the 'progressive' indian was one who renounced alcohol consumption, bare feet and adobe houses and learnt Spanish. Following *mestizaje*'s expectation that ethnicity would be replaced by (national) culture, three censuses between 1950 and 1980 contained no question about ethnic affiliations, prefering to measure language use in the home. *Mestizaje* was the route to key modernist reforms in diverse areas, including the economy (a growing internal market for the 'right' goods), gender relations (ending female tradition which blocked progress), cultural difference (with the aim of a homogeneous society) and hygiene (personal and public) (see Radcliffe 1996). Luna Tamayo's 'Modernización' illustrates such views:

> The day we see Indian women wearing patent leather shoes, silk stockings, elegant dresses and hats, strolling the streets of Ambato on the arm of well-dressed Indian [men], that day will be a blessing for the history of the national economy, because we will have obtained for our industry a million and a half or two million new consumers.
>
> (quoted in Clark 1998: 10)

From the early to mid-1980s, a major transformation occurred in the ways in which the state represented, communicated with and legislated for indigenous populations (Ibarra 1992; Crain 1990). The emergence of the new indigenism marked the move away from paternalistic and white-*mestizo* centred notions of indigenous cultures, towards a form of multiculturalism in which diverse cultural traditions were to be acknowledged and celebrated. In 1984 the Oficina Nacional de Asuntos Indígenas (National Office of Indigenous Affairs) was founded to encourage dialogue, and rural development was handed over to the Fondo de Desarrollo Rural Marginal (FODERUMA), although policy was 'assistentialist' rather than empowering, with the exception of bilingual education. Additionally, the increasing role of tourism – linked to the geophysical, natural and social

Figure 7.2 Maria
Source: Sarah A. Radcliffe

features of the country – was an additional incentive to re-evaluate the 'value' of indigenous-ness in national society.

An Andean woman in her thirties, active in extension work and popular education, Maria (Figure 7.2) is representative of the non-politician social group adopting indigenous clothing in the 1990s.[3] Growing up in a small farm of under 5 hectares that was formerly part of a hacienda estate, Maria and her siblings had not spoken Quichua although it was her parents' first language, as they wished to denote social advancement through speaking Spanish and wearing 'urban' *mestizo* clothes. Her brothers combine agricultural work with consultancies, moving between urban and rural areas with ease, while her sister studies at university. Maria passed through the first year at secondary school before starting work, although she has since resumed education. She grew up wearing 'western' clothes which signalled her willingness to enter into the *mestizo*, national and development mentalist agenda of social change. However, in 1990, Maria changed to wearing 'peasant' or 'indigenous' clothes, framing it in terms of memories of what her grandmother had worn. Nowadays she wears a set of clothes immediately recognisable to other Ecuadoreans as signifying an Andean indigenous woman: a black wrap-around straight skirt, flesh-coloured nylon tights, black velvet shoes with gold embroidery, a blue synthetic long-sleeved jumper, a blue woollen shawl knotted at the front, red bead bracelets and necklaces, a black fedora hat and glasses.

When Maria first started wearing 'indigenous' clothes, people were surprised, and commented 'you're *gente blanca* [white], why give up that?' Yet to her – and to other women and men – wearing indigenous clothes expresses pride in indigenous practices and representations, a pride which reflects up and amplifies the indigenous political presence, especially after the 1990 Uprising. She is particularly proud of the respect accorded her by older villagers (interviews 1994, 1996).

Rather than see this latest cultural twist merely as a moment of resistance to Hispanicising dominance of the *mestizo* state, such re-indigenising elements can lead to a reflection upon the geographies and gendering of domination/resistance engaged in new hybridisations. An interaction between the 'modernity' of the nation and the seeming tradition of the costume entails the unpacking of the process by which such hybridisation of identity occurs, and how power works through dress, nation, gender and tradition.

Unpacking hybridity I: geographies and power

Widespread in discussions of post-colonial radicalised minorities, the concept of hybridity offers much to research on cultural issues and ethnic resistance. Drawing on work by such writers as Bhabha on colonialism and hybridity (Bhabha 1990, 1994),[4] and Taussig on mimesis (Taussig 1993),[5] this approach suggests that hybrid cultures are those where form as well as content has been shaped by the coloniser/powerful. The power invested in mimetic objects is the power transferred from the hegemonic force (coloniser) into objects produced by the colonised (Taussig 1993). Hybridity rightly points up the complexities of power engaged in cultural contact and social change, highlighting the deeply entangled and impure, non-essential nature of societies and identities. The process of hybridisation denotes a question of ambivalence, suggesting heterogeneous social fields in which identities are formed through the simultaneous processes of resistance and hegemony, identity and difference.

Additional to the ambivalence of hybridity, it is argued here that hybridity contains within it a complex *geographical* process of ongoing, and closely inter-related, relations of domination and resistance. In order to understand how certain hybridities arise and not others, the unpacking of hybridity within a specific geographical and social context offers a means of understanding how, in this case, the relations of domination and resistance are so profoundly interconnected and overlapping. While the notion of hybridity undercuts the idea that domination/ resistance are mutually exclusive, its unquestioned inbuilt spatial assumption risks reinscribing a particular – global – geography. Hybridity, in writers such as Bhabha and Taussig, refers implicitly to a geography of coloniser/colonised, an international spatial economy. While the workings of modernity *are* grounded in a global spatial field of contact and colonisation, this is not the only space of operation; in Latin America, the 'inbetween-ness' of colonial and then republican society – which looks to both metropolitan culture and elements of 'local' culture (Moore-Gilbert 1997) – is crucial to an understanding of creole/*mestizo* elites' responses to Europe and 'global' practices and images as well as the making of their own 'counter mimicry' (Moore-Gilbert 1997: 149; also Poole 1997; Muratorio 1994b). Understanding the relations of domination/resistance in this context thus means opening up – and analysing – these other, multi-scalar and overlapping geographies, including the space of the clothed body. Geographies of hybridity can then be defined as operating via continuously transformed social relations in a diverse range of places (locations, sites) and via spaces (flows of images, goods, people), in which the separation of domination from resistance is often complex and in many respects, analytically unsatisfactory.

Unpacking hybridity II: the space of the body

> [The body is] the literal site on which resistance and oppression have struggled, with the weapons being in both cases the physical signs of cultural difference . . . for the possession of control and identity.
>
> (Ashcroft *et al.* 1995: 322)

While foregrounding a violent or panic-stricken moment (often under colonialism) as one in which hybridity emerges as a strategy of displacement, Bhabha argues that hybridity 'entertains difference without an assumed or imposed hierarchy' (Bhabha 1994: 4; cf. Young 1995). Nevertheless, hybridity's ambivalence – which according to Bhabha inheres in its instability between mimesis and mimicry of the coloniser – is, crucially related to the relations of domination which cause hybrids, and the entangled resistances that give rise to mimicry. Hybridity refers as equally to the 'effects of the politics of exclusion within key modernist institutions' (Harvey 1996: 25) – such as the nation – as to playful mimicry, in which the outcomes of domination are not pre-given.

In this context, the bodily space – among other spaces – articulates and reflects back issues of identity and positioning (Duncan 1996). Gilroy's (1996) argument about post-colonial agency dwindling down to the family and then to 'the exterior surface of the body', belies the workings of power relations in more ambivalent and spatially variable ways. Personal body space can be seen as a valid arena for colonial and national control, yet in many situations the body becomes an arena for diverse 'stories', a space over which contests are (re-)made, intertwining power-to, power-over and power-within. The following account focuses on the ways in which the bodily space – and its re-presentation through clothing – articulates geographical, power-invested relations of identity and place in contemporary Ecuador.

During the colonial period, indigenous clothing varied from colonial district to hacienda estate, as a means of distinguishing indians and preventing escape from onerous hacienda work. Historically, in Andean provincial and rural areas the type of clothing one wore indicated status and the treatment to be accorded in social interaction between unequals. Criteria for social distinction rested on dress, with violence and discrimination meted out to those transgressing such boundaries (Stark 1981). In southern Peru, indigenous peoples wearing 'white' people's shirts were punished by *mestizo* elites (Weismantel 1988: 81). Clothing was used by landowners and officials to reinforce their own social status, a visible indicator of their dominance.

Complex new geographies affect the positioning of indigenous groups today and the power relations around definitions of nation, gender and tradition. Formerly clear-cut boundaries between rural and urban, between peasants and informal sector workers, have been profoundly reconfigured during economic restructuring and modern development. Although the state has been active in this period reinforcing national-based identities, there is in fact a new ambivalence around ethnic, national and class boundaries. All of these factors are reflected in a new politicisation of dress, as a surface upon which dynamic relations of power are worked through gender relations, meanings of nation and tradition, and the imagined community of nationhood.

During the 1980s, there was a reinvention of rural social relations in non-state spaces in a discursive and organisational project that impacted widely on public culture. With the strengthening of the indigenous movement (Zamosc 1994; Bebbington *et al.* 1992), the so-called 'indian problem' of the 1950s and 1960s irreversibly became an 'ethnic-national question' both at the level of demands made by indigenous nationalities, and in the manner in which the state reorientated its discourses (Wray 1989). Moreover, CONAIE's formal political power increased with Congressional elections and ministerial appointments.[6]

Such organisational strengths rest upon the creation of Andean rural identities in addition to the use of cultural politics as a means to highlight indigenous presence. An Andean identity and practice emerged through grassroots and community-based processes, despite increasing linkages to labour and product markets, and despite the official value placed on urban *mestizo* identities. By contrast, Andean rural society was reformulated not only in terms of political economy, but also through cultural and social organisation. Permanent and semi-permanent urban migrants remained in close contact with rural areas of origin. Overall, the boundaries around social groups gained a new ambiguity, and the boundaries between indigenous and *mestizo* became more ambivalent. During the 1980s, as land distribution in the Sierra showed equal proportions of land among small, medium and large landowners (Zamosc 1994: 43), the ethnic identification of these groups became less predictable. The 1990 Uprising revealed widespread support among '*mestizo*' communities who historically would not have identified with these demands.

With the rise of an indigenous political presence, cultural politics are now used to amplify a new political culture around nation and indian. One key indigenous demand is the ending of ethnic discrimination, and the view that indians are second-class citizens. As CONAIE leader Macas states, 'we [indigenous] were, have been and are still [seen as] inferior; we are second class people' (Macas 1993: 116). In this context, one politicising strategy is to stress the widespread existence of indian backgrounds among the Ecuadorian population, as well as the political (rather than 'scientific') choice of acknowledging an indigenous heritage.

To summarise the argument so far, while the possibility of hybridity has emerged within Ecuadorian society, it is contextualised by the historical hegemony of a move from indigenous to *mestizo* in which the presentation of subjects' affiliation to place has occured through the layering of clothing on a racialised, colonised body. Moreover, today's affirmation of 'indigenous' clothing (and pro-indigenous affiliations which it represents) plays into a geography of social relations of domination in which the ethnic signs ascribed by dominant colonial society to indigenous groups were forcibly, and then willingly, put on. In other words, the 'ironic compromise' (Bhabha 1994: 85) of mimicry rests on the ways in which clothing has come to represent a reconfigured geography of identity.

Gendered bodies

In the gendered politics of culture, women more frequently retain 'traditional' clothes than men in the process of national integration and development (Crain 1996; De la Cadena 1995). From Guatemala to Mexico, female indigenous clothing represents the officially endorsed essence of nationhood, the mimesis of nation in female clothed bodies (Hendrikson 1991). Under the complex power relations of colonialism too, heterosexuality – what Young calls the 'implicit politics of heterosexuality' (Young 1995: 25) – and gender relations have long provided a template for the organisation of relations of power and difference. In republican nation-states, the hybrid feminine symbol carries with it a recognition of the gendered, sexualised relationship of colonial power as well as the enabling power of the state to mould its national population (Stoler 1995; Young 1995). Indigenous political culture, in which the female indigenous figure carries so much ideological baggage, is thus to be understood in the light of longer trajectories of entangled powers, ones that stretch across numerous spaces as well as engaging profoundly contested notions of what symbolises nation and belonging.

Tracing the history of the figure of the indigenous woman in Ecuador reveals the entangled webs of power around its construction today as an icon of political and national identity. Part of a budding transatlantic trade of images and texts about the Andes, the *Relación histórica del Viaje a la América Meridional* (Narrative of a Journey to Meridional America) was published in 1747/8 by the Spaniards Jorge Juan and Antonio de Ulloa. Europe's fascination with Andean others was sparked by La Condamine's 1736 French Academy of Sciences expedition, while Juan and Ulloa's account of that journey was widely read in Europe (see, among others, Pratt 1992; Poole 1997). Being associated with La Condamine, the narrative and representational dimensions of the report entered into later republican self-identities, having been mediated by extensive European circuits of knowledge about the Andes, particularly concerning the emerging modernist narrative of 'types', hygiene and gender.

The purpose of indigenous femininity within such representations was to mark the lower limit of 'colonial' society. As Fitzell (1994) explains in her discussion of one print, the indian woman's placing to the far right and background of the scene indicates her marginality not primarily in racial terms but in geographical distinctions between rural/urban, capital/provinces and in gendered relations. Fitzell highlights the non-racialised but highly gendered nature of difference in this image, an emphasis that echoes a wider French imaginary of the time (Poole 1997). In marking the gender contrast, the rural male indian accompanying her stands to her left in a separate 'class', yet closer to those of urban residence and noble birth. In other words, the figure of the indigenous woman had become profoundly marginal yet crucial in the reinforcement of ethnic and class boundaries, later to be reworked into national imaginaries.

For a brief moment at independence in Andean America, a discontinuity in the positioning of indigenous femininity appeared, with the widespread use of representations of *La India* (a generic indigenous female icon, kitted out with feathers) on shields, flags, medals, coins and monuments (Muratorio 1994a: 13; Taussig 1993). Framed within a transatlantic circuit of imagery, the *India* was part of the making of meaning by new creole elites, staking a claim to their own regional authenticity and their difference from the Spanish ex-colonial masters. Temporarily, the indian woman became a symbol of both Latin America and Liberty, a powerful figure reflecting the entangled power relations between Europe and former Spanish colonies.

As the nineteenth century progressed, with rising economic and political power among *mestizo* elites, and their marginalisation of liberal agendas that would have permitted indigenous (and black) political participation in nation-building, so too the iconography of a feminised nation increasingly moved away from indigenous representations (see Guerrero 1994 for a general discussion). Ecuador's public iconography shifted, as the *India* was replaced by more European-oriented imaginaries. The main public symbol of Ecuadorian republicanism – Quito's Independence statue – is of a white woman. To commemorate early struggles for independence against the Spanish, the 'Diez de Agosto' (Tenth of August) committee was founded in the early 1890s, to judge on proposals for a monument (fieldnotes, 1996). The design by Fransisco Durini de Cacerez and company was finally accepted, and in 1894, Juan Batista Minghetti was commissioned by the municipality to make a design. A special tax was raised for the monument and paid for its prefabrication and transport from Italy. By the time the monument was inaugurated in Independence Square, Quito, in 1906 (in time for the new liberal constitution), the indigenous *América* had become an Italian-made maid, with a thin, European face, short wavy hair and a laurel crown. In other words, during the course of a century, the *India* had transformed herself into an Old World goddess refracted through French Revolution iconography and draped in fluid Grecian fabrics, her appearance giving no risk of identification with indigenous femininity. As an equivalence was increasingly drawn between racial hybridity and 'degeneracy' in late-nineteenth-century social Darwinist notions, so the 'whitening' of public female icons occurred, as in Quito's monument (cf. Muratorio 1994b).

During the twentieth century, the figure of the indigenous woman came to represent a new 'limit', another marker of the extent of nationhood, under the idea of *mestizaje* as a route to modernity. In continuity with early-nineteenth-century ideas of a marginal yet crucial position, indigenous women embodied a 'foil' to identity, in the new context of modernisation-oriented development. In the 1960s, the Andean mission explicitly recognised indigenous women as the main group to be targeted in development programmes, perceiving them as where 'tradition' was strongest. 'Women in rural areas tend to a greater degree of conservatism and to the retention of local customs, dress and manners' reported the Andean Mission (1960: 5). Following decades saw development programmes aimed at incorporating rural, largely indigenous, women into the national project (see Radcliffe 1996).[7]

As tourism and market economies have expanded, so too has the commercialisation of indigenously dressed female bodies entered into the dynamics of representation. Migrating into Quito, young indigenous women are hired for hotel employment and are dressed to symbolise an exotic alterity to domestic and foreign businessmen (Crain 1996). Their work clothes are in fact too expensive for general use, yet represent the powers (power over, as well as power within) of modernity's hygiene, simultaneous with the representation of the hetero-sexualisation of client–maid interactions and a mimesis of ethnicised relations of servility on the hacienda (cf. McClintock 1995).

Hybridity in indigenous eyes

Andean groups historically made sharp distinctions between *mestizo* and indigenous dress such that boundaries between these two categories were reinforced. While migration permitted the dropping of indigenous clothing in favour of 'western'/ modern clothing, moves from 'indigenous' to 'white' were perceived by indians as 'inauthentic'. Changing clothing meant affiliation to the *mestizo* national sphere, a 'putting on' of a non-indigenous identity, a disguise or mimicry of what it meant to be 'Ecuadorian'. The incomparable categories of indigenous and *mestizo* were frequently reinforced by dress codes (Stark 1981). Among indigenous Otavaleños, there are marked distinctions between the 'authenticity' – demonstrated through bodily presentation of clothing and hair – of themselves and *mishu/mestizos*, and the inauthenticity of the *chaupi-mishu* ('half-mestizos'). With their cut hair or residence in Quito, *chaupi-mishus* were seen by other indigenous to be strongly identified with modern nationhood (Frank 1991: 516).

Simultaneous with mounting indigenous political presence from the 1970s, the adoption of indian dress signals the entangling of powerful homogenising national cultures and their claims to tradition, together with the power of authenticity for indigenous claims to national belonging. Indigenous groups have endowed mimetic power on to clothing within the mirror dance that is Hispanic-indigenous encounters over time. However, by inscribing 'indigenous-ness' into the national political project, it gained a new cultural meaning, reiterating a positive ethnic difference, out of a previously negative inscription. The 'double inscription' (Hall 1996) of post-colonial identities thus gives rise to ambiguities, just at the time when a claim to authenticity is politically charged. The mimetic power embodied in such items is not easily relinquished, as a shift to (another) apparently less 'authentic' position represents a potential instability in identity. As Alonso says 'any departure from a mimetic performance of an invented past can be construed as a loss of original substance' (Alonso 1994: 398). Yet gender relations step in here to provide a grounding for 'original' substance, and for claims to indigenous tradition, at the time when highly succesful indigenous claims are also being made on the state.

From the perspective of indigenous communities, the transformation of women into *mestizo* subjects is a more ambiguous and threatening process than for men. While men in the largely 'indigenous' area of highland Zumbagua could adopt 'white' aspects such as jeans or short hair without threatening their masculinity or indigenous identity (Weismantel 1988: 82), women from the same region could not do so. If the women moved away from strictly demarcated categories of indian and white, they upturned gender relations, being accused of non-conforming sexual behaviour (not marrying, or being lesbian). From this perspective, the move to *chola* status is deeply inscribed in the power relations around heterosexuality. It would appear that it is not only the Victorians whose notions of hybridity engaged very directly with sexuality; in Ecuadorian indigenous communities, the politics of heterosexuality is explicit. However, in this case the destabilising element adheres not primarily to race (the linchpin for Victorian debates about degeneracy) but to gender.

Yet the domination of a heterosexual visual economy cannot be the only side of the question, for if, as Mallon (1996: 176) notes, 'the women are supposed to be the *tabula rasa*, the ground on which men inscribe ethnicity or national identity in their struggles for power, what happens when the ground moves and speaks?' For Maria and other women, there now exists a 'new' authenticity and community due to their adoption of indigenous dress. From Maria's perspective, wearing indigenous clothes has confirmed her political values while also providing a basis for social interaction with a distinctive, 'other' community with which she can identify and socialise. Since wearing 'indian' clothes, Maria has found she can 'laugh with and have fun with people who wear traditional clothes' even if they are not personal friends (interview 1994). The adoption of female indigenous dress by women inserts itself in that space where mimesis can offer the possibility of a non-essentialised femininity, alive to the instability of that sign.

Conclusion

With the rise of the Ecuadorian indigenous movement, the figure of the clothed, indigenous woman has been remade – once again – in its (dis)continuous history of power and spatial relations. In this new context, how can indigenous femininity stand in for indigenous resistance? At the present time, the figure is appropriated by Andean indigenous groups as a rural claim, away from the spatial concentration of power in cities. Additionally, the 'pluri-national' claims speak to a pan-American indigenous renaissance, as well as a constitutional reconfiguring of the Ecuadorian nation-state, providing more autonomy to indigenous-majority regions. Yet it also speaks to the biopower of modernist hygiene; the acceptable indigenous figure is clean, sober and ordered, the twentieth century equivalent to Otavaleño 'acceptable hybrids' sent to the 1892 Madrid exhibition (Muratorio 1994b: 125, 134).

Thinking back to the case study of Maria, we can reinterpret her situation and the complex, overlapping issues surrounding readoption of indigenous clothing. Maria represents part of the new Andeanity, the economic and social circuits linking rural and urban areas and crosscutting ethnic boundaries. Her insertion into the circuits of modernity's cultural geographies have made her put on certain items of clothing as symbols of indigenous resistance. The relations of domination associated with modernist notions of hygiene are behind her appearance, as are the heterosexual relations of gender power which reinforce her embodiment as a summation of the 'essence' of identity. The items of clothing she wears are what the consumer culture of market relations offer, while laying claim to the reciprocity and non-market relations of Andean solidarity. In one respect, she epitomises Luna Tamayo's dream of modern development, although the circuits which create meaning for her are social, economic and political linkages operating across the terrain of national space. A circle is thus completed: the hybrid indigenous with which Ecuadorian nationhood imposed an alter-nation/native has now rearticulated an imagined community on the basis of that overdetermined elision of femininity and nationhood. In the interconnected relations of power, the indigenous figure is both a produced symbol and a natural given, a complex mimesis of that which was forged at the boundary of imagination and practice.

In the adoption of 'indigenous' clothing by men and indigenous political leaders, there is an aspect of what Bhabha terms the 'contestation of the given symbols of authority which shift the terrains of antagonism', and which are 'displaced in a supplementary movement' (Bhabha 1994: 193). Although malleability in power relations is analytically useful, it has been argued that analysis of hybridisation needs to be more attentive to the *spatial* relations and power relations which lie behind them. The post-colonial discussion of hybridity has been largely constructed within an a-spatial paradigm, in which the global is read as the (former or indirect) coloniser and the colonised is overdetermined as the 'local'. Even if agency is seen to coincide with the family or body surface, such a geography disregards the interrelated scales and *multiple* geographies around which agency, power and resistance are articulated. This chapter has argued that the spatial relations present in the formation of hybrids are operating at a variety of different scales, as well as constituted by a number of spatial relations organised around place (specific locations) and spaces (flows of imagery, populations, ideas and meanings). In the Ecuadorian case, the complex multi-level and diversely located geographies and social relations behind the use of indigenous dress speak to the (gendered) body and additionally to other spaces and places.

Moreover, by recognising the gendering of the bodily space, this chapter has argued for a sustained *gender* focus in discussions of hybridity. Perceived as a complex interplay between relations of domination and resistance, the ambivalence of the hybrid with relation to unequal gender relations points to the domination within hybridity (Bhabha 1994: 86). In the Ecuadorian case, such domination rests largely upon the politics of heterosexuality. Andean patriarchal popular imaginations link femininity with the risk of loss of essence. Whereas the male adoption of indigenous clothes is couched within a context of masculine (albeit subaltern indigenous) agency, the readoption of indian clothing by women speaks to (but is not encompassed by) gender relations of heterosexuality.

In conclusion, bringing together the facets of spatial and social powers by which hybrids are constituted, the Ecuadorian material illustrates the role played by these dimensions in constituting what at first view appears to be an act of resistance, and what turns out to be a story about complex entanglements of resistance/domination. What is resistance in one facet (ethnic resistance) is in fact entangled with republican elite relations of resistance to a former colonial power, while also being founded on the ambivalent relations of heterosexualised gender relations in which female bodies represent an essence of identity, while offering community to women positioned in this way. By focusing on the space of the gendered/racialised female body, it has been argued that the indigenous woman is not solely to be understood in terms of her located, ethnic community but equally within a transatlantic and inter-regional set of powers. The Ecuadorian 'new indigenous' are not the 'unreliable moderns' of the Mexican *naco*, with their inability to distinguish between high and low culture (Lomnitz 1996: 57), but rather the assimilation of an imitation with highly politicised regard to an 'original'. Ecuadorian indian presence is reinforced through the adoption of indigenous clothing, yet given the baggage with which such signs come weighted – nation, gender, authenticity – the putting on of these clothes is also a story about hybridity. Written from a gender and spatially sensitive perspective on hybridity as a route into understanding relations of domination/resistance, the chapter shows the complexities of power as they work through gender, nation, tradition and dress.

Acknowledgements

This chapter was published in revised form in *European Review of Latin American and Caribbean Studies* (1997) 11: 9–27 and I wish to thank the editors for permission to re-present the material here. I am also grateful for comments from the editors of this collection, as well as their original invitation to give a paper to the conference Geographies of Domination/Resistance in Glasgow.

Notes

1 In Latin America, the predominant analytical framework for indigenous culture and politics has long been 'resistance' (Field 1994), whereby the process of identity definition and social practice takes place in the struggle of indigenous groups *vis-à-vis* a dominant acculturating state or class.
2 In discussion of Zumbahua, Cotopaxi, Weismantel (1988) notes that while white rice (a symbol of nationhood) is consumed every two weeks by Quichua-speaking peasants, they continue to produce and consume barley soup, which symbolises Quichua-inflected reciprocity and history.

3 All names have been changed.
4 Young (1995: 23) defines a hybrid as a 'transmutation of . . . culture into a compounded, composite mode . . . While hybridity denotes a fusion, it also describes a dialectical articulation.'
5 Taussig (1993: xiii) discusses mimesis in terms of 'the copy drawing on the character and power of the original, to the point where the representation may even assume that character and power.'
6 In the 1996 national elections, eight indigenous representatives of the Movimiento Nuevo-Pais Pachacutic were voted into Congress, and a number subsequently gained key ministerial positions.
7 Yet bringing indigenous women into the white, urban and modern nation rests upon the ambivalence of the resulting hybrids. In migrating to cities to undertake domestic work, indigenous women became *cholas*, and to paraphrase Bhabha, almost white but not quite (Bhabha 1994: 85). Despite cutting hair, wearing make-up and changing clothes, *cholas* remained 'unmoderns', ambivalent mimics of the goal of *mestizaje* nationhood.

References

Abercrombie, T. (1991) 'To be Indian, to be Bolivian: "ethnic" and "national" disourses of identity', in G. Urban and J. Sherzer (eds) *Nation-States and Indians in Latin America*, Austin: University of Texas Press.

Ahmad, A. (1992) *In Theory: Classes, Nations, Literatures*, London: Verso.

Albán Gomez, E., Andrango, A. and Bustamente, T. (comps) (1993) *Los indios y el estado-pais: pluriculturalidad y multi-etnicidad en el Ecuador*, Quito: Abya-Yala.

Alonso, A. (1994) 'The politics of space, time and substance: state formation, nationalism and ethnicity', *Annual Review of Anthropology* 23: 379–405.

Andean Mission of the United Nations (1960) *Integration of the native indian population in Ecuador*, Andean Mission/Government of Ecuador.

Ashcroft, B., Griffiths, G. and Tiffin, H. (eds) (1995) *The Postcolonial Studies Reader*, London: Routledge.

Bebbington, A. *et al.* (1992) *Una década ganada*, Quito: Abya-Yala.

Bhabha, H. (1990) *Nation and Narration*, London: Routledge.

Bhabha, H. (1994) 'Of mimicry and man: the ambivalence of colonial discourse', in *Location of Culture*, London: Routledge.

Breuilly, J. (1982) *Nationalism and the State*, Manchester: Manchester University Press.

Clark, K. (1998) 'Race, culture and mestizaje: the statistical construction of the Ecuadorian nation, 1930–1950', *Journal of Historical Sociology* 11: 1–20.

Crain, M. (1980) 'The social construction of national identity in Highland Ecuador', *Anthropological Quarterly* 15(3): 43–59.

Crain, M. (1996) 'La interpenetración de género y etnicidad: nuevas auto-representaciones de la mujer indígena en el contexto urbano de Quito', in L. Luna (ed.) *Desde las orillas de la política: género y poder en América Latina*, Barcelona: Universitat de Barcelona.

De la Cadena, M. (1995) '"Women are more indian": ethnicity and gender in a community near Cuzco', in B. Larson and O. Harris (eds) *Ethnicity, Markets and Migration in the Andes*, London: Duke University Press.

Duncan, N. (ed.) (1996) *Bodyspace: Destabilising Geographies of Gender and Sexuality*, London: Routledge.

Field, L. (1994) 'Who are the Indians? Reconceptualizing indigenous identity, resistance and the role of social science in Latin America', *Latin American Research Review* 21(3): 237–248.

Fitzell, J. (1994) 'Teorizando la diferencia en los Andes del Ecuador: viajeros europeos, la ciencia de exotismo y las imágenes de los indios', in B. Muratorio (ed.) *Imágenes y imagineros: representaciones de los indígenas ecuatorianos, siglos XIX y XX*, Quito: FLACSO.

Frank, E. (1991) 'Movimiento indigena, identidad étnica y el Levantamiento: un proyecto político alternativo', in I. Almedia (comp.) *Indios: reflexiones sobre el Levantamiento indígena*, Quito: ILDIS.

Gillis, J. (ed.) (1994) *Commemorations: The Politics of National Identity*, Princeton, NJ: Princeton University Press.

Gilroy, P. (1996) 'Route work: the black Atlantic and the politics of exile', in I. Chambers and L. Curtis (eds) *The Post-Colonial Question*, London: Routledge.

Guerrero, A. (1994) 'Una imagen ventrilocua: el discurso liberal de la "desgraciada raza indígena" a fines del siglo XIX', in B. Muratorio (ed.) *Imágenes y imagineros*, Quito: FLACSO.

Hall, S. (1996) 'When was the "postcolonial"? Thinking at the limit', in I. Chambers and L. Curtis (eds) *The Post-Colonial Question*, London: Routledge.

Harvey, P. (1996) *Hybrids of Modernity: Anthropology, the Museum and the Nation-State*, London: Routledge.

Hendrikson, C. (1991) 'Images of the indian in Guatemala: the role of indigenous dress in Indian and Ladino constructions', in G. Urban and J. Scherzer (eds) *Nation-States and Indians in Latin America*, Austin: University of Texas Press.

Ibarra, A. (1992) *Los indígenas y el estado en el Ecuador*, Quito: Abya-Yala.

Lomnitz, C. (1996) 'Fissures in contemporary Mexican nationalism', *Public Culture* 9: 55–68.

Macas, L. (1993) 'Tenemos alma desde 1637', in E. Albán Gomez (comp.) *Los indios y el estado-pais*, Quito: Abya-Yala.

McClintock, C. (1995) *Imperial Leather: Race, Gender and Sexuality in the Colonial Contest*, London: Routledge.

Mallon, F. (1996) 'Constructing mestizaje in Latin America: authenticity, marginality and gender in the claiming of ethnic identities', *Journal of Latin American Anthropology* 2(1).

Moore-Gilbert, B. (1997) *Postcolonial Theory: Contexts, Practices, Politics*, London: Verso.

Muratorio, B. (1994a) 'Introducción: discursos y silencios sobre el indio en la conciencia nacional', in B. Muratorio (ed.) *Imágenes y imagineros*, Quito: FLACSO.

Muratorio, B. (1994b) 'Nación, identidad y etnicidad: imágenes de los Indios ecuatorianos y sus imagineros a fines del siglo XIX', in B. Muratorio (ed.) *Imágenes y imagineros: representaciones de los indígenas ecuatorianos, siglos XIX y XX*, Quito: FLACSO.

Poole, D. (1997) *Vision, Race and Modernity: A Visual Economy of the Andean Image World*, Princeton, NJ: Princeton University Press.

Prakash, G. (ed.) (1995) *After Colonialism: Imperial Histories and Postcolonial Displacements*, Princeton, NJ: Princeton University Press.

Pratt, M. L. (1992) *Imperial Eyes: Travel Writing and Transculturation*, London: Routledge.

Quintero, L. and Silva, E. (1991) *Ecuador: una nación en ciernes*, Quito: FLACSO.

Radcliffe, S. A. (1996) 'Gendered nations: nostalgia, development and territory in Ecuador', *Gender, Place and Culture* 3(1): 5–21.

Radcliffe, S. A. (1998) 'Frontiers and popular nationhood: geographies of identities in the 1995 Ecuador–Peru border dispute', *Political Geography* 17(3): 273–293

Radcliffe, S. A. (1999) 'Popular and state discourses of power', in J. Allen and D. Massey (eds) *Human Geography Today*, Cambridge: Polity.

Radcliffe, S. A. and Westwood, S. (1996) *Remaking the Nation: Place, Identity and Politics in Latin America*, London: Routledge.

Santana, R. (1995) *Los indios en la política, la política de los indios*, Quito: Abya-Yala.

Silva, E. (1991) *Los mitos de la ecuadorianidad: ensayo sobre la identidad nacional*, Quito: Abya-Yala.

Stark, L. (1981) 'Folk models of stratification and ethnicity in the highlands of Northern Ecuador', in N. Whitten (ed.) *Cultural Transformations and Ethnicity in Modern Ecuador*, Chicago: University of Illinois Press.

Stoler, A. L. (1995) *Race and the Education of Desire: Foucault's 'History of Sexuality' and the Colonial Order of Things*, Durham, NC: Duke University Press.

Stutzman, R. (1981) '*El mestizaje*: an all-inclusive ideology of exclusion', in N. Whitten (ed.) *Cultural Transformations and Ethnicity in Modern Ecuador*, Chicago: University of Illinois Press.

Taussig, M. (1993) *Mimesis and Alterity: A Particular History of the Senses*, London: Routledge.

Urban, G. and Scherzer, J. (eds) (1991) *Nation-States and Indians in Latin America*, Austin: University of Texas Press.

Weismantel, M. (1988) *Food, Gender and Poverty in the Andes*, Pittsburgh, PA: Pittsburgh University Press.

Whitten, N. (ed.) (1981) *Cultural Transformations and Ethnicity in Modern Ecuador*, Chicago: University of Illinois Press.

Wray, N. (1989) 'La consitución del movimiento étnico-nacional indio en Ecuador', *América Indígena* 49(1): 77–99.

Young, R. J. (1995) *Colonial Desire: Hybridity in Theory, Culture and Race*, London: Routledge.

Zamosc, L. (1994) 'Agrarian protest and the Indian movement in the Ecuadorean highlands', *Latin American Research Review* 21(3): 37–69.

8

JAMAICAN YARDIES ON BRITISH TELEVISION

Dominant representations, spaces for resistance?

Tracey Skelton

The media create space through which representations are both constructed and deconstructed; these representations are part of the highly dynamic and entangled power plays between domination and resistance. Processes of interpretation are multifaceted and engage several actors in the 'writing' and 'reading' of the media text. Those involved in the creation of a particular media space will have multiple understandings of what it is they are trying to convey, while the audience of such a space will have multiple understandings of what it is they are trying to perceive. Hence the media are highly diverse, dynamic and many layered spatialities and a knitting-wool tangle of power. Visual media broadcast on television are particular formations of complex power relations of domination and resistance. The media are useful examples of Pile's (1997: 2) discussion of the fact that 'geographies of resistance do not necessarily (or even ever) mirror geographies of domination, as an upside-down or back-to-front or face-down map of the world'. In this chapter I consider two television programmes which have a similar focus of attention, Jamaican Yardies, but they are far from mirror images of each other and both offer dominant and resistive elements. There are several ways of reading and inter-preting both films, different ways of giving them meaning; what I offer here are just some parts of those possibilities.

The making of television programmes (re)presents and interprets subjects through a myriad of images. The viewing of television programmes (re)interprets representations through complex interrelationships of the subjects we view and our own subject identities. The processes of making and watching are linked with power: the economic power to make and to have access to watch; the social power to be behind the camera, to write the script, to travel and observe; the political power to construct the programme or to be at liberty to view it; and the cultural power to represent, symbolise, interpret and give meaning to what is shown. Making and watching television programmes are part of the processes of domination and resistance. An individual programme can at once be dominant and resistive. Dominant, as for a period of time it can present us with a particular interpretation of a situation, an event, a process and as we watch that is the dominant image we receive. Resistive, because that same programme which dominates the screen for a period of time may be providing a narrative which resists the status quo, critiques the 'official version', and so allows hidden voices and visions to be heard and seen.

The visual media actively construct spaces both real and imagined, created and filmed by the camera. While there are multiple meanings at play around a television programme there are also what Hall (1987c) calls 'preferred meanings'. That is the meanings which those who construct the images and match the narrative to those pictures would like the audience to receive. Hence many people will read a particular programme in a way very similar to the 'preferred meaning' of the producer/director, but there will always be those who resist such 'preferred meanings' because of their own subject knowledges, their own life experiences and/or their own social status. Consequently the meanings and interpretations taken from television programmes can be as entangled as the lives and experiences of those who make and view them. Television programmes are a material practice, and in this case the two programmes are firmly located within UK material practices, although the *Yardies* programme (Zabihyan 1995) was also broadcast on Jamaican national television.

This chapter, therefore, focuses on two television programmes (self-identified as documentaries) broadcast in 1995 that constitute part of a 1990s fascination with the so-called Jamaican Yardies. The first one considered is *The Informer* (World in Action 1995) and the second is *Yardies* (Network First 1995). The chapter demonstrates the dominant power of the creation and reinforcing of racist representations in the first film, and investigates the resistive tendencies against the racist stereotyping of Jamaicans, and especially young Jamaican men, in the second. The chapter also examines the ways in which there could be resistive readings from the first programme and perceptions of dominant stereotypes from the second. The chapter unravels the threads of domination/resistance knitted through the two programmes and demonstrates the active roles of power played through the spatialities of visual media.

Both documentaries are essentially about two places: the UK and Jamaica. Hence they represent facets of uneven geographies within the world economy. In *The Informer* the UK is represented as 'First World', sanitised, wealthy, a fighter of crime, working to ensure security for its citizens, mighty and on the side of right. Jamaica is a 'Third World' country, chaotic, poor, creator of crime, failing to provide security and safety, weak and riddled with wrong. However, it also offers a compelling critique of the ways in which some members of the London Metropolitan Police Force have been operating in relation to the so-called Yardies. So while the programme uses extensive racialised stereotypes, it also offers a resistive critical take on some aspects of the British police. In *Yardies* we see Jamaicans struggling to make a living within the context of severe economic depravation and we see the Jamaican police trying to control crime, to control young Jamaican men, and to protect the wealthy of the nation. Yet this programme also contains dominant meanings because it exploits preconceived binaries: Yardie youth/police officers; poor downtown Kingston/rich uptown Kingston. The relationship between Jamaica and the UK is de-historicised in both programmes, although some of the viewing public will know that the connections were established through slavery, plantation economics and empire. In both programmes the dominant nation, the UK, through its media might 'represents' certain aspects of the sub-ordinated and yet resistive nation, Jamaica.

In this chapter I offer a particular interpretation of the two television programmes which, together with print media coverage of the Yardies phenomenon (Skelton 1998b), have created a cultural space in which the British public are encouraged to perceive young black men, not necessarily just Jamaican, as violent, ruthless and criminal. I wish to demonstrate the ways in which the power of such a cultural space is racist, violent, harmful and constitutes a process of excluded geography (Sibley 1995). I also wish to consider a particular way in which such negativity might be challenged and resisted within the same cultural spaces. The chapter is therefore an example of a discursive approach because it looks at the ways in which meanings and representations are constitutive of particular constructions of knowledge and understanding. To quote Hall:

> The 'discursive approach' is concerned with the effects and consequences of representation – its politics – how language and representations produce meaning but [also] how the knowledge which a particular dis-course produces connects with power, regulates conduct, makes up or constructs certain identities and subjectivities, and defines the way certain things are represented, thought about, practised and studied.
>
> (Hall 1987a: 6)

In the next section I consider the theoretical and conceptual debates around representation with a specific focus on the media and issues of 'race'. The third section demonstrates the ways in which the British media have constructed particular representations and meanings around black people based in the UK. I also show that such stereotypical images have always been contested and that there have been significant shifts made in the ways in which television interacts with issues of 'race'.

'Practices of representation': 'Regimes of truth'

These are not my phrases but placing them together is my invention because I have been struck by the aptness of both concepts for a discussion around the media, 'race' and the entanglements of power.[1] Before considering these phrases and in particular the concept of 'representation' it is important to backtrack a little and consider the complex term 'culture'.

Culture means very different things and is a highly complex concept yet is used somewhat glibly and in a myriad of contexts. Here I wish it to be understood (yes, it would be my preferred meaning!) as a process, a set of cultural practices, as the production and exchange of meanings between members of a society or group who can then build upon shared meanings (Hall 1997a). In cultural practice, things, people, events and places have meanings because we give them meaning through the ways in which we think about them, how we use them, how we experience them and so forth. However, we need to ask where and how are the meanings produced?

Meaning is constantly interpreted and is produced through different forms of media; one such form is television programmes, and in this particular case, documentaries. How sure are we that people share the same meanings given to the things, people, events and places that are interpreted through, in this case, television programmes? Hall argues that people who are part of a culture must share concepts, images and ideas to enable them to think and feel about the world, so they then experience that world in broadly similar ways. In order to do this they use the same 'cultural codes' (Hall 1997a: 4). Hence there has to be a common language of cultural understanding and this includes being able to read visual media in the same ways. Such a language works through representations in that there are elements which we use to express or communicate something – a thought, concept, idea or feeling – which others will understand in a similar way to that which we mean. Thus meaning is conveyed through representation, but meaning is rather like frog spawn – slidy, sticky, disguised, at times transparent, translucent and opaque, but still one part of the various formations of a frog.[2]

> Meaning is not straightforward or transparent, and does not survive intact the passage through representation. It is a slippery customer, changing and shifting with context, usage and historical circumstances. It is never therefore finally fixed. It is always putting off or 'deferring' its rendezvous with Absolute Truth. It is always being negotiated and inflected, to resonate with new situations. It is sometimes contested and bitterly fought over. There are always different circuits of meaning circulating in any culture at the same time, overlapping discursive formations, from which we draw to create meanings or express what we think.
>
> (Hall 1997a: 9–10)

This means that while the programme makers of the two films under discussion might be using shared cultural codes and representations of meanings in relation to young Jamaican men in particular (young black men in general), there are inevitably going to be multiple meanings attached to the programmes. A few will come through within the programmes themselves, but most with the different readings that the audiences provide as they receive the visual images and the language of the narration. Yet it is also the case that because the programmes are part of the construct of British media culture, there are shared codes and strong 'preferred meanings' which will present dominant representations of the subjects of the filming. Hence there is likely to be a majority audience in the British context who receive the dominant meaning much as it was intended, but resistive readings and interpretations are always possible because there is always contestation over representation. Nevertheless both films are an example of 'The embodying of concepts, ideas and emotions in symbolic form which can be transmitted and meaningfully interpreted [and is what is meant] by the "*practices of representation*"' (Hall 1997a: 10, added emphasis).

The two programmes demonstrate 'practices of representation' and present 'preferred meanings' in relation to Jamaican Yardies. They are in a position to do this because, although as Foucault (1980) argues meanings and knowledges, and the practices based upon them, change over time, there is a relationship connected with meaning and representation which remains constant and that is the relation-ship of power. How do 'practices of representation' build up some form of control over the way in which we perceive and interpret the notion of Jamaican Yardies, and by extension young black men? Foucault would argue that it hinges on how knowledge and power are interlinked. When knowledge and power intersect there is an assumption of 'truth', and also there are the means/practices to make something true. Knowledge can be used to control the behaviour of others, to determine how we think about others, to legitimise practices which restrict and constrain the behaviour of others. As Hall (1997b: 49) points out, drawing on Foucault, 'what we think we "know" in a particular period . . . about crime has a bearing on how we regulate, control and punish criminals'. Hence once knowledge and meanings of the Yardies are constructed through a particular discourse ('practices of representation'), such as TV documentaries, print media stories, police training videos, TV news stories, 'gangster' novels and music, then relations of power come into play around control and punishment. We might not be very fussy about who and how we represent – all young black men can therefore be constructed by the discourse as Yardies or potential Yardies. In such a way we construct a 'regime of truth' (Foucault 1980), one in which not all black young men are criminals but dominant meaning systems suggest this to be so. We might punish them for what we believe to be true, and then it might come true as those that are assumed to be criminals and punished for it might decide that they should in fact become criminals. Consequently a 'regime of truth' may lead to actual reality. However, power is never monopolised by one group all of the time, since there is always resistance and opposing 'regimes of truth' which can be very productive and develop new 'practices of representation'.

Television documentaries are perceived to be 'truthful', to present both sides of an argument and to leave the viewer to make decisions for themselves about what they think of the situation represented. We expect dramas to 'have a side' to them, we know that soap operas are not 'real' but we are led to believe that we can trust documentaries to 'tell it as it is'. In fact what they do through their particular 'practices of representation' – interviews, narratives, edited footage of 'real' events – is to construct regimes of truth. Invariably we get a dominant perspective threaded through the programme and the resistive elements might be weak. In such cases the resistance might be located outside of the programme and in the knowledges constructed by the viewers. The two programmes discussed here are described as documentaries but there are 'preferred meanings' attached to them which play around with the concepts of domination and resistance. The next section summarises the key 'moral panics' and 'folk devils' which have been constructed in the media around black people in the UK over the post-war decades and how this is located in a longer history of 'racialising the Other' (Hall 1997c). I also comment on the ways in which television represented black people through these years and how black people have worked towards changing such representations of themselves.

British media dominant representations of black Caribbean people

In order to dominate or resist through material spatialities we tend to construct social and cultural geographies of exclusion or inclusion (Sibley 1995). The construction of 'others' is part of the process of domination and resistance. In order to construct dominant social and cultural meanings of 'others' we have to establish boundaries and identify those we render as 'other' with all that we loathe and fear. The 'other' is associated with defilement, pollution, danger, dirt and incomprehension. In order to resist such processes of 'othering' there are attempts to lower the boundaries, to create spaces within which, through actual contact or particular representations, we learn of similarities rather than differences. Later I will show how the two television programmes follow these patterns of domination and resistance – The Informer constructs the Jamaican Yardie as 'Other' and desires boundaries to keep 'them' out, while Yardies is at pains both to deconstruct and contextualise the notion of Yardie in order to illustrate the similarities between 'us' (the white UK viewing audience) and 'them' (the so-called Jamaican Yardies). However, as has been discussed, there are resistive elements of The Informer and dominant elements of *Yardies*. For now, let us consider how the media have represented black young men and alongside that comment specifically on other representations around black people which have been constructed by television programmes.

Hall (1981) uses an interesting metaphor to describe the ways in which racial ideologies work. He talks of a reservoir of words and images which can be drawn upon at certain key points, one which all those raised within the reach of such a reservoir can readily identify. Let me pursue the watery similes. Within racist society stereotypes and negative representations do not have to be constantly redrawn and recreated because there is a deep reservoir within which all those growing up in a racist society are submerged. We absorb the images and the negativities almost by osmosis, and when required, the memories of what we have imbibed, bubble to the surface and become understood as common-sense, natural 'truths' that take a considerable amount of energy and political awareness to question. The British hegemonic constructions of Caribbean migrants and resi-dents, media obsessions with immigration, and the ways in which relations between the police and black communities have been represented, are examples of these reservoirs. When we are encouraged either to lightly sip (as in the form of racist jokes) or to drink a hefty draught (as during the times of urban unrest, often dubbed 'race' riots) then the mind and soul are flooded with the predefined prejudices and bigotries. The damage is done. In some people the waters of hate subside, but slowly, they may come to feel prejudice but not necessarily act upon it; in others they are drained away rapidly, dried up almost completely, as they resist the racism and counter it with anti-racist strategies; but in still others the waters boil over and physical harm is meted out without care or thought, and the racism becomes part of their dominant positionality, part of their identities.

Below I draw upon some of the key features of a media reservoir of racist stereo-typing, racial bigotries and exclusionary/inclusionary practices – exclusionary against black Caribbean people, inclusionary through a definition of white British people as everything that the Caribbean population is not. The two programmes under detailed analysis here are therefore the recent inpourings of a well-established and constantly maintained source of racist representations. These representations of the post-war era have provided repetitive and pervasive cultural representations of young, black men as dangerous, criminal, out of control and violent.

Hall (1997c) argues that there were three significant moments when the 'West' encountered black people and from that developed multiple representations which were part of the marking of racial difference. First, in the sixteenth century European traders had contact with West African kingdoms; slaves became the commodity of exchange for the next three hundred years. Second, the late-nineteenth-century 'scramble' for Africa saw European powers falling over each other in order to carve up the continent through the processes of imperialism and colonialism. It is the third period of Commonwealth immigration in the mid-twentieth century which I focus on here because it also matches the time of the growing availability of television, which has become a central arena of representation. Consequently 'racialising the Other' (Hall 1997c: 239) has a long, long history which is why the reservoir is apparently bottomless.

During the immediate post-war period Britain suffered a major and damaging labour shortage and attempts were made to secure new labour from novel sources, but, as Harris (1993) demonstrates, there was enormous resistance to the employment of black labour, despite the fact that people of the West Indies were effectively British citizens and many of them had worked in the UK during the Second World War. The same government which was so determined not to recruit labour from the colonies was at the same time recruiting from eastern European countries. The UK position had to be explained to the governors and populations of the West Indies, a representative visited the region and gave several arguments as to why Britain could not employ people from the Caribbean.[3] However, by the late 1940s and early 1950s it was becoming harder to recruit East Europeans so the UK had to resort to black and Asian labour from its colonies and former colonies. The government was far from happy about this, having never wanted to encourage the migration of black people to the UK. Its policies and debates began to fill the reservoirs of resentment which washed out into the streets where many everyday folk held the same sentiments.

Jamaicans were the first to arrive on the Empire Windrush (492 of them in 1948) and they constituted the largest single Caribbean group. Throughout the history of the colonisation of the region, Jamaica had probably caused the most 'trouble'; the various slave uprisings and the existence of the Maroons had been a source of enormous concern, anxiety and fury for the British (Craton 1997). Yet the Jamaicans were also admired, in particular for their musical production and their dancing skills. Such a love–hate relationship is part of the complex process of 'othering': we are reviled by the 'other' at the same time as being attracted to it (Said 1978; Hall 1997c). However, inevitably, the impact on Caribbean migrants was overwhelmingly negative with accommodation in the 1940s and 1950s extremely hard to find and experiences of racism in the workplace and on the street being common. The scenes were set for the persistently negative representations of the black population in general, Jamaicans in particular.

By the late 1950s it was becoming clear that the black workers from the colonies and Commonwealth were likely to remain in the UK. They now had jobs, had managed to get better and more secure housing, and many had been in the UK for over ten years. What ensued each decade was a specific media focus on, and public attention to, a particular 'feature' of supposed black 'natures'. These socially and politically constructed so-called 'natural' characteristics identified black people, especially black men, as lawless, violent, uncontrollable and unpredictable.

The media coverage of the 1958 'race' riots that took place in Nottingham and Notting Hill (London) is probably the earliest example of the process of domination through representation in this era (see Gutzmore 1993). In both cases white men (many of them Mosleyite fascists) began to attack black people (mostly men) on the streets, who then eventually responded in self-defence (Gutzmore 1993: 212). However, media descriptions twisted the story around and stated that the people to fear, because of their aggression, were the black men. The media talked of savage black men running amok in the streets, brandishing machetes and showing no mercy. Press coverage of what the white protagonists had done and how they behaved was minimal. What it firmly established as a representation, however, was the violent and uncontrolled 'nature' of black men which reinforced the notion that not only are they highly dangerous but that they also do not belong; they are not like 'us', they are not British, they should not be here. These two strands of the representations – violent and 'other' – have been resurrected with consistent regularity since 1958. Representations which ran alongside the 'violent' versions of black masculinities were studio discussions, dramas and documentaries which examined the 'troubled state of race relations' and attempted to explain black people to white (Daniels 1994). One of television's responses to the problematic 'race' situation was to provide a service for the immigrants and the 1960s was a time for language programmes with the emphasis that cultural differences were the problem, implying that if black migrants could learn to be more like 'us' the problems would subside. Black people were the 'troubled social subjects' and were acceptable on television only in light entertainment and sport genres (Malik 1998: 313).

During the 1970s the media, police and consequently the public, obsession was with the 'mugger'. A mugger was a young man (almost always black) who approached people in the public space of the street and threatened to harm them, usually with a knife, unless all their money was handed over. The media and political representations were so pervasive and powerful that young black men were increasingly associated in the public's mind with the aggressive act of mugging. By the late 1970s and early 1980s it was not uncommon to see young black men being taken to one side of public pavements and forced to empty their pockets by two or three police officers at a time. The fear of 'mugging' had enabled the police to push the government to pass legislation which allowed them to stop and search anyone they thought was behaving in a 'suspicious' manner, the so-called 'sus laws'. As the black communities and anti-racist organisations feared, being black, male and young appeared to be 'suspicious' enough to be stopped and searched (Gordon 1983, 1986). Throughout the 1970s 'mugging' era the dominant representation of the dangerous, criminal, violent, young, black man was becoming a common-sense, naturalised understanding of what young black men were like. Counter-representations of young black men working hard, succeeding and contributing positively to British society were, and remain, almost non-existent. However, there was growing anger among the black population about the ways in which television both represented them and excluded them; black viewers, media workers, academics and campaign groups made their dissatisfaction clear and independent television companies began to experiment in response with programmes in the off-peak slots. Yet the programmes were still about explaining black people to the white audience; black programme makers had very little access to production and black actors remained type cast (Barry 1988; Daniels 1994; Malik 1998)

When the urban disturbances of the early and mid-1980s were reported, similar narratives related to the muggers came forth. Britain was becoming a riot-torn society and an 'alien disease' was spreading through the towns and cities of the whole country. Young black men did not share the values of 'law-abiding society'; they were outsiders who wished to undermine the basis of society through violence against people, property and the police. There was a real dread that the police could not contain such violence and that white British people would be left to fend for themselves against a group of men with no morality or social sense of responsibility and who were capable of excessive violence. Certain place names – Brixton (London), Toxteth (Liverpool), Handsworth (Birmingham) – became racialised and associated with danger, destruction and lawlessness. In reality the so-called 'race' riots were moments of urban unrest which involved as many white people as they did black, but throughout this period it was black city communities which experienced police harassment and violence (Mama 1993). Just as the early 1980s was a time of urban unrest it was also a time of change in relation to television media. Channel 4 Television was launched in 1982. Part of the remit of Channel 4 was to emphasise diversity and new ideas and to commission programmes from both existing television companies and new independent ones, hence providing opportunities for black media workers. This was the first genuine recognition that there was a significant non-white audience. Channel 4 employed someone explicitly to commission multicultural programmes (Daniels 1994; Malik 1998).

In the 1990s the enemies within 'our' midst are the 'Jamaican Yardies'. These are the latest in a line of 'folk-devils and moral panics' (Sibley 1995). As Sibley comments:

> One of the most remarkable features of moral panics is their recurrence in different guises with no obvious connection with economic crises or periods of social upheaval, as if societies frequently need to define their boundaries . . . The resonance of historical panics in modern crises is

worth noting because it demonstrates the continuing need to define the contours of normality and to eliminate difference.

(Sibley 1995: 39–40)

However, the 1990s are also a time of considerable growth in black media production and programmes which represent the heterogeneity of the black population. In 1998 (at the time of writing this chapter) there was a full, rich and varied season of programmes associated with the fifty years' anniversary of the docking of the *Empire Windrush*. The programmes fit genres which are celebratory, analytical and/or entertaining. The diversities of black Caribbean experiences have been investigated. Consequently through the persistent resistance of black people against dominant 'practices of representation' and 'regimes of truth' we can all enjoy diverse and professional television programmes.

I have demonstrated the longevity of certain representations and the ways in which these are both racialised and gendered. Young black men have been persistently cast in the role of criminal, and while there are inevitably black men who are criminals, the insidious ways in which representations work mean that there are no distinctions made. If you are young and black, if you are young and Jamaican, then if you are not already a criminal your 'natural' inclination will be to become one or to behave in a violent and uncontrolled way. Through such representations the media are instrumental in the construction of both boundaries and excluded geographies and constitutes a cultural space through which processes of domination, and to a lesser extent within the contexts described above, of resistance are practised. What it has also shown is that television broadcasting consistently rendered black people as the social problem and produced and perpetuated negative stereotypes about them. However, through concerted resistance and hard work the face of television programming is being transformed.

World in Action *The Informer*

The Informer was investigated by Duncan Stuff and Nick Davies,[4] produced by Granada TV and broadcast on 6 November 1995 on the independent commercial channel (ITV). The basic story line of the documentary is this. Eaton Green, a Jamaican Yardie, not only became an informer to Scotland Yard but also continued his own criminal activities, most infamously being part of the largest armed robbery in British history in Nottingham where 150 people were robbed. When Nottingham police tried to investigate further they found their work hampered by Scotland Yard, who appeared to be protecting Green. The programme interviewed people involved with 'handling' Green (the Metropolitan Police), representing him (solicitors) and trying to catch and arrest him (the Nottingham Police). It also used tape recordings of Eaton Green himself. The programme traced two Yardie associates of Green back to Kingston, Jamaica, where they too were interviewed. The television programme aimed to tell the story of possible police mistakes in their use of a Yardie informer. It appeared to be an exposé of police activities and questioned their tactics and problematic relations with a Jamaican Yardie. This material allowed the space for a resistive reading of a critical analysis of the Metropolitan Police and for raising troubling questions about why the police, in conjunction with the immigration officials, allowed known Jamaican criminals to enter the UK and worked with them. The programme therefore resisted the dominant notions of the police as upholders of the law and protectors of the innocent. However, the programme also did much more than that. Coexistent with its resistive quality it also, and highly effectively, constructed and perpetuated dominant racist and harmful representations of young black men, in particular those of Jamaican origin or descent.

The Informer opened with the narrator talking of extreme violence, drug trafficking and a Yardie who 'went on a sustained spree of crack-dealing, armed robberies and shootings'. A part of Kingston, Jamaica, the ghetto, was 'a place run by criminal gangs so powerful they're linked to the main political parties'. The narrator set the scene, geographically placed the conflicts, established the extreme dangers and talked of the inability of the British police to cope:

> From the streets of Kingston to the streets of London and New York, Jamaican criminals are pushing crack-cocaine with unprecedented violence. It's a world police officers have found almost impossible to penetrate until they found a top informant. Scotland Yard thought they could fight fire with fire.

Kingston, Jamaica, was firmly in the picture and repeatedly identified as the source of this evil, the 'Yardie'. Again to quote from the narration heard over images of streets in Kingston filmed from a moving vehicle with a Bob Marley soundtrack:

> Eaton Green's road to the dock had begun here in the ghetto of Kingston, Jamaica . . . Green became a gun man at the age of 12, by 24 he'd served five years in prison. He told Nottingham police that he had no choice but to become a gun man . . . Green's story is all too common for a boy from the ghettos of Kingston.

This commentary placed trouble within Kingston and implied the criminal 'natures' of all those who grow up there. Eaton Green began his criminal activity as a child of 12 and the last sentence implied that all young men from Kingston are, or will become, 'Yardies'. Hence the discussion of Kingston was provided as proof that in those circumstances true criminal 'natures' will come forward. Without saying the words 'black Jamaican men' they can be immediately recognised by the viewer – the reservoir was beginning to be tapped.

The programme used the words of top police officers to construct the 'other', the folk-devil, the group which must be excluded. Commander John Grieve of the Metropolitan Police described the Yardies:

> They are a group of travelling drugs-traffickers who are willing to resort to extreme violence, they are difficult to penetrate, they are paranoid, treacherous, violent, unstable and they operate in a culture which is difficult for other cultures to understand.

This is quite an impressive construction of the 'other'. The link with travelling should not be missed, another deeply ingrained British construction of an 'other' given its own reservoir of hatred and loathing of the gypsies and other travelling people (Okely 1996; Sibley 1995). Here the terms 'violence' and 'violent' were repeated and the concept of 'difference' from the white norm was identified. The uncontrollable nature of the Yardies was very clear and hence served as justification for the use of 'one of them' as an informer.

The central themes of this programme were that the Yardies dominate and cause anarchy and chaos in Jamaica, that they get into Britain with ease, that once here on British streets they disregard all laws and perpetrate violence unknown in Britain before. The central shots throughout the film were of the spaces of the respectable side of law enforcement, police and solicitors' offices, a court house; of a derelict warehouse where the robbery took place; of an actor playing Eaton Green; of the arrivals screen at an international airport. An imaginary place was constructed to show the phone calls made to the Yard – a red telephone box was used, the quintessential icon of Englishness – and an old-style desk and Bakelite phone represented the office where the calls were received. 'Real life' consisted of the police cars on London streets, of people and police in a gun battle on the main streets of the Kingston downtown area, of people arriving at Gatwick airport.

My interpretation of the images and their meanings is that they reinforce the sense that places of safety for law-abiding and respectable people (inside the police station, the court room, the solicitors' rooms) are under severe threat from the outside, the marginals, the dangerous and the violent. They, the Yardies, come into Britain on planes, they use our cherished red phone boxes to peddle lies and contaminate our police force, they come from places where there is no law, just dirt, squalor and destruction. The discourse of this programme is that all we hold dear to keep us safe is endangered by Yardies.

The Yardie in question, Eaton Green, was played by an actor who never becomes a 'real person' – we never see his whole face, we never hear his voice, he is set against a bright white background and is preoccupied with putting on gold rings and loading and firing a large and powerful gun. Eaton Green is not a 'person' but a Yardie, a folk-devil, a myth, a terror; he is also any young black man. The only time we see Eaton Green himself is from a security camera picture in a bank or building society – reinforcing the sense that the Yardie is everywhere, that he could be the black man in front of you at the bank. This image was used in repeated shots interspersed with gun shot sounds and a closing camera shot, as if someone were running with the camera, of a telephone box, the place where Green was finally arrested by Nottingham Police. The particular use of this grainy, blurred image of Green's face in a repeated way reminded us that here is a criminal of the highest order in a 'safe' everyday space. In a similar way the airport becomes a dangerous border zone where such 'others' are allowed in – we were repeatedly shown an arrivals board showing a flight from Montego Bay, Jamaica with the word 'landed' blinking its presence. These dangers are coming into Britain everyday, walking through the arrivals hall and on to the streets of London. The programme told the story of Eaton Green being allowed to bring two Jamaicans into the UK even though it was thought that they were multiple murderers. This story reinforces the sense that once one of 'them' crosses the boundary, more and more of 'them' come in with ease. The programme constructed a 'regime of truth' which leaves us (the white British for whom the programme is effectively made) feeling vulnerable, unsure of who we can trust – how many other police officers have come under the spell of Yardies? All we know is that the danger is black, male, armed and on our streets. Throughout the programme we are manipulated by a dominant racialisation which over years has constructed the close connections of:

black youth = criminal = violent = 'other' = to be excluded and contained

Although *The Informer* can be read as presenting resistive tendencies   exposing some aspects of policing, questioning immigration practice and informing us of a specific criminal group – I have read it as an example of dominant racist representation. While it can be read as a legitimate documentary about criminal practice it is also contributing to longstanding 'regimes of truth' about young black men. Depending on one's political positioning *vis-à-vis* race one either believes its 'regime of truth', rejects its representation totally or accepts some parts of the regime but resists others. However, it is important to see this programme in the wider context of television and the media in general. When we consider who controls and has access to the mainstream media then we have to ask where is the space for resistance? How can young black men counter these stereotypes? Faced with the inaccessibility of the mechanisms of media production for the majority of marginalised and excluded groups then we have to consider other forms of resistance which are not necessarily the mirror image of the dominant forms. We have to consider media spaces in which other more honest, reflective and positive representations might emerge. In the final section I consider the second documentary and analyse the ways in which, imperfect as they might be, there are resistive elements which can offer alternative representations. However, as my analysis of a particular audience discussion about the programme will show what I, placed within a UK context and through a comparison with *The Informer*, read as resistance can be part of dominant representation for an audience from a Jamaican context.

Network First *Yardies*

Network First is a forum in which new directors' work is broadcast. It too is shown on the commercial channel of ITV and this programme directed by Kimi Zabihyan was broadcast on 21 February 1995. The programme was filmed in Kingston, Jamaica, and attempted to deconstruct the notion of 'Yardie': 'a Yardie in Jamaica does not mean gangster, simply a Jamaican'. The programme showed the levels of poverty, hardship and struggle that people in the 'ghettos' of Kingston experience (for a more detailed discussion of the 'ghettos', see Skelton 1998a). The film followed a special police squad at work in the ghetto communities of downtown Kingston but also showed us 'real' people, a brother and sister, Peter and Rosie – their daily lives, their vulnerability to crime and attack within the ghetto, and the ways in which they earn a meagre living. What this film achieves, and what I think allows it to be categorised as a form of resistance culture, is that the Jamaicans of Kingston become in some ways 'real'; we, the viewers get to know something about them, we can identify with the difficulties that they face and in some way understand the pull towards crime and the desire to 'escape' through migration. The film's narrator told us:

> Peter lives in Southside and has tried six times to stow away to the United States, each time he was deported back. If he had succeeded he might have been described as a Yardie, but Peter is not part of an organised crime ring, he's just a typical youth dodging poverty, the police and the tensions of an overcrowded ghetto . . . His sister Rosie has to hustle for money but everyone in the ghetto has a cash-flow problem . . . she lives in a tiny room with her baby.

The young people we followed through the film were 'real', they have voices, they have dreams and hopes and they have genuine fears, many of which were connected with their social, economic and political environments. We see Jamaicans at home, at work, with their children, in church and in the streets. The Jamaican Yardies of Zabihyan's film were not folk-devils, they were not to be feared, but were people 'like us' and as Peter told us, their only crime was being poor.

The other focus of attention throughout the film was the Special Anti-Crime Task Force (SACTF) set up to attack crime (and, on occasion, the people) in the ghettos. They have far-reaching powers of random stop and search, and of arrest; they can hold 'suspects' for hours without charge or notice of what they have been arrested for. The SACTF raided an open-air dancehall and loaded young men into closed lorries. A policewoman repeatedly slapped a young woman who spoke back to her. Although no identification was asked for, the men were accused of being deportees from the USA or the UK. We rejoined the young men where they were still being held in a large room the following morning. In a macabre reminder of slavery and the sale of the fittest the men who claimed to have been at the dancehall for the talent contest had to 'prove' their ability and show their bodies as evidence that they were not part of gangs (bodies with bullet wounds, knife or acid scars were seen as proof positive of gang involvement and consequently crime). If the men performed to the police officers' liking and their bodies were 'clean', then they were released. When a young man at the back of the room complained, a police officer threatened to hit him with a wooden stool, but he was reminded of the camera and satisfied himself with a hard kick into the side of the seated youth.

The Network First film showed us the grinding poverty and the complex social problems that arise from having to live in such close confinement with other people, all of whom are equally poor, have so little to lose, and may use violence to defend what little they have. We were introduced to Lloyd, a construction worker, whose face was severely scarred by acid, a bucket of which was thrown at him in a case of mistaken identity. Lloyd told us of people's shock when they see him and how he kept his face covered to protect them and himself. Lloyd was confident he knew what would reduce the violence:

> It needs to get the youth to work, if they do that then the kids will work and you won't find youth on the corner, hold up no-one . . . find something to give the youth, I know that you give them work and they will work good, work – honest bread.

The brother and sister both had work of a kind. Peter made his living videoing weddings and parties, but mostly funerals, especially funerals of his own peer group, many of them gunned down in drug-related crimes. Rosie earned a meagre income for her and her baby by selling foam and risking her life by working on the ships as a prostitute. The film provided happy endings for some. With his money from the video films, Peter bought a new suit, 'gangster style', to make himself look and feel good as he went to the major source of entertainment, the dancehall. Rosie finally had enough money to buy a sewing machine and set herself up as a seamstress and was so happy because she did not have to work the boats any more. At the end of the film we were told that Lloyd was fatally stabbed in the ghetto.

The programme was not totally located within the downtown areas. A short part of the film followed the SACTF as they investigated a burglary up in the cool hills of Jamaica's uptown suburbs. The people told their story, and retold stories of their neighbours' burglaries, to the police and the cameras. These are the wealthy elite of the country and it was at this point, through one of the interviewees, that there was some recognition that the economy might be a significant factor in the levels of criminal activity. However such political-economic context to the films was quickly bypassed by the narrator: 'but the economy is a tough nut to crack, in the meantime the police keep up a constant pressure in downtown areas. [We follow the police into the downtown dancehall venue described above.]'

When I first watched *Yardies* after viewing *The Informer* I read it as highly resistive and as a welcome intervention in the Yardies discourse. It gave a sense of what life was like in the ghettos of Kingston and placed the so-called Yardies in their socio-economic context. It did not seek explicitly to excuse Yardies and the level of violent crime, but through implication offered some explanatory factors. It also showed us aspects of lived realities in Jamaica. Yet over the times of viewing and in discussion with young Jamaicans living in Kingston I began to question my initial readings. It was clear that while resisting *The Informer* style representation of Jamaican young men as criminals, *Yardies* in fact was part of a discourse about 'Third World' countries and poverty which comprises an equally problematic 'regime of truth'.

While the *Yardies* film is in part resistive there are problems with its representation of the people of Kingston, and by implication, all Jamaicans. *Yardies* draws upon another reservoir of prejudice, that of pity for those in the 'Third World'. Yes, we come to identify and sympathise with the very difficult situations in the Kingston ghettos, and yes, there is some analysis of the ways in which the extremes of poverty and wealth in the country might contribute to this, but how resistive is understanding through pity? It is a close line between pity and cultural explanations, such as, 'people are criminals because they are poor, people are poor because they do not work like 'us', they do not have 'our' efficiency'.

When I showed this programme to Jamaican students at the University of the West Indies (UWI) their comments and discussion surprised me because they focused on how negative the representations were.[5] They argued that it showed Jamaica in a negative light and while some aspects of the programme were realistic it was not balanced. Some of the students had either grown up or worked in some of the areas filmed and one mature student actually knew some of the shop-keepers and traders shown in the programme; these students therefore spoke from their own experiences. They were vociferous in their arguments that the Kingston ghettos were much more differentiated than had been shown. They resented the homogeneity of the representations. They stated that the positive side of ghetto living, the close connections within the community, the mutual help and support that people gave each other were not fully reflected. Indeed, they said that the community as an entity was not seen at all, the focus always being on individuals (very much a western obsession). Many of the students believed that a lot of the material was contrived for the camera. This was especially true of the police activities. One young man commented that the young men held overnight in the large room were probably just as impatient with the camera crew as with the police! The students wanted to know how Rosie had suddenly got so much money, asking 'Was she paid for participating?' Most thought that she had been, which somehow confirmed their sense of a staged performance for the cameras. Several women commented that Rosie would not be able to resist going back into prostitution. There are hundreds and hundreds of seamstresses in the ghetto, many from the rural areas where as part of small-scale development projects they are taught to sew. Once they can do so they leave for the city and find that almost every woman has those same skills and so they have to find other ways of earning a living. The students pointed out the stark economic realities – a woman could earn as much in a night of prostitution as she could for a full week of sewing.

I learned a great deal from these discussions, which were candid and confident. It developed into a broader debate about the ways in which people in the UK learned about Jamaica and a genuine concern was expressed that programmes such as *Yardies* were the only way that people in the UK could learn something about Jamaica. As one student stated:

> You see, we get to see a lot of your television and we see lots of different things, so we know that one programme isn't the whole story about Britain. But if you only see one thing like that then really you don't know anything at all about Jamaica.

What these discussions illustrate is that meanings and interpretations of representations are highly dependent on the audience's social, cultural, political and economic position, their identity politics and their own positionality in relation to 'race' and geography. Whatever the intentions of the makers of *Yardies* the readings of the film cannot be predicted. In one geographical context the film can be read as resistive and as trying to right the balance, but in another it is highly problematic and fails to be balanced enough. This means that while the media have power, viewers, through their own resistive readings have power too. This implies that power is not a chain but rather circulates; the discourse that develops in reaction to dominant power can be significant and positive in its effects (Foucault 1980; Hall 1997b).

Conclusion

Television is a very powerful tool in the construction of dominant representations, and yet at the same time can provide an equally powerful form of resistance. The media are therefore both wrapped in the entanglements of power and themselves are active entanglers. Documentary makers have a message to present and a 'preferred meaning' that they wish to convey, but once they have made their film and it is broadcast it falls out of their control. If they have woven a tight 'regime of truth' there is a high chance that their preferred meaning will be received by the majority of viewers. However, if there are internal contradictions, knots and weaknesses in the threads, then the stories might be easier to unravel and multiple readings developed which knit new patterns. We might reject the dominant representation and draw out the resistive qualities, we might construct our own 'regimes of truth' which draw upon our own knowledges and interpretations. Hence the media, and television programming in particular, always carry multiple meanings and offer these through a myriad of representations. What we have to ensure is that we critically engage with such representations and pattern our own meanings from them which allow space for inclusivity, fairness and balance.

Through a consideration of the material spaces of power produced by two television documentaries this chapter has shown the ease with which a reservoir of representations of young black men can be tapped to construct geographies of exclusion, racialised stereotypes and constructions of the Other. It has also shown, however, that there are ways in which dominant representations can be resisted, not necessarily by those most marginalised, but by intercedents who wish either to directly resist harmful images and symbolisations, or to offer more nuanced and sophisticated representations which engage and counter the dominant norms. However, there is always a danger that intercedents and those who resist processes of domination on *behalf* of those most affected by, in this case, damaging representations, might create new patterns of domination. In some ways this chapter could be identified as having fallen into this same trap. I am not Jamaican, I am not black and I am not male; however, I am a geographer, I have worked in Jamaica and I fight racism in a range of ways. I am also very privileged to be able to learn from Jamaicans themselves, just as television controllers have learned from black people that black audiences are responsive and sophisticated and that they too require and can produce positive representations. Within my capacity as an academic – through my writing and my teaching – I am in a position to challenge dominant representations of young black Jamaican men, lend support to others who resist in a multitude of ways, and begin to question the ways in which we can resist racism and the harm it does. In such a manner, I can be part of the process of ensuring there are spaces of, and for, resistance.

Acknowledgements

I would like to thank, once again, all the Jamaican UWI students who gave their time to watch *Yardies* and discuss their reactions with me. This would not have been possible without the support of Dr Patricia Mohammed, the course director and the seminar tutors whose teaching time I stole! I would additionally like to thank the editors, who included me in the conference programme and then gave me helpful advice on an earlier draft of this chapter. My time at the UWI Mona Campus, Jamaica was funded by a Commonwealth Universities Development Fellowship, and I gratefully acknowledge this support.

Notes

1 'Practices of representation' is taken from Hall (1987a); 'Regimes of truth' comes from Foucault (1980).

2 I should perhaps explain this metaphor. It was stimulated by the quote from Hall which follows in the chapter – the idea that meaning is a 'slippery customer, changing and fixing'. I always enjoy Hall's metaphors and was inspired to think of my own. In my family the discovery of frog spawn in our small pond has always had different meanings for us: I loved to find it because I found the life cycle of the frog fascinating, especially its ambiguous status as water/land dwelling, and I envied its ability to hold its breath for so long underwater; my youngest sister saw it as something of a scientific discovery; my middle sister loathed the way it felt but had no aversion to frogs; my mother hated the lot – spawn, tadpoles and frogs. My mother was in the position of dominant power in the household and so the frog spawn had to go, but there was resistance from me and my sisters. So the spawn was carefully transported to large lakes on the outskirts of the city where no doubt happy, healthy and multiply-meaning frogs lived ever after.

3 Harris (1993: 22) lists the following 'explanations': the jobs being advertised were only temporary vacancies not real jobs; there were serious problems of accommodation; Caribbean workers would not be able to work outside and would catch colds and develop lung infections, but underground work in mining would be too hot for them; also the men tended to be lazy and prone to fighting among themselves! One wonders where these excuses were when Caribbean men and women were called in to work and fight during the Second World War.

4 Davies also wrote a lengthy article for the *Guardian* (6 November 1995: 2–5) under the headline 'The Yards Yardie' detailing the story line followed in the documentary. This article followed in the well-established line of 1990s coverage in the same newspaper of the so-called Jamaican Yardies. Examples include articles by Bowcott (1990); Campbell (1991, 1993); Nevin (1991); Campbell and Chaudhary (1993); Johnson (1993); Sharrock (1993); Travis (1993); Chaudhary and Campbell (1995) and Thompson (1995). Nick Davies himself later narrated a follow-up programme by World in Action (broadcast in 1997) which presented a very similar story line and repeated much of the footage used in the 1995 programme.

5 I was a visiting Commonwealth Fellow at the Mona Campus of UWI during September 1997. Part of my fellowship involved teaching third year students and a group of them agreed to watch the Yardies video and then discuss their reactions to the programme.

The discussions were taped but unfortunately the tapes were stolen in transit between Trinidad and Grenada. The above commentary on their responses was taken from notes made at the time of the discussion.

References

Barry, A. (1988) 'Black mythologies: representation of black people on British TV', in J. Twitchim (ed.) *The Black and White Media Book*, Stoke-on-Trent: Trentham.

Bates, S. and Bunting, M. (1993) 'Expulsion row grows', *Guardian* 27 December: 1.

Bowcott, O. (1990) 'Suppliers and addicts urge moderation', *Guardian* 14 November: 5.

Campbell, D. (1991) 'Crime PLC', *Guardian* 14 December: 4–5.

Campbell, D. (1993) 'Record seizures as drugs sales rise', *Guardian* 4 September: 5.

Campbell, D. and Chaudhary, V. (1993) 'Four arrested over PC killing', *Guardian* 22 October: 1.

Chaudhary, V. and Campbell, D. (1995) 'Traditional gang rivalries take more sinister turn', *Guardian* 11 December: 3.

Craton, M. (1997) 'Forms of resistance to slavery', in F. W. Knight (ed.) *General History of the Caribbean, Vol. III, The Slave Societies of the Caribbean*, London: UNESCO/Macmillan Education.

Daniels, T. (1994) 'Programmes for black audiences', in S. Hood (ed.) *Behind the Screens: The Structure of British Television in the Nineties*, London: Lawrence and Wishart.

Davies, N. (1995) 'The Yards Yardie', *Guardian* 6 November: 2–5.

Foucault, M. (1980) *Power/Knowledge*, Brighton: Harvester.

Gordon, P. (1983) *White Law: Racism in the Police, Courts and Prisons*, London: Pluto.

Gordon, P. (1986) *Racial Violence and Harassment*, London: Runnymede Trust.

Gutzmore, C. (1993) 'Carnival, the state and the black masses in the United Kingdom', in W. James and C. Harris (eds) *Inside Babylon: The Caribbean Diaspora in Britain*, London: Verso.

Hall, S. (1981) 'Teaching race', in A. James and R. Jeffcoate (eds) *The School in the Multicultural Society*, London: Harper and Row.

Hall, S. (1987a) 'Introduction', in S. Hall (ed.) *Representation: Cultural Representation and Signifying Practices*, London: Sage/Open University.

Hall, S. (1987b) 'The work of representation', in S. Hall (ed.) *Representation: Cultural Representation and Signifying Practices*, London: Sage/Open University.

Hall, S. (1987c) 'The spectacle of the "other"', in S. Hall (ed.) *Representation: Cultural Representation and Signifying Practices*, London: Sage/Open University.

Harris, J. (1993) 'Post-war migration and the Industrial Reserve Army', in W. James and C. Harris (eds) *Inside Babylon: The Caribbean Diaspora in Britain*, London: Verso.

Johnson, A. (1993) 'The harder they come the harder they fall', *Guardian* 13 November: 27.

Malik, S. (1998) 'The construction of Black and Asian ethnicities in British film and television', in A. Briggs and P. Cobley (eds) *The Media: An Introduction*, London: Longman.

Mama, A. (1993) 'Black women and the police: a place where the law is not upheld', in W. James and C. Harris (eds) *Inside Babylon: The Caribbean Diaspora in Britain*, London: Verso.

Nevin, C. (1991) 'Northern soul searchers', *Guardian* 15 June: 4–5.

Okely, J. (1996) *Own or Other Culture*, London: Routledge.

Pile, S. (1997) 'Introduction', in S. Pile and M. Keith (eds) *Geographies of Resistance*, London: Routledge.

Said, E. W. (1978) *Orientalism*, London: Routledge and Kegan Paul.

Sharrock, D. (1993) 'Drugs linked to gun murder', *Guardian* 1 June: 2.

Sibley, D. (1995) *Geographies of Exclusion*, London: Routledge.

Skelton, T. (1998a) 'Ghetto girls/urban music: Jamaican ragga music and female performance', in R. Ainley (ed.) *New Frontiers of Space, Bodies and Gender*, London: Routledge.

Skelton, T. (1998b) 'Doing violence/doing harm: an analysis of British media representations of Jamaican Yardies', *Small Axe: A Journal of Criticism* 3 (March).

Thompson, T. (1995) 'Violence in Britain: Yardies, myth and reality', *Guardian* 19 September: 5.

Travis, A. (1993) 'Fear of return to old refusal rate', *Guardian* 24 December: 3.

World in Action (1995) *The Informer*, broadcast 6 November, Granada TV for ITV.

Zabihyan, K. (dir.) (1995) *Yardies*, broadcast 21 February, Network First, ITV.

9

ORGANISATIONAL
GEOGRAPHIES

Surveillance, display and the spaces of power
in business organisation

Philip Crang

This chapter examines the entangled spaces of power in business organisation. It is probably true to say that, having been the subject of some of the best and most influential work in human geography in the 1980s, as in writing about the links between different forms of capitalist production and the spatial dynamics of economic activity (Scott and Storper 1986), questions of organisational geography are no longer the intellectual flavour of the month. To reproduce a rather crude disciplinary historiography, for much of the 1990s organisational geographies have been rather sidelined by a number of 'cultural turns' (Cook *et al.* 1999) which have stimulated more interest in the realms of both identity politics and consumption than in questions of production, the social division of labour, and work. Indeed, the *passé* status of questions to do with business organisation has recently been officially confirmed by having pretty much a whole Economic and Social Research Council (ESRC) *thematic priority* devoted to them. However, there is now an increasing body of writings attempting to bridge this gap: writings which look to develop insights from cultural studies and cultural geography while applying them in the context of workplaces and organisations. Despite occasionally giving in to the temptation simply to adorn unremarkable empirical findings with a new set of theoretical lapel badges – thereby writing about 'organisational cultures and rituals', 'hybrid workplaces' and so on – on the whole this slice of theoretical mobility has productively recast debates concerning the social relations in and of production. It is this positive outcome that is the starting-point for my chapter.

In particular, I want to sketch out something of how an understanding of the power*ful* nature of business organisation is fostered, or at least hinted at, in this literature. More particularly still, I want to reflect on the kinds of organisational *geographies* that these relations of power might be seen as both constitutive of and constituted by. To that end, the chapter is divided into three main sections, each of which pivots around discussing a couple of significant books rather than drawing directly upon my own empirical research (Crang 1994, 1997).

I want to begin by thinking about how we might view business organisation as marked by geographies of *surveillance*. To do this, I briefly revisit an absolute, and deserved, classic of the literature on work, Braverman's (1974) *Labor and Monopoly Capital*. Published in the 1970s, this book established the terms of what has now become known as the 'labour process debate', a debate which dominated the social science literature on work up until the end of the 1980s. I wish to pull out from Braverman his understanding of organisational politics, an understanding that highlights the logical compulsion in a capitalist mode of production for managers to monitor and to control workers ever more tightly so as to maximise productivity and diminish labour costs. I also want to see how the many criticisms heaped on Braverman might lead us to a rather different portrait of organisational surveillance, one in which processes of surveillance are reframed as inherent to processes of modern organisation. Here I focus in particular on the work of Dandeker, and on what he calls his 'neo-Machiavellian' approach to organisation, as set out in his *Surveillance, Power and Modernity* (Dandeker 1990). I then want to argue that analyses of organisational geographies need to supplement their interest in surveillance with an explicit examination of practices and technologies of *display*. Drawing from work in service management studies – rather sadly I often find myself hiding in the corner of the library avidly devouring the latest issue of the *International Journal of Hospitality Management* – I outline both the potential significance of these geographies of display and some analytical dimensions along which they might be unpacked. Finally, I discuss how questions of organisational surveillance and display might be extended by linking into more recent work on both the ordering activities of 'actor-networks', in the process skimming across Law's (1994) *Organizing Modernity*, and the discursive constitution of organisational identities, as pursued for example in du Gay's (1996) *Consumption and Identity at Work*.

Throughout I shall pay attention to the entanglements of power characterising organisational geographies, stressing how the spaces of business organisation are always shot through with complex intersections between what is 'taken' (appropriated both overtly and covertly) by those in charge and what is 'given' (wittingly and otherwise) by those doing the productive work. In moving to a conclusion, I endeavour to make more explicit the implications of seeing organisation as a process of networking and government, one marked by definite geographies of surveillance and display, for attempts at rethinking the geographies of domination and resistance.

Organisational geographies of surveillance

Dandeker (1990) defines surveillance in the following terms:

> In a general sense, surveillance activities are features of all social relationships. The exercise of surveillance involves one or more of the following activities: (1) the collection and storage of information (presumed to be useful) about people or objects; (2) the supervision of the activities of people or objects through the issuing of instructions or the physical design of the natural and built environments . . .; (3) the application of information gathering activities to the business of monitoring the behaviour of those under supervision, and, in the case of subject persons, their compliance with instructions.
>
> (Dandeker 1990: 37)

Surveillance hence involves both information collection and, usually, to quote Dandeker again, a 'certain capacity to supervise and manage behaviour' (Dandeker 1990: 39). It also involves questions of spatial organisation, insofar as organisational geographies are a central resource in, as well as a barrier to, both the control and the monitoring of activities.

Much of the literature on organisational surveillance concerns the spaces of labour control, especially the workplace (Crang 1994). In part, workplaces organise labour through their potential, as places, for being encoded with cultural meanings and embodying organisational power and ethoses. However, the organisational power of workplaces stems at least as much, if not more, from their ability to structure social and spatial practices (Baldry 1997). In particular, and to use de Certeau's (1984) formula of contrasting dominating strategies and resisting tactics, workplaces have the potential to operate as 'spatial strategies' facilitating 'the calculus of force-relationships which becomes possible when a subject of will and power (a proprietor, an enterprise, a city, a scientific institution) can be isolated from an "environment"' (de Certeau 1984: xix). Through a bounded definition of a place of employment, workplaces allow a clear symbolic and contractual demarcation of paid work, and of the social practices expected in paid work, from other social arenas. Thus Sayer and Walker (1992) talk generally of how:

> The basic organising principle of the workplace is containment within a limited area. Direct connection and immediate access are its chief integrative effects . . . The workplace is also a place of confinement, a piece of turf where the boss rules, a symbolic and social world in which the capitalist's hegemony is normally reinforced. In short, the workplace is a system of labour control.
>
> (Sayer and Walker 1992: 120)

In a similar vein, but more specifically, a number of writers have discussed the role of the nineteenth-century factory in disciplining labour into the routines of industrial capitalism. For Pollard (1965) and for Thompson (1967), of particular importance was the imposition of new forms of work time, as enforced through clocks, bells, bureaucratic procedures for monitoring arrival and departure along with fines for lateness or unauthorised breaks, and close personal supervision. For Stein (1995), in his study of factory life in Ontario in the last quarter of the nineteenth century, of rather more significance were new forms of spatial practice and consciousness: 'Sealed off from the outside world, split up into departments and work-rooms, assigned specific tasks and work stations, workers experienced a discipline that was acutely spatial' (Stein 1995: 289).

While rarely raising the issue of workplace geographies at all explicitly, Braverman's (1974) classic account of the nature of labour in the period of what he, following Baran and Sweezy (1966), termed 'monopoly capitalism' was an attempt to take these questions of time-space and labour surveillance into the twentieth century. Braverman developed two main arguments. In the first instance he empha- sised how modern management has developed as both an organisational function and a set of occupations because of the need to control the labour force. Through the wage system capitalists buy only the 'potential' of labour time, which means that labour needs to be carefully managed in order to convert the time spent at work into actual work. As Braverman put it: 'Like a rider who uses reins, bridle, spurs, carrot, whip and training from birth to impose his [sic] will, the capitalist strives, through management, to control' (Braverman 1974: 66). Braverman's second main argument was that a particular form of management, based in the 'deskilling' of workers through the socio-spatial separation of conception and execution alongside the tight monitoring of pre-established work routines, becomes the ultimate expression of this functional need to control. For Braverman, capitalist organisations are thereby structured around a socio-spatial 'separation of con- ception and execution', such that 'in one location the physical processes of production are executed . . . [and] in another are concentrated the design, plan- ning, calculation, and record-keeping' (Braverman 1974: 68). All of these locations are marked by an impulse to pre-establish the time-space routines of workers in the greatest possible detail, and then to monitor and to enforce these routines as closely as possible through various managerial technologies of work design (such as the assembly line) and also by the ensuring of a managerial 'panoptical' vision (whether from supervisors on the line or from electronic point of sale terminals in a supermarket). (See Chapter 1 for a discussion of Foucault on the technologies of 'panopticism' and dominating power.) In consequence, so Braverman believed, capitalist workplaces are designed not only to bracket out work time-space from the rest of everyday life, but also actively to purify that bracketed time-space of all deviations from concerns with corporate need and profitability.

Braverman had little to say about how such managerial ambitions were resisted by workers, but his analysis does open up 'space' for others to document the tactical incursions into corporate workplaces made by employees. One example, Beynon's (1984) account of *Working for Ford*, highlights not only the colonisations of the self demanded by Ford, and the spies and stool pigeons used to enforce these strictures, but also how workers on the early assembly lines developed a variety of tactics to reinsert their rounded humanity into a dehumanised workplace. For example, quoting the work of Sward, he notes how:

> Ford men, before long, became noted for their ingenuity in circum-venting the ironclad law against talking at their work. They developed an art of covert speech known as the 'Ford Whisper'. Masters of this language, like inmates in a penal institution, could communicate in undertones without taking their eyes off their work . . . Ford's tool and die men used to be governed by the same taboo, though their work by its very nature compelled them to move about . . . Their technique was to exchange small talk while gesticulating in mock earnestness at parts of a lathe, or while feigning interest in a blueprint. One highly intelligent artisan in this department . . . learned to talk like a ventriloquist. After spending ten years in Ford's service, this man became the laughing stock of his wife and friends, for the habit of talking out of the side of his mouth finally became ungovernable; he began to talk that way . . . at home or in the most casual conversation with someone outside after working hours.
>
> (quoted in Beynon 1984: 43–44)

Moreover, Beynon goes on to see how in the Halewood plant of the 1960s the shop stewards movement set about opening up cracks in the surveying gaze of management so as to challenge the entire factory regime. Efforts were made to 'crush' the foremen responsible for supervising the line. Places on the line with less tight surveillance were targeted for disruptive acts (for example, scratches made to cars in the wet deck of the paint shop would not be seen until three hours later, so this was a major area of activity). Humour was used as a weapon to mock the claimed authority of managers and, to quote Beynon, to respond to managerial 'assaults with guile rather than brawn' (Beynon 1975: 95). Most famously, this saw the plant convenor, Eddie Roberts, accepting a decision to discipline him 'verbally', but then enacting a wonderfully tellable story by going on the sick so that he had to be asked several times to come into work in order to get his warning (which he eventually received while grinning and looking out the window). Here, then, we have an account of workplace geographies that sets up a powerful picture of managerial domination through tight surveillance. In turn, worker resistance is seen as enacted through: (a) the sneaking into the workplace of undetectable moments of illicit everyday life; (b) identifying gaps in the surveillance systems so that one can escape managerial gazes; and/or (c) trying to get away with staging mocking performances or poses to be seen by those gazing managers and, ideally, by fellow workers as well.

Such geographies of domination and resistance still have a strong resonance in more recent portraits (and experiences) of working lives. They are there, for example, in Allen and Pryke's (1994; see also Massey and Allen 1995) fascinating account of security guard work, policed and indeed comprised, as it is, by the requirement to log one's walks through pointing a laser reader at the bar codes situated around a building. Yet such accounts also have a number of well-documented analytical limits, and these are of three main sorts. First, there is the understanding of management offered by Braverman (1974). This is both thoroughly (if refreshingly) functionalist, since for Braverman management simply does capital's bidding with no independent agency, and decidedly narrow, since management is seen as only concerned with the control of labour. Second, there is the picture of labour offered which is based on the argument that capital aims to reduce workers to an abstract and replicable deliverer of labour power, and so *either* little attention is paid to worker subjectivity *or* any incursion of worker subjectivity tends to be cast as inherently transgressive. Third, there is the mono-theistic quality suggested for managerial strategies and workplace geographies, implying that there is really *one* true best way to organise a business and its labour force. Participants in the labour process debate have long dined out on examples of alternative, less direct forms of management control (for example, strategies of responsible autonomy in which workers are encouraged to buy into company values: see Friedman 1977; also Fox 1974). Paralleling such examples, moreover, one could discuss different constructions of the workplace to the highly regimented and purified work space outlined by Braverman. Think of the hi-tech companies that construct a flexible time-space in which work and home become blurred (as described in Henry and Massey 1995). Or consider the customer contact jobs where workers are expected to join in with the consuming pleasures being offered, an extreme case of which would be the individuals hired by many bars in Mediterranean resorts to attract customers, whose job is to work at chatting, flirting, drinking and dancing to make themselves and those around them feel good (Crang 1994: 151). Or contemplate the visits of mobile sales executives, as they work at organising other people's spaces into their own projects (Laurier and Philo 1999). The point is that the closed factory, or factory-like office, is a particular workplace form, and is not the model that all workplaces are destined to follow (see also remarks on the 'nomadic workplace' and the home as workplace in Sayer and Walker 1992: 120–121).

One way to deal with these limits to Braverman's analysis is therefore to frame management and organisation in rather broader terms, and to accept a greater variety in configurations of workplace geography. An account that begins this process is, as mentioned earlier, Dandeker's (1990) portrayal of surveillance in modern organisations. Dandeker develops what he calls a 'neo-Machiavellian' perspective on organisation which is based on three main arguments. First, management is understood to originate in more than just the control of wage labour. Instead, stress is placed on the technical necessities for managerial and bureaucratic work in socially and spatially complex organisations, and on the multiple concerns of management (including labour control, but also revolving around product markets, competitive environments, state regulation, constructions of self-interest, and so on). Second and relatedly, management is cast as being concerned both with processes internal to an organisation *and* with relations to 'external' environments. Third, both the logistical requirements for organisational work and the constructions of internal and external environments are seen as shaped by questions of geography. In consequence, organisational geographies are seen as impacting on both managerial strategies and organisational forms, as well as being partly determined by them. Dandeker illustrates this last argument with reference to the nineteenth-century American railroad industry (see also Revill 1991, 1994). Dandeker (1990) stresses how the progressive adoption of divisional and regional organisational forms by this industry from 1875 onwards – forms that replaced earlier centralised functional departments, and ones that were later taken up by other organisations – was intrinsically entwined with the growing geography of the railroad companies: with increasingly extensive railway networks being established which then posed new problems of coordination and demanded new forms of administrative space.

Dandeker duly argues for an understanding of management as about more than simply controlling employees, such that other managerial concerns (for instance product quality, or relations with regulatory agencies, or the defence of one's own managerial fraction) are seen as shaping the precise means of labour control. He regards organisation as a process of connecting people and things across time and space in order to meet a range of varied and often conflicting objectives, a perspective which also resonates with the 'actor-network' claims of Latour and others (see both Hinchliffe and Wilbert, Chapters 10 and 11 in this volume). Such a view leads Dandeker to suppose that the spatial character of the connections being organised then impacts backs on the forms of organisation themselves.

Organisational geographies of display

Dandeker (1990) pursues these arguments primarily with reference to forms of organisational surveillance, but shadowing much of what he says are also questions of organisational display: that is, questions about the active presentation of organisational identities to individuals and institutions understood to be external to the presenting body. Such self-presentations have been paid particular attention in studies of the rhetorical construction called 'service employment' (Adkins and Lury 1996; Crang 1997; Hochschild 1983; Sturdy 1998). Indeed, accounts of service employment tend to shift attention away from organisational geographies of surveillance and toward the socio-spatial relationships between producers and consumers. This shift is very apparent in the established concern of service management literatures with the co-presence of employees and their customers in 'service encounters'. It is implicit in the discursive association of service employments with moral dilemmas of honesty/dishonesty, trust/distrust, and the seen/unseen (Kramer and Tyler 1996; Thrift 1996). Yet above all, it is reflected in the dramaturgical or performative metaphors commonly used to represent and to structure much service employment. In recruitment for service jobs, the emphasis is often on personality and talents, rather than on supposedly impersonal technical skills, as managements search for the right kind of service-oriented people. Once recruited, workers have to be directed into their roles. This may occur in the context of clearly delineated workplace environments or sets, and it is evident that '[m]anaging the elements of the physical environment of the service organisation is one means to characterise services and, in addition, to affect the way in which they are delivered and perceived' (Upah and Fulton 1985: 255). It may involve the 'costumes' worn by employees, for a 'person who is not dressed for the part will not be able to play that part effectively', as Solomon (1985: 69) argues when discussing the 'packaging of the service provider'. Emphasis is also placed on the coaching of staff in the surface acting skills of communicational interaction, while some managements even train their staff in a 'deeper' Stanislavskian level of performance in which the surface effects of the body (involving the likes of eye contact, smiling, open gestures and so on) are established and controlled through deliberate forms of 'emotional management'. The implication is that such bodily actions then stem from 'genuine' states of feeling: I smile because I am pleased to see you and so on (Sturdy 1998; Wouters 1992).

These fairly familiar and well-rehearsed understandings of service products, in the guise of the performances of the service worker, are productive for managerial writers because they impact back on debates over the appropriate character of managerial surveillance in service industries. In particular, they are often used to throw into question an emphasis on work task predetermination and worker deskilling. If service products are largely made up of indeterminate and interactive social encounters, and if these affective encounters are service labour processes, then such labour processes cannot be predetermined by management because there is actually an inherent requirement for the 'indeterminate skills' possessed by the employees working in them. Thus, the discursive construction of service employments is used to legitimise forms of managerial surveillance that set themselves up expressly in opposition to those of routinisation and tight monitoring.

Pre-eminent are a range of human resource management strategies which emphasise a different kind of control over more autonomous jobs and workers through the latter's adherence to organisational values and 'culture'. They lay emphasis on how the value-adding intangibles and indeterminates of the generic service product — *'that special something extra'* — require flexible interactional resources on the part of service workers, ones that management cannot conceptualise in advance nor deliver in detail. A claim is made, then, for a need to manage autonomy in inter-personal work if value-adding elements of the product are to be maintained and developed. In turn this state of affairs is seen as giving such service work a particular character. Here, for example, is populist management guru Peters (1994) eulogising generally about customer contact work under the heading of 'theatre on the retail stage':

> Retail, whether in the classroom or the showroom, is a performing art
> . . . That's why I love retail. Sure, I have to count on the organiser of
> the seminar to bring in a good crowd, to select an adequate conference
> facility and to get a hundred logistical details just so . . . But after that, it's
> my show or your show. The conference hall opens, the Body Shop's door
> clanks upward, the school bell rings and we are absolutely, positively in
> charge. It's our stage . . . Retailing also allows you — no, it requires you
> — to reinvent . . . For a great actor or actress, each day is a golden
> opportunity to experiment with a new approach — in fact, with nothing
> less than a new persona. What are you going to be today? . . . what's your
> new twist?
>
> (Peters 1994)

In the service management literature, emphasis is therefore laid on how the geographies of product manufacture and provision, as understood through notions of display and performance, interrelate with both product quality and forms of organisational surveillance.

What must also be considered is the extent to which this new emphasis on display and performance within many service organisations may also generate myriad new opportunities for workers to rebel against what is expected of them. On the one hand, it is evident that there is here an even more thoroughgoing colonisation of the worker's self, a process that sees the dominating power of the corporate concern taking over not just the body (its capacity for manual labour power) and parts of the mind (its capacity for mental labour), but also the very 'presentation' of an individual's self-identity (how people present themselves to others, chiefly customers, in their everyday work practices).

On the other hand, it may be that, in the very process of trying to cajole from workers a range of displays which can never be completely scripted, the result is to leave these workers a measure of autonomy wherein they can act in ways that tread a fine line between selling the company's product and anarchic subordination. For instance, it is argued that when workers such as waiters are smiling without meaning it or feeling it, this lack can readily be detected by both team-mates and consumers. Moreover, in some situations it can be used as a form of disruptive performance on the service assembly workfloor, something perhaps akin to Beynon's (1984) worker actions at Ford. The difficulty for management is to be able to spot exactly when such disruption is occurring.

Once in this 'grey area' of service workers putting on displays and performances, there is room for wilful resistance, a little like Scott's (1985) foot-dragging, which cannot be simply detected and countered by those in authority (unlike in the stark polarised organisational politics outlined by Braverman). A specific case was retold to me, concerning a cut-price flight between two British cities run by a new airline company which prides itself on having young, easy-going, have-a-laugh flight crews. As the airplane was taxiing down the runway ready to get airborne, the pilot announced that, 'because [the airline] is such a low-cost company, we will be turning off the cabin lights for take-off', to which everyone laughed heartily. Such a piece of performance could be interpreted as the pilot happily fitting in with the company's image, but it could also be an act of defiance by a pilot who thought that he was being paid far too little and having to work in unreasonable conditions imposed by a 'cut-throat' operation. This may be only a trivial case, but it does indicate the particularly entangled moments of power, notably of domination leaking into resistance, which may accompany the new organisational geographies of display.

Organising and governing

In broad terms what I have been arguing is that Braverman's powerful portrait of organisations as marked by managerial domination and (on occasions) worker resistance needs to be reworked within a broader understanding of organisational geographies of surveillance and display. Law's (1994) account of *Organizing Modernity* is, so I would suggest, a fascinating attempt at just such an understanding. Spinning off wildly and wonderfully from a rather thin ethnography on the organ-isation of 'Daresbury' Laboratory, Law makes an argument for understanding modern life not in terms of the social order, as given, but in terms of plural and partial 'modes of ordering'. These modes of ordering are partly a matter of the stories and conceptual apparatuses through which we construct a meaningful environment for our own and others' actions. Yet they are not only narrative structures, for they are performed and embodied in networks of relations and interactions between different materials whose very differences are themselves constructs of these orderings. Thus, in Law's 'actor-network' approach, organisation is all about ordering and reordering materials so as to construct, recruit and arrange various actors (including employees but also raw materials, consumers, products, financial institutions, allies and competitors, retailers and manufacturers, and so on through and beyond the circuits of capital) into productive and temporarily sustainable networks of relations. Surveillance and display are socio-spatial practices that are centrally involved in the construction and reconstruction of these networks. They help both to 'constitute' and to 'recruit' materials into organising webs. They operate to establish the terms of involvement within those networks, and they exist as interrelated technologies and practices of interaction.

I fully admit that all of this, at least in my telling of it, may remain desperately vague, so let me look at one particular mode of ordering in a little more depth, that of 'enterprise'. Law (1994) spends some time outlining the impact of enterprise discourses on Daresbury in particular, and on contemporary forms of organisation more generally. Enterprise, rather like service, tends to deploy surveillance as a *post-facto* monitoring of a space of discretion and self-management. In so doing, it makes certain kinds of clearly demarcated personal and organisational displays especially important: think of quality audits and assessments, complete with their textual and interpersonal stagings of organisational achievement. Clearly, these enterprising technologies, practices and spatialities of surveillance and display are markedly different to those constructed within ordering discourses of worker deskilling and work routinisation, a difference which is reflected in distinct constitutions of the entities of manager, worker and consumer. An elaboration of this claim can be found in du Gay's inquiries into the discursive production of working subjects. What du Gay (1996) does is to contextualise technologies of managerial surveillance within broader strategies of organisational government, again paying particular attention to recent concerns with enterprising organisations and workers. Rather than dismissing such a shift as a rather feeble attempt to disguise the real power of managers to control workers, du Gay argues that such concerns amount to a thoroughgoing reconstruction of the very entities of management and labour, and of the terms of their interrelations. As he puts it:

> Changes in the way of conceptualising, documenting and acting upon the internal world of the business organisation actively transform the meaning and reality of work . . . In other words, the identity of the 'worker' has been differentially constituted in the changing practices of governing economic life. 'Workers' and 'managers' have been 'made up' in different ways – discursively re-imagined and conceptualised – at different times through their positioning in a variety of discourses of work reform.
>
> (du Gay 1996: 53, 55)

Hence, via Dandeker and a gaggle of service management writers, Law and du Gay, we are now returning to questions of management and labour in a rather different way to how we left them when in the company of Braverman. Now they are not so much about surveillance as a form of labour control, but rather are more about surveillance, display and the making up of organisational subjects. They are no longer so much about the physical spaces of the workplace and its tactical inhabitation, but are more about the always material *and* mental discursive construction and inhabitation of the organisation and its spaces. They cease to be so much about the *one* best way to extract surplus value as about the many ways in which to enlist materials, including subjectified and embodied actors, into networks productive of economic, cultural and psychic surpluses.

Conclusion

Let me draw things to a close with an attempt to pull out four threads from the preceding discussions. In the first place, while I began with a nod to the so-called 'cultural turn', noting the possibilities for economic geographers and sociologists when responding to its seductions, I hope that the examples given here demonstrate that such responses have to go beyond merely appealing to something like 'organisational culture' as an explanatory entity. More useful will be intensive inquiry into the processes of fashioning, organising and governing networks and entities, processes that are in part about the construction, negotiation and contestation of meanings, values and feelings. Nonetheless, these are processes which cannot be reduced to a simple scripting of organisational form, and instead they are firmly embedded in a whole range of organising practices. In the second place, I have tried to suggest that examining the socio-spatial relations of surveillance and display provides a convincing way into studying these processes of organising and governing. If I had had more words here, I could have said a little more on how these socio-spatial relations might be conceptualised (but see my more extended accounts: Crang 1994, 1997). In the third place, I have suggested that analysing these geographies provides a means of going beyond the narrow conceptions of both labour and management developed by Braverman.

Finally, I have also highlighted the potential dangers of this conceptual broadening, in particular as the clear organisational politics of domination and sometime resistance, as so starkly analysed by Braverman, disappear into a fudge of network negotiations and managements. Power in this new Machiavellian depiction is so entangled and so entangling that just how to analyse its operations becomes far from clear. Above all, to quote Cockburn (1992: 42), power in this new vision becomes a matter of 'capacity', of the ability to influence the network, to 'mould materials or sway opinions', something which is 'circulated, employed and exercised through a net-like organisation'. All too easily, questions of domination and resistance, or of oppression and liberation, are replaced here by 'an enthusiasm for the minutiae of organisational decision-making as an intellectual puzzle and human drama' (Cockburn 1992: 43). Having said this, two potential ways of analysing these operations of power are perhaps worth further consideration, in part because they sharpen up a sense of the politics of organisation. First, notions of performance strike me as particularly useful in approaching the socio-spatial relations *in* organisation: that is, the organising powers that construct and reconstruct organisational situations. A dramaturgical analysis very helpfully provides a set of concepts which bring together questions of surveillance and display, scripting and interactionally developed improvisations, actors and settings, and also the terms of inclusion and exclusion (Goffman 1956). Second, though, such an emphasis has to be combined with a grasp of the socio-spatial relations *of* organisation: that is, an understanding of the kinds of surpluses extracted through organisational networks. Whereas for Braverman the forms of surplus (surplus value) and the sites of their extraction (the labour process in the fixed workplace) are very clear, I have tried to argue that modern organisations are characterised by more multiple surplus productions and extractions. Much more is being 'taken' from workers than just their physical labour power, often across a variety of sites and situations, but at the same time the opportunities for workers to cut back on what is being 'given' to the company are arguably multiplying as well (for example, in their unscripted performances workers may resist directing their looks, words and movements wholly toward corporate goals). Even so, the mutiplicity of possibilities opened up by the approach advocated here is certainly no excuse for failing to go beyond the techniques of organisation in order to analyse their broader social productivities. Indeed, it is vital to remain attuned to the overall, and continuing, threads of domination and resistance that are always entangling the spaces of power in business organisation.

Acknowledgements

Thanks to Eric Laurier and Chris Philo for their kind and generous help.

References

Adkins, L. and Lury, C. (1996) 'The cultural, the sexual and the gendering of the labour market', in L. Adkins and V. Merchant (eds) *Sexualising the Social*, London: Macmillan.

Allen, J. and Pryke, M. (1994) 'The production of service space', *Environment and Planning D: Society and Space* 12: 453–476.

Baldry, C. (1997) 'The social construction of office space', *International Labour Review* 136: 365–378.

Baran, P. A. and Sweezy, P. M. (1966) *Monopoly Capital*, New York: Monthly Review Press.

Beynon, H. (1984) *Working for Ford*, Harmondsworth: Penguin.

Braverman, H. (1974) *Labor and Monopoly Capital: The Degradation of Work in the Twentieth Century*, New York: Monthly Review Press.

Cockburn, C. (1992) 'The circuit of technology: gender, identity and power', in R. Silverstone and E. Hirsch (eds) *Consuming Technologies: Media and Information in Domestic Spaces*, London: Routledge.

Cook, I., Crouch, D., Naylor, S. and Ryan, J. (eds) (1999) *Cultural Turns/Geographical Turns: Perspectives on Cultural Geography*, London: Longman.

Crang, P. (1994) 'It's showtime: on the workplace geographies of display in a restaurant in South East England', *Environment and Planning D: Society and Space* 12: 675–704.

Crang, P. (1997) 'Performing the tourist product', in C. Rojek and J. Urry (eds) *Touring Culture: Transformations of Travel and Theory*, London: Routledge.

Dandeker, C. (1990) *Surveillance, Power and Modernity: Bureaucracy and Discipline from 1700 to the Present Day*, Cambridge: Polity.

de Certeau, M. (1984) *The Practice of Everyday Life*, London: University of California Press.

du Gay, P. (1996) *Consumption and Identity at Work*, London: Sage.

Fox, A. (1974) *Beyond Contract: Work, Power and Trust Relations*, London: Faber and Faber.

Friedman, A. (1977) *Industry and Labour: Class Struggle at Work and Monopoly Capitalism*, London: Macmillan.

Goffman, E. (1956) *The Presentation of Self in Everyday Life*, Edinburgh: Edinburgh University Press.

Henry, N. and Massey, D. (1995) 'Competitive times in high/tech', *Geoforum* 26: 49–64.

Hochschild, A. (1983) *The Managed Heart: The Commercialisation of Human Feeling*, Berkeley: University of California Press.

Kramer, R. M. and Tyler, T. R. (1996) *Trust in Organisations*, London: Sage.

Laurier, E. and Philo, C. (1999) 'Meet you at junction 17: a socio-technical and spatial study of the mobile office' (Project Report to the ESRC), Glasgow: Department of Geography and Topographic Science, University of Glasgow.

Law, J. (1994) *Organizing Modernity*, Oxford: Blackwell.

Massey, D. and Allen, J. (1995) 'High-tech places: poverty in the midst of growth', in C. Philo (ed.) *Off the Map: The Social Geography of Poverty in the UK*, London: Child Poverty Action Group.

Peters, T. (1994) 'Theatre on the retail stage', *The Independent on Sunday* 6 March.

Pollard, S. (1965) *The Genesis of Modern Management*, London: Edward Arnold.

Revill, G. (1991) 'Trained for life: personal identity and the meaning of work in the nineteenth-century railway industry', in C. Philo (ed.) *New Words, New Worlds: Reconceptualising Social and Cultural Geography*, Lampeter: Social and Cultural Geography Study Group.

Revill, G. (1994) 'Working the system: journeys through corporate culture in the railway age', *Environment and Planning D: Society and Space* 12: 705–725.

Sayer, A. and Walker, R. (1992) *The New Social Economy: Reworking the Division of Labour*, Oxford: Blackwell.

Scott, A. J. and Storper, M. (eds) (1986) *Production, Work, Territory: The Geographical Anatomy of Capitalism*, London: Allen and Unwin.

Scott, J. (1985) *Weapons of the Weak: Everyday Forms of Peasant Resistance*, New Haven, CT: Yale University Press.

Solomon, M. R. (1985) 'Packaging the service provider', *Service Industries Journal* 5: 64–72.

Stein, J. (1995) 'Time, space and social discipline: factory life in Cornwall, Ontario, 1867–1893', *Journal of Historical Geography* 21: 278–299.

Sturdy, A. (1998) 'Customer care in a consumer society: smiling and sometimes meaning it?', *Service Organization* 5: 27–53.

Thompson, E. P. (1967) 'Time, work discipline and industrial capitalism', *Past and Present* 38: 56–97.

Thrift, N. (1996) *Spatial Formations*, London: Sage.

Upah, G. D. and Fulton, J. W. (1985) 'Situation creating in service marketing', in M. R. Solomon and S. Solomon (eds) *The Service Encounter*, Lexington, MA: Lexington Books.

Wouters, C. (1992) 'On status competition and emotion management: the study of emotion as a new field', *Theory, Culture and Society* 9: 229–252.

10

ENTANGLED HUMANS

Specifying powers and their spatialities

Steve Hinchliffe

> It is difficult to deny that human bodies have certain powers within the plane of practice which have real consequences for cultural extension.
>
> (Thrift 1997: 138)

Powering up the human

In this chapter I develop the argument that the attribution of power to human bodies requires extended qualification. In order to do so I critically engage with various accounts of power. First, I draw up a caricature of causal power. I argue that if power is seen as emanating from powerful bodies then this is conditional upon a particular spatial imagination. Furthermore, this imagination closes down more political options than it manages to open. Second, I contrast causal power with a version of relational power. Again, I emphasise the spatial consequences and conditions for thinking power in relational ways. The specific form of relational thinking that I take up is most readily associated with actor-network theory, and in particular the work of Latour. The latter's interest in the powers of association is taken up in the third part of the chapter. Here I develop a number of related problems with viewing power as associative. The first of these takes issue with the role of associations in the construction of power. The argument is made that power can be performed through dissociations as well as associations. This leads to a refiguring of the topology of power. Third, the chapter turns to the conduct and performance of power. In seeking to understand power as a practical and expressive effect of conduct, the chapter closes with the suggestion that the distinctions between human body-subjects and non-human body-subjects are local, and not generalisable, achievements.

Causal power and its geometry

As a crude summary, or a caricature, causal versions of power picture somebody getting somebody else to do something that they would not normally do. The superior body can do this by virtue of its ability to tap into a resource that is somehow lacking or less abundant in the other body. Typically, in accounts where the bodies in question are human bodies, the powerful actor draws on intelligence, charisma, bodily strength, access to a gun, the backing of state apparatus or whatever in order to ensure that the other person does what the actor says. Here agency is neatly divided. It is an internal and/or incorporated property of the powerful and something that is externally visited upon the powerless or the less powerful.

This version of power is also often characterised as a zero-sum game. As the powerful body accumulates more power, the subjugated body simultaneously loses power. The total power available remains the same, only redistributions of power can be initiated. Redistributions are gravitational and thus one-way – power moves from the weak to the strong. Almost inevitably (or so it would seem), stored blocks of power become invincible, and can only grow more powerful as they subject larger numbers of other bodies to their scheme. As Clegg (1989) puts it, an inescapable conclusion inhabiting causal versions of power is that, as power gets stored up by one faction, it is possible for that particular faction to continue to marginalise resistance in an ever more one-dimensional universe. In other words, power, in this account, rests with the victors (the dominant). Resistance is largely excluded from the power regime (a questionable assumption). This geography of dominance and resistance is straightforward, but the simplifications involve costs.

Viewed in terms of a progressive politics, there are unattractive and attractive aspects to this model. Causal accounts of power are unattractive because the world seems destined for more of the same. The central block can only grow and become more powerful. Those endowed with more power than the rest will climb their way to the top of the pile, and in doing so collect more power as they ascend. Any resistance seems to grow less and less potent and more and more unlikely to succeed as each battle is won by the powerful, who in turn can store up the equivalent power exerted by the failed resistance. This inevitability is clearly disheartening. But there is an attractive aspect to this model of power; this attraction stems from an ability on the part of the resisters to locate the central power-holders. We know where you are, goes the chant. Power, agency and responsibility map on to each other in a neat fashion. The enemy is identifiable and the political aims are relatively straightforward. There is a daunting, but nonetheless comprehensible, map of power to be drawn; this mapping seems to involve a straightforward representation of pre-constituted objects in two (sometimes three) dimensions. Differently sized and positioned regions of stored power are drawn on a plane, or sometimes are sketched as volumes in three dimensions. This might be called a geometry of power. To be sure, the causal account of power that underlies this geometry is far removed from the account of power offered by Massey, which is also sometimes called a geometry (see Massey 1992, 1993, 1999). Massey's spatial politics are counterposed to this kind of surface mapping. But I use the geometrical metaphor here to emphasise the topology that often underpins causal accounts of power. This is a map of objects, concentrations and geometric distances. Representing power in this way confirms a simple spatial politics. Revolutionary politics says 'turn the map of power on its head', make it revolve in the mechanical world that it constitutes. The geometrical, if not the practical, simplicity of an inversion, and the taking of sides, has helped to reproduce a geographical imagination of centres and margins.

But this attractive simplicity has its drawbacks. This is the fatal side to the attraction. One consequence of drawing up satisfying battle-lines has been to treat the resistant margins as somehow (still) beyond the core. They are treated as distinct regions, as independent entities fighting for prolonged independence. Again there is the attractiveness of a simple map and a potentially romantic story to be told of defeating the enemy despite the odds. But this is a world of pre-constituted objects, and of background contexts that magically help to explain the ensuing orders; the shortcomings of such imaginings are at their most apparent when it is cultures and humans that are rendered through this grid of causal powers. 'The local culture resists the global culture'. We say this as if cultures were entities, objects that transcend representations and relations. This is a world where questions of difference and similarities 'appear to refer to self-evident manifestations of lifestyle and community . . . Such a context, in turn, re-creates an essentialist interpretation of cultural uniqueness' (Strathern 1995: 57). The relational processes of connection and disconnection are effaced to produce a spatio-temporally impoverished account of cultural change and stability (for a critique, see Massey 1999). Cutting a long story short, progressive politics are made even more difficult as cultures are fixed in form.

A similar move is made when it is individual human beings that resist the dominant scheme of things. The romance often lies in the suggestion that certain human beings manage to hold out against the march of the inhuman regime by protecting their humanness, preserving a nugget of their true selves, and this human nugget sounds as though it exists prior to social practices. But this act of purification, based upon an essential quality of humanity, is difficult to effect in practice. Moreover, maintaining a pure space to be human is, to paraphrase N. Rose (1996), part of and not external to the stream of everyday life which calls upon people in a multiplicity of ways. In other words, human agency does not, in Rose's invocation of Foucault, emanate from within, but is an effect of a complex of relations and practices. The effect of the human essence is but one manifestation of social practice.

In sum, causal accounts of power might provide some of the sharpest imagery, in terms of satisfying maps of power and straightforward responsibilities. These images enable an ability to judge (see Lee 1995). But, the very same maps can result in some very blunt tools for producing conditions for change. Essentialising human and social relations is one example of this desensitised or numbed social theory of power. So what of other models of power and their spatialities? The brief mention of Rose's work brings me to relational theories of power.

Relational power and its topologies

There is, as Latour, Foucault and others have been at pains to point out, a paradox in causal, zero-sum formulations of power. Latour (1986: 264–265) puts it like this: 'when you simply have power – in potentia – nothing happens; when you exert power – in actu – others are performing the action and not you'. In simple terms, rather than seeing power as the cause of A's domination over B, Latour would have it that power is the consequence of A's abilities to bring B into its programme of action. It is through action, and in this case through B acting in accordance with the aims of A, that power is composed. Although A might enjoy a power effect greater than B, there is no necessary relationship whereby A's so-called power is at the expense of B. In the process of connecting, both parties may enjoy a variety of power effects, which will probably not be equal, but are unlikely to be unidirectional. In other words, power and indeed 'objects' (like the cultures and persons referred to above) are constituted through relations.

In a relational view of power, the power map takes on a different guise. Or rather, as Latour (1997) puts it, there is a change of topology. The spheres, volumes, concentrations and metric distances are no longer positioned and traced as self-evident objects that can be used to account for or explain power. Rather, these 'objects' and the power that they wield are treated as outcomes or effects. Where they are traced or drawn together, in everyday practice as well as in academic accounting, this itself is an act of power. In other words, if we say that something is powerful, we are both commenting on an outcome of a collection of practices that go to produce the effect of power (and the something), and, in our own small way, we may be adding to that effect of power. Tracing power is always active. It is a drawing together and at the same time an attribution. The method-ological point is that we move from self-evident blocks to tracings, drawings and acts of pulling and drawing together (see Latour 1990). This is one aspect of an approach to working actively that is sometimes called actor-network theory (ANT) – a theory that is seemingly powerful, but of course, like many powerful things, it is crafted heterogeneously. This is not the place to rehearse, or draw together, yet another account of ANT – there are more than enough representations and far fewer performances of active tracings (but see Law 1997); however, it is worth emphasising a point about topology. Latour (1997) suggests ANT marks a topological shift from surfaces to filaments.

> Instead of thinking in terms of surfaces – two dimensions – or spheres – three dimensions – one is asked to think in terms of nodes that have as many dimensions as they have connections . . . modern societies cannot be described without recognizing them as having fibrous, thread-like, wiry, stringy, ropy, capillary character that is never captured by the notions of levels, layers, territories, spheres, categories, structure, systems. It [ANT] aims at explaining the effects accounted for by those traditional words without having to buy the ontology, topology and politics that goes with them.
>
> (Latour 1997: 2)

In terms of explicating domination and resistance, the tautological tendency to explain power by power is sidestepped in favour of practices that confer strength. Power is now distributed across the neat dividing lines that characterised the causal account. Indeed, the neat lines have started to dissolve. The threads and strings evoke a much more entangled geography of power. The metaphor is characteristically physical, drawing on an imagery of string bags and other kinds of woven or latticed 'objects'. But this is no ready-made ontology. The importance lies in that the stringing together is achieved through a variety of means, practices and materialities. For Latour (1997: 2): 'Strength does not come from concentration, purity and unity, but from dissemination, heterogeneity and the careful plaiting of weak ties'. It is active, it is always in process.

Power in this sense is about, and constituted from, acts of arranging, ordering, organising and delegating. Meanwhile these acts seem not to have their rationale or cause rooted in some guiding architectonic of action. This focus on arrangements demonstrates the indebtedness of Latour and others who are associated with ANT to Foucault's project on power, and in particular his account of the emergent disciplinary power in the eighteenth and nineteenth centuries in Europe (Foucault 1979). The micro-physics of disciplinary power share a focus upon arrangements of objects (and, importantly, on the simultaneous constitution of those objects). They also provide a sense that these arrangements of power are not made for power itself (as in sovereign models) but for a host of other organisational aims which are the outcome of a diversity of linguistic and material practices. Again, then, power is not the cause of action, but an effect of action.

Taken to its limits, it is clear that power cannot, in this scheme of things, be possessed (to 'have' power results in immobility, but to exercise power is to distribute power across a network). The concentrations, storages, volumes and surfaces that are often used to account for or map the location of power have, as I have suggested, largely (or at least momentarily) evaporated under an analytical style of tracing associations. As a result, Latour (1986) has suggested that the word power be dropped as an analytical tool. Given that power, like society, capitalism and so on, is the precarious or sometimes seemingly durable outcome of associations, then analysts should pay attention to associative work and not simply the forms that can result (see also Pels 1997: 704). As Law (1994) puts it, this is the distinction between a sociology of order, and a sociology of ordering. Let me now summarise something of this method of tracing associations as a means of accounting for the operations of power. In doing so I shall emphasise a series of points that will be taken up and developed in the next section.

Here is the crux. For Latour, the world is ordered, made stable and the effect of power is thereby produced, through processes of extension; this extension is materially heterogeneous, involving combinations of humans, non-humans, technologies and so on. Extension is not simple, it is not an extension through a context. An idea, or statement (and the statement can be a collection of words as well as a collection of materials, and will probably be a combination) does not simply travel or diffuse through a context. Rather statements and contexts are related in more complex ways than would be inferred by suggesting that the context was peculiarly amenable to the transmission of a statement, or it was resistant. Indeed, statements and contexts are co-produced in the network tracing activities. In order to avoid the term 'context', Latour follows Serres (1987) in adopting the term 'quasi-object' as a way of emphasising this co-determinacy. More recently, Latour has put it in these terms.

> A ball going from hand to hand is a poor example of a quasi-object since, although it does trace the collective and that the playing team would not exist without the moving token, the latter is not modified by the passings.
>
> (Latour 1997: 10)

The point here is to emphasise that for a statement (be it a fact or an artefact) to exert influence it needs to move (in both senses of that term), and in order to move it will change, as will the networks which constitute its conditions of possibility. Extensions of network statements involve varied translations.

In earlier work on the practice of science, Latour (1987) makes this extension–translation relationship, or tension, the basis of his method for following emergent scientists/engineers/economists – whoever – as they succeeded or failed, through a chain of statements, to produce a power effect of truth, agency or efficiency. Latour called this 'metrology' – a gigantic enterprise of contextual and textual construction – making the outside world, in Latour's (1987: 251) terms, a world inside which facts and machines can survive. In a similar vein, Latour (1987: 248) strikes out a naive universalism sometimes attached to facts and machines by stating that 'no one has ever observed a fact, a theory or a machine that could survive outside of the networks that gave birth to them'. Clearly, in this scheme of things, a successful or powerful statement (machine, theory etc.) can be described as such only if it can be drawn together with a network that is long and durable. In metrology, network length becomes the measure of success, the more there is enrolment, the longer the chain, the more successful the statement. Length and durability of heterogeneous associations becomes the metric by which power is explicated.

Relating and restoring power

As a method and a theory of power, metrology can be effective, but there are doubts. I shall develop three concerns. They are related and I shall tell them as a progression. The first concern questions the privilege afforded to connections, associations and network length. The second seeks to extend the topological possibilities of ordering practices. The third dwells on the conduct of power, and in doing so opens up a politics of practice.

Association and dissociation

The constructionist claims that something does not hold because it is true or efficient, rather something is true/efficient because it holds is a neat way of capturing the relational topology of power sketched previously. If a fact, organisation, person or machine is to be successful, the longer and more durable the network or the hold, the more they are true (for facts), stable (for organisations), accomplished (for people) and efficient (for machines). But Munro (1997a) is concerned that we are managing only to dispose of one set of divisions (between truth and falsity in technoscience studies for example) by erecting others. The division here is between durable and non-durable networks – and the division is located somewhere along the length of a chain of statements. Munro goes on:

> We can learn to 'see' the long networks when they are long, as he [Latour] helps us to do, yes: but we need to be careful not to be captured by these in ways that prevent sight of the exact moments when a dissociation is being mobilized.
>
> (Munro 1997a: 9)

Dissociation and disconnection are not always signs of weakness. Munro reminds us that the durability/network length/power equivalencies that Latour tends to trace can miss the ways in which networks might retreat (in order to recoup the detail), and even diminish themselves in order that they can be overlooked. In a separate paper, Munro (1997b) draws partly upon Barnes (1988) and Law (1991, 1994) to emphasise the role of the 'overlook' as an integral part of organisational work. In doing so, he provides a useful insight into the power of management.

The management technologies that interest Munro are records and accounts. Being called to account, and the making accountable through written inscriptions, records and audits, is all part of 'a proliferation of technologies of control' (Munro 1997b: 8). Drawing things together in an audit or through record keeping can give the impression that discretion is reduced as people are ever more firmly called to account. In other words, all this surveillance and tracing produces people who act in line and are reduced to relays on a switchboard, or, in Barnes's (1988) language, 'authorities' (people without discretion). But this version of power and ordering would confuse durability with rigidity, and would invoke a network that looked more like a 'thing' than a process. Fixing the relations, making the network an object wherein all the elements play their roles to perfection, without missing a line, can be effective – some of the time. This can form an important part of power stories. But there are other parts and other tricks. One way of bringing these to the fore is to insist on reading 'network' as a verb (Law 1994). In doing so, the actions that enable networking practices and organising can at last be seen. As Munro suggests, there will be activities that connect and extend as well as activities that cut, perform stops and waits, and withdraw. Without a cut, for example, it is impossible to make decisions or judgements (Lee 1995; Law and Mol 1996). Similarly, cuts are necessary in order to assign responsibility – another important ingredient to the task of going on. As Lee (1995) argues, responsibility engineering is every bit as complicated as, and not always isomorphic with, power engineering. The cartoon (Figure 10.1) is a useful illustration – the fat cat enjoys his power effects by being associated with all parts of the organisation, and at the same time does not suffer responsibility effects because he cuts himself off from various particulars (events, people, things: see Lee 1995).

The point is this. When discretion is exercised in bypassing the account or by cutting certain ties, the effect of power is not necessarily diminished. Indeed, it can be quite the reverse. In some circumstances, working by the book, and carrying things out to the letter, might be less effective in management terms than predominantly bypassing the formal network. For one thing, there are simply not enough hours in the day for organisations to spend following up every recorded infringement of designated norms (see Munro 1997b). In addition, it 'is not the surveillance effect of durable records that makes the figure of the manager more "powerful", so much as they help facilitate the movement of moments of un-decidability over time and space' (Munro 1997b: 12). In other words, deferring a decision and thereby overlooking an account can be a way in which power *is* stored

Figure 10.1 Power and responsibility engineering in large corporations
Source: S. Bell (1988) *If . . . Breezes In*, London: Methuen, p. 57

up. As Munro suggests, deferral does not mean that the slate is wiped clean. It can mean that the weight of the matter that has been overlooked is left dangling, waiting for the moment when the manager decides to call the employee to account (see Munro 1997b, where a suggestive play on waiting and weighting is extended).

Munro is not alone in seeking to develop a distributive and discretionary approach to power. Law (1991, 1994) and Barnes (1988) before him have both sought to describe ways in which power can be stored (see also Pels 1997). These authors have all attempted to combine discretionary approaches with a radical relationalism. In other words, the ability of a manager to act with discretion is not an inversion of Latour's disavowal of powerful entities. Above all, 'this is not to slide back into an emanationist or reificatory conception of power' (Pels 1997: 708). As Law (1991: 170) has noted, '"power to" and "power over" may be stored and treated by social analysts as a potential or a set of conditions so long as we do not forget that they are also an effect, a product of more or less precariously structured relations'. Law's question was then to ask how are relations maintained long enough to generate power effects? His provisional answer is to follow strategies which embody those relations in durable materials (including, of course, the textual inscription, the tracing activity). But it is only a provisional answer: in the very same paper, and more clearly in his later work, Law adds to this relational materialism by considering performativity and further topological possibilities (see for example Law 1994, 1997; Mol and Law 1994).

Before considering these topologies and performances in a little more detail, here is a summary of the arguments made so far regarding relational power. I have looked at two possibilities. The first is suggested by a surface reading of ANT, one that tends towards a totalising account of power and a paradoxically limited spatial imagination (see Lee and Brown 1994, for a critique of this rendition of ANT). The second opens up a more heterogeneous account of power. Here is the caricatured version of these two possibilities.

First, order and power are cemented through an extension of network practices, bonded more or less precariously by the toing and froing of durable materials (like records or some other inscription or statement, including machines). Power is distributed through relays, or authorities, which ideally have no discretion (machines are supposedly very good at acting as authorities). As the network is established through the implementation of durable materials, so the capacity for discretion is reduced.

Second, order and power are precarious outcomes of a wealth of activities, some of which involve associations, some dissociations. Some increase the predictability of events (punch cards tend to produce workers that turn up on time). Others increase the undecidability (management makes decisions if or when to act on the occasional late arrival). Importantly, discretion is redistributed, rather than eliminated (Law 1991; Munro 1997b).

Running with this second possibility, I want to consider two implications. First, there are some important implications for the ways in which power is spatially and temporally constituted that are not necessarily captured[1] by the geometries and topologies that have been sketched so far in this chapter. Second, there is a need for a closer understanding of conduct that can cope with the indeterminacy and undecidability that haunts any ordering of power and action. These are related concerns, but I shall deal with each in turn. I start with a brief discussion of the space-time of power.

Network topologies and suspended sentences

The topology of the thread, the rope and the filament that Latour has used to destabilise the surfaces and spheres that have inhabited mainstream social theory has much to recommend it. This network-like ontology is attractive, not least because it resonates with many aspects of organisational activity (see Lee and Brown 1994). But there may be other topological possibilities that are ignored when the network dominates our vision (see Mol and Law 1994). The network metaphor is partly derived from semiotics, which is translated as 'path-building, order-making' (Latour 1997: 6). This order-making is viewed as parallel to the building of meaning through written and other forms of text. But the blindness to the dissociations that are played out in ordering activities suggests that the topological metaphor of the network and the semiotic structure of the sentence are sometimes, but by no means all the time, a deficient means to trace power effects. In Munro's management example, the organisational network does not hold itself together through invariant relations. Neither do dissociations always bring the walls tumbling down. They may even be made more robust. And this is the key: order and power need not be made up of a pure origin or homogeneous entity (this much has been taken from Latour and Foucault and has helped to destabilise the volumetric and regional thinking of much social theory: see Mol and Law 1994). But neither need power always be made up of a durable set of associations held by threads and relations. Power can be conducted through a mixing of space-times, so that a deferral of the decision by management amounts to a waiting that disturbs the normal pattern of rules. The (semiotic) sentence is suspended – not broken or disrupted. And it does not mean that there is now a free for all. Individuals might know that they have displeased management, or they might not. But even if they are fairly sure they have got away with it, there is the possibility that they will not next time, and that someone will decide to notice the record and/or implement the penalty. These are complex time-spaces of conduct. The point here is not to name a new topology (although sentences in suspension, mixed time-spaces and so on make fluid metaphors very attractive). Rather it is to signal that any analysis of power might benefit from being topologically open at the outset. That means not only being open (or, in Latour's terms, agnostic) about the shapes of the actors ('any shape is possible provided it is obsessively coded as heterogeneous associations through translations': Latour 1997: 6), but also being open about the make-up (and breakdown) of relations. Relations are not all of a kind. They will produce differing temporalities and spatialities, different weights and waits. As Latour has conceded:

> ANT is a powerful tool to destroy spheres and domains, to regain the sense of heterogeneity and to bring interobjectivity back into the centre of attention . . . Yet, it is an extremely bad tool for differentiating associations . . . It is thus necessary, after having traced the actor-networks, to specify the types of trajectories that are obtained through highly different mediations.
>
> (Latour 1997: 10)

Conduct and performance

Understanding power in this more open temporal and spatial sense presents us with an age-old difficulty. Quite rightly, causal or emanationist powers have been dissolved and power has emerged as a relational effect of action. But we have also noted instances when and where power and fairly well defined agents are re-stored. If we are too quick with the analysis, this can seem to be a means of smuggling naturalised versions of power and (more often than not) an unproblematised human agency back in through the open door of relational word-play. A manager's discretion can all too quickly provide a ticket to an uncritical singling out of human actors' abilities to direct the play. They, or some of the luckier or better equipped individuals, are the ones who flick the switches that make other people and things behave as authorities. Discretion sounds too much like an inherently and purely human quality. In other words, we are in danger of returning to a people view of power, where knowledge is internalised as thoughts, beliefs, intentions and values, and where action is painted as the transmission of knowledge into the outside world (Munro 1997b: 1). Power in this sense is the source and the result of calculative actions by the powerful.

In fact, in treating power and knowledge in this calculative fashion, we are simply back to a causal version of power where to those that have, more shall be given. Yet, as we have seen, command and control versions of power are as unpopular in social theory as they are in economic policy. The question I want to pose at this stage is how can other understandings of action open up alternative politics of power? I do this by briefly sketching out some theories of practice and by doing so lead into a discussion of the possibilities that are opened up for progressive politics by working with understandings of the conduct of power.

I start by returning to Munro. Like most good ethnographers, Munro makes an important though often unnoticed switch in the politics of explanation. This is the switch: do not dig deep to explain action. Start with it.

> So, instead of attempting to go forward, adding narrative onto narrative in order to arrive at a suitable story of social action, it seems more apposite to take one step back and begin with conduct. Organizing is made possible, less through action being elicited by commands, and more through conduct.
>
> (Munro 1997b: 5)

Conduct is a useful term here. In the first place, it is suggestive of the directions and momentum that, to greater or lesser extents, frame action. To conduct oneself appropriately is to follow some rules, discourses and codes and to engage in mimesis (Munro 1997b: Taussig 1993). But it is also less determined than this. As Munro argues, conduct is ongoing. It is neither automatic behaviour characteristic of the authority or the cultural dope, nor is it a calculative redirection of events characteristic of the power or agent. While the spatial metaphor might not be useful, Law (1991: 171–172) was surely right to ask 'is there not, in fact, a large territory between explicit calculation on receipt of signs on the one hand, and "automatic" response to the input of signs on the other?' His answer is that there is 'a continuum where some are (relational) powers, others authorities, but most are strung out between the two, borrowing and so embodying more or less explicit strategies of calculation' (Law 1991: 172).

The term 'embodying' is crucial: I take it that embodying here shares something with the notion of translation in ANT. It is the translation of a more-or-less specified programme of action into the action. It is the making real, embodying a statement into an artefact, a text, a bodily movement, a more-or-less durable relation. The term embodying should not be read as a process that completes the programme. Like any translation it has the potential to supplement the rule or programme, it is both traduction and trahision (Law 1997) – transmission and treason. Nor should embodying, at this moment, be read as always or necessarily synonymous with what we often take to be the human body. I shall be coming back to the notion of human embodiment in a moment.

Let me return this sociology of translation to the question of discretion and conduct. As I have suggested, the temptation is to infer a specifically human agency from acts of discretion. That is, there is a rush to reach the satisfying conclusion that despite all the audits, records and bureaucracy, it is humans, after all, that direct the show. This is an example of what Law and Mol (1996) have called a 'framing assumption'. It is the will to show singularity in descriptions of conduct. It is to single out an event, a decision, a calculative action, an agent from the stream of conduct. Wielding the knife in this way is to contribute to the effect of power storage. The idea that there is a burden of discretion is derived from a singularistion of the decision. But this is too clean. We need a more serviceable understanding of conduct than this in order to grasp at something of the way in which decisions are multiple (Law and Mol 1996), and how it is that responsibilities fail to stick to the power containers (Lee 1995; Law and Mol 1996).

One potential source for the understanding of conduct and its role in organising power effects lies in those streams of social psychology (Shotter 1993, 1996; Newman and Holzman 1997), geography (Thrift 1996, 1997; G. Rose 1997) and sociology (Law 1994) which have as their concern the emergent qualities and properties of social action. Shotter's work, in particular, is worthy of attention at this point. Drawing on Wittgenstein's later works, Shotter offers an understanding of practice that firmly locates action within the stream of conduct.

The dialogical or joint nature of action, and the non-cognitive assertion that actions are not simply expressions of prior intentions but mark the partial completion of thoughts (Newman and Holzman 1997), provide a means for working through the problematics of power and agency. One outcome of attempting to understand power in this way is the attention that is afforded to the situations within which conduct develops.[2]

> Joint activities have a dialogical or mixed character to them. In such circumstances, outcomes cannot be attributed to the desires or plans of any of the individuals involved, neither can they be attributed to outside agencies. It is as if the particular situation itself were a third agency in the exchange with 'its' own unique requirements.
>
> (Shotter 1996: 296)

The agency of the situation is a collective production that arises, not from the unproblematic pooling of minds or resources, but from the openings and the partial understandings that ensue. Again, it is the connections and the disconnections that matter. As Shotter tends to suggest, social scientists are prone to seeing a proliferation of gaps as the noise that must be eliminated for there to be clear communication (see also G. Rose 1997). But Shotter's reading of Wittgenstein reminds us that it is the gaps that enable conduct to keep going. Indeed, ordering activities are predicated on a third presence that often goes unnoticed.[3] There are striking similarities here to Michel Serres's invocation of a third presence in information exchange. Serres has allocated various names to this figure, including the angel, the demon, the third man and the parasite (Serres 1982; Harari and Bell 1982; Hetherington and Lee 1997). For Serres, the parasite is too often characterised as an operator that disrupts a system of exchange. But the parasite, 'by virtue of its power to perturb . . . ultimately constitutes like the *clinamen* and the demon, *the condition of possibility of the system*' (Harari and Bell 1982: xxvi, original emphasis). What we, as social scientists, have formerly called noise turns out to be the very condition for making sense, for ordering worlds through dialogical actions.

So where does this leave our theories of power and of human agency? First of all, it suggests that power and agency are noisy. But to put the question in these terms, in the terms of theories of power and human agency, is to miss the more radical possibilities that are opened up by these writers. The second, more crucial, point is that understanding actions as co-productions wherein there lie no pure thoughts or deep-seated blueprints makes some very different demands of social scientists. These understandings

> 'move' us toward a new way of 'looking over' the 'play' of appearances unfolding before us, such that instead of seeing the events concerned in terms of theories as to what they supposedly 'represent', we see them 'relationally' – that is, we see them practically as embedded in a network of possible connections and relations with their surroundings 'pointing toward' the roles they might actually play in our lives.
>
> (Shotter 1996: 305)

For Shotter (1993), this requires a knowing of the third kind. We are concerned neither with *episteme* (knowing that), nor with *techne* (knowing how). The third kind of knowing is a practical activity, a knowing from within (*sui generis*) that generates its own conditions for change. It is 'an ethical know-how, to do with our way of being in the world, our stance in relating ourselves to our surroundings' (Shotter 1996: 309). It is a practical, tool-and-result method (Newman and Holzman 1997) that seeks to work with rather than refuse the noisiness of action.

Its ethics are modern. The cacophony of conduct is not acceptable at face value. But there is no vain quest for an ideal speech situation, free from the parasites of communication. The power to change stems from the possibilities for constructing other than already existing relations from the resources available in the gaps that open up in the course of gesturing and acting (Shotter 1996: 301). The question might now be who or what gets to twist or re-twist the words and actions?

I shall now draw this chapter to a close. I do so by questioning the implicit reattribution of a human form of agency that can occur in this Wittgensteinian world.

Conclusion: powering down the human

Latour (1997: 6) states that in itself ANT is not a theory of action. The relational approach developed by Shotter and others as a means of gesturing towards a practical method of coping with conduct offers a preliminary means of establishing action within ANT's own practical method. The attractiveness of this kind of move stems from the similarities that exist within these writings with respect to their formulations of the openness of conduct and the emergent properties of ordering activities. Likewise, both sets of writing suggest that the concern of the social scientist should be ontological rather than epistemological. They offer points of engagement and possibilities for new practices rather than 'new knowledge' about the world.

But there are problems in grafting together these writings: for the most part these centre on the role and status ascribed to humans in this practical method (see also Thrift 1997: 136ff). As the opening quote to this chapter suggests, Thrift would be unhappy with what he calls the hard-line Foucauldian move to allow for the possibility of agency to be distributed evenly across any situation. The rationale for this disquiet is practical and ontological:

> The attempt to make humans and non-humans into simply different locations in networks or assemblages, with humans becoming the effects performed by them is not as simple (to carry through) as it may at first seem.
>
> (Thrift 1997: 137–138)

Thrift is referring to the common move in ANT and other relational theories to mark out the 'human' as an achievement, a precarious effect of networking practices (see Callon and Law 1995 for a rehearsal of this kind of move). But for Thrift there is a special case to be made for humans. In doing so, humans are attributed a particular role in the ordering of worlds. This special case is unlike most other attributions. The latter are often based upon human intentionality and language use (see Callon and Law 1995) or specifically on human cognition.

Thrift's case centres on human embodiment. In particular it is the body-subject's tacit nature, its expressive side that endows it with the peculiar capacities for turning programmes of action into sometimes quite other ends. For Thrift (1997) and for Shotter (1996), the gesturing, joint actions of body-subjects, with their inherent ambiguity and ambivalence, provide the key expressive medium to configure and turn life worlds. It is embodied action, with its incompleteness and its becoming, that offers the most potent sources for a Wittgensteinian poetics of practices. The body-subject in Thrift's writings provides a sophisticated means of avoiding an understanding of human conduct as that which is entirely entrapped within discourse.

My point here is not to disagree. I have no doubt that processes of human embodiment are elusory and opaque (Thrift 1997; Radley 1995). Nor do I question the assertion that a common, or perhaps the most common, attributions of agency, discretion and 'power to' are afforded to human bodies. This, after all, is what many forms of organising are set up to achieve. It would be problematic for a manager in an organisation to find that agency was always being attributed elsewhere. Indeed, the technologies of managing, the tracing activities, the record keeping and so on can be regarded as creating, again precariously, a space for discretion. This is a mode of ordering that 'defines and distributes the character of persons and papers' (Callon and Law 1995: 494). So the point is not to suggest that Thrift's metaphysics are wrong or right. But it is to ask whether or not there is something to be gained in remaining agnostic for a moment longer with respect to the necessary attribution of a special status to human embodiment? In other words, do we need to make the distinction (to paraphrase Callon and Law 1995: 499) that people are Wittgensteinian, and everything else is not? For Callon and Law (1995: 502): 'Machines, persons, texts, all are opaque. All resist, or have the potential to resist, re-presentation'. Like human embodiment, other embodiments can signify in ways that are not reducible to representational forms. There is play in the so-called material world.

To be sure, the attributions drawn up by Thrift and Shotter go a long way to refiguring agencies as other than intentional, linguistically complex beings. They are also vital in offering ways of understanding that refuse to view bodies as entrapped within discourse. It is no doubt useful to follow Wittgenstein as a means of opening up other practices in those circumstances where human bodies are performed as prime movers (an empirical example can be found in Hinchliffe forthcoming). But these human-centred treatments of power speak to specific, if common, configurations, and they should not rule out other possibilities. Strategic embodiments will always involve translation and treason in the stream of conduct, no matter what kinds of bodies they go on to form. These, and non-strategic embodiments, will always emerge in the stream of conduct in ways which are not wholly determined by discourse.

Let me return to the opening quotation taken from Thrift (1997: 138): 'It is difficult to deny that human bodies have certain powers within the plane of practice which have real consequences for cultural extension'. Thrift's writing offers a useful means to heed Latour's call for ANT to differentiate associations. It is a means of avoiding the tendency to reduce the world to an 'undiscriminated field of wills, point forces and resistances' (Lee and Brown 1994: 774). But humans are not alone in the world, they are entangled with other bodies that have their own, emergent specificities. They too can turn worlds and may have a vital role in offering alternative courses and other than already existing relations. Indeed, it would be unfortunate to lose sight of this ability, offered most readily by ANT, to attribute powers and agencies in ways that are orthogonal to the conventional understandings of conduct and action (Callon and Law 1995: 497). For to do so, to lose ANT's methodological commitment to a-humanism, would risk too much unravelling of wordly affairs. It would risk a return to the binary maps of power with which we started. More than this, it would only serve to impoverish our theories of practice. At their best the latter can offer ways and means of intervening in affairs. To work, this practical politics must always remain entangled in the world. This is not to admit defeat and suggest that actions remain entrapped within discourse. It is to insist on the importance of embodiment (human, non-human, mixtures thereof) and the connections and cuts that make this a far from complete exercise of power.

Acknowledgements

Thanks to Doreen Massey, Dave Featherstone and the editors for reading and commenting on an earlier draft of this chapter.

Notes

1 'Captured' is the wrong word. The aim is not capture some absolute sense of reality. It is to provide tools for understanding how reality is built. It is not the shapes of extant actors that is the concern. It is the means of recording those shapes that is the focus of attention (see Latour 1997: 6).
2 Talk of situations should not be confused with a 'presentist' or isolationist geography of action (see Thrift 1996: 7). Situations are extended through the kinds of practices that Shotter is characterising.
3 This presence is far from straightforward – it may well entail a presence of absence, as in the silences that punctuate everyday talk, or the blank figures that offer the conditions of possibility for ordering in a whole variety of actions, including modern mathematics (see Hetherington and Lee 1997).

References

Barnes, B. (1988) *The Nature of Power*, Cambridge: Polity.

Callon, M. and Law, J. (1995) Agency and the hybrid *collectif*, *South Atlantic Quarterly* 94: 481–507.

Clegg, S. (1989) *Frameworks of Power*, London: Sage.

Foucault, M. (1979) *Discipline and Punish: The Birth of the Prison*, London: Peregrine.

Harari, J. V. and Bell, D. F. (1982) 'Introduction: journal à plusieurs voies', in M. Serres (ed.) *Hermes: Literature, Science, Philosophy*, Baltimore, MD: Johns Hopkins University Press.

Hetherington, K. and Lee, N. (1997) Social order and the blank figure, Mimeo.

Hinchliffe, S. (forthcoming) 'Performance and experimental knowledge', *Society & Space*.

Latour, B. (1986) 'The powers of association', in J. Law (ed.) *Power, Action, Belief*, London: Routledge and Kegan Paul.

Latour, B. (1987) *Science in Action*, Milton Keynes: Open University Press.

Latour, B. (1990) 'Drawing things together', in M. Lynch and S. Woolgar (eds) *Representation in Scientific Practice*, Cambridge, MA: MIT Press.

Latour, B. (1997) 'On actor-network theory: a few clarifications', Centre for Social Theory and Technology, Keele University, available at: http://www.keele.ac.uk/depts/stt/staff/jl/pubs-jl2.htm

Law, J. (1991) 'Power, discretion and strategy', in J. Law (ed.) *The Sociology of Monsters*, London: Routledge.

Law, J. (1994) *Organizing Modernity*, Oxford: Blackwell.

Law, J. (1997) 'Traduction/Trahison – notes on ANT', Centre for Social Theory and Technology, Keele University, available at http://www.keele.ac.uk/depts/stt/staff/jl/pubs-jl2.htm

Law, J. and Mol, A. (1996) 'Decision/s', paper presented to Centre for Social Theory and Technology seminar on Complexity, Keele University, November.

Lee, N. (1995) 'Judgement, responsibility and generalized constructivism', paper presented at the Centre for Social Theory and Technology annual workshop on The Labour of Division, Keele University, December.

Lee, N. and Brown, S. (1994) 'Otherness and the actor network: the undiscovered continent', *American Behavioural Scientist* 37: 772–790.

Massey, D. (1992) 'Politics and space/time', *New Left Review* 196: 65–84.

Massey, D. (1993) 'Power-geometry and a progressive sense of place', in J. Bird, B. Curtis, T. Putnam, G. Robertson and L. Tickner (eds) *Mapping the Futures: Local Cultures, Global Change*, London: Routledge.

Massey, D. (1999) 'Spaces of politics', in D. Massey, J. Allen and P. Sarre (eds) *Human Geography Today Polity*, Cambridge: Polity.

Mol, A. and Law, J. (1994) 'Regions, networks and fluids: anaemia and social topology', *Social Studies of Science* 24: 641–671.

Munro, R. (1997a) 'Ideas of difference: stability, social spaces and the labour of division', in K. Hetherington and R. Munro (eds) *Ideas of Difference*, Oxford: Blackwell.

Munro, R. (1997b) 'Power, conduct and accountability: re-distributing discretion and the new technologies of managing', paper presented at the Actor-Network Theory and After conference, Keele University, July.

Newman, F. and Holzman, L. (1997) *The End of Knowing: A New Developmental Way of Learning*, London: Routledge.

Pels, D. (1997) 'Mixing metaphors: politics or economics of knowledge?', *Theory and Society* 26: 685–717.

Radley, A. (1995) 'The elusory body and social constructionist theory', *Body and Society* 1(2): 3–23.

Rose, G. (1997) 'Situating knowledges: positionality, reflexivities and other tactics', *Progress in Human Geography* 21(3): 305–320.

Rose, N. (1996) *Inventing Ourselves: Psychology, Power and Personhood*, Cambridge: Cambridge University Press.

Serres, M. (1982) *Hermes: Literature, Science, Philosophy*, Baltimore, MD: Johns Hopkins University Press.

Serres, M. (1987) *Statues*, Paris: François Bourin.

Shotter, J. (1993) *Cultural Politics of Everyday Life: Social Constructionism, Rhetoric and Knowing of the Third Kind*, Buckingham: Open University Press.

Shotter, J. (1996) 'Living in a Wittgensteinian world: beyond theory to a poetics of practices', *Journal for the Theory of Social Behaviour* 26(3): 293–311.

Strathern, M. (1995) 'The nice thing about culture is that everyone has it', in M. Strathern (ed.) *Shifting Contexts: Transformations in Anthropological Knowledge*, London: Routledge.

Taussig, M. (1993) *Mimesis and Alterity*, New York: Routledge.

Thrift, N. (1996) *Spatial Formations*, London: Sage.

Thrift, N. (1997) 'The still point: resistance, expressive embodiment and dance', in S. Pile and M. Keith (eds) *Geographies of Resistance*, London: Routledge.

11

ANTI-THIS – AGAINST-THAT

Resistances along a human–non-human axis

Chris Wilbert

In 1996 two of the many protesters seeking to protect fields and woodland areas at an anti-roads demonstration in Newbury, southern England, while dressed as a pantomime horse, were arrested for attempting to prevent site clearance work (see Figure 11.1). Such scenes are part of a long tradition of pranks, of dressing up in ways associated with 'different places' in order to bring a kind of carnivalesque air to resistant practices, partly in order to ridicule the networks of forces being opposed. Strathern (1992: 174) suggests that the fields and woodlands, as parts of 'nature' dominated and seemingly passified by economic and cultural practices, appear to need such human inputs to ensure their continued existence. However, there are other ecological, ethical and some scientific arguments – such as Gaia theory – which stress the activity of 'nature' and of non-humans operating *alone*, and, in some more forthright cases, the possibility of this 'nature' somehow wreaking revenge upon those humans who dominate it.

In fact we find a whole range of such views running together through many ecological forms of politics regarding the agency of both humans and non-humans. This points to differing questions of what human/non-human material and symbolic relations are, or could be, as well as to forms of asymmetrical alliances occurring between some humans and non-human animals in ecological and animal rights protests. This raises political and ethical questions about who, or what, can or cannot act purposively to bring about change, and one of the newer aspects of some ecological struggles is the perceived role not only of people, but also of 'nature' and non-human animals. Broader questions are also raised about the relations between culture and nature itself.

Here I wish to move briefly through some of the openings of debate about human and non-human animal agency in terms of what has been termed the 'modern constitution', the historical work of producing a distribution of powers over, as well as a separation of, the social and natural worlds, and how this constitution is currently being challenged.

238

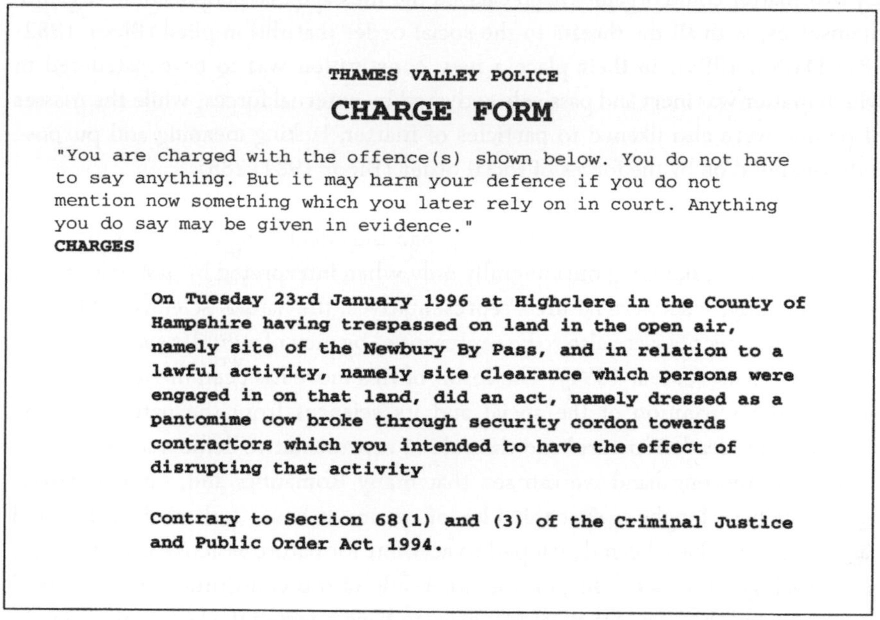

THAMES VALLEY POLICE

CHARGE FORM

"You are charged with the offence(s) shown below. You do not have to say anything. But it may harm your defence if you do not mention now something which you later rely on in court. Anything you do say may be given in evidence."

CHARGES

On Tuesday 23rd January 1996 at Highclere in the County of Hampshire having trespassed on land in the open air, namely site of the Newbury By Pass, and in relation to a lawful activity, namely site clearance which persons were engaged in on that land, did an act, namely dressed as a pantomime cow broke through security cordon towards contractors which you intended to have the effect of disrupting that activity

Contrary to Section 68(1) and (3) of the Criminal Justice and Public Order Act 1994.

Figure 11.1 Thames Valley Police charge form
Source: Reproduced by CoMedia from *Squall* 12 (spring 1996)

The modern constitution

It has become increasingly conventional for some historians and sociologists of science to argue that it was in the seventeenth century that the modern constitution, separating the social world from the natural and a western world from other cultures, was developed. Latour (1991, 1993) has argued that this modern constitution is summed up in the dispute between Robert Boyle and Thomas Hobbes against a background of revolutionary ferment and hermetic and cabalistic philosophies of nature.[1] According to Latour, what Boyle and Hobbes sought to do was to distribute and organise a whole series of powers and rights of who, or what, was able to speak or meaningfully act (Latour, 1991: 12–13). This same point was made somewhat earlier by Bloor (1982) in seeking to show how Durkeim and Mauss's theory of 'primitive classification' could indeed show how 'the classification of things reproduced the classification of men [sic]' (Bloor 1982: 290–291).[2] To put this point very simply, the aim of this constitution was to deny intentionality and agency to nature (as was being advocated by Hermeticists and Levellers) and to deny self-organisation and meaningful action to certain people (revolutionaries, the peasants, the poor and women). The argument is that these ideas of intentionality and agency had to be denied in both these cases because to

239

say that matter could organise itself carried the message that people could organise themselves, with all the threats to the social order that this implied (Bloor 1982: 287; Daston 1995). In their place a new constitution was to be constructed in which matter was inert and passively activated by external forces; while the masses of people were also likened to particles of matter, lacking meaning and purpose until operated on by the forces of social order (Bloor 1982: 288).

The eventual dominance of these doctrines (and the work that had to go into making them dominant is a massive story) had the effect of nature becoming mute and mechanistic, behaving meaningfully only when interpreted by *men* of a certain social standing, who were nature's representatives, the natural scientists; likewise the masses of people were also to be represented by men of a similar social standing – the politicians (Latour 1991). The result of this move has been the long modern project, the separation of the social and its sciences from the natural and its sciences, the 'death of nature' and the rule of representative democracies.

Now on the one hand we can see that many Romantics and, later, environmentalists have fought against this idea of nature as brute and mechanistic, and various theories have been developed to account for nature as active and affective, as we shall go on to see.[3] But, on the other side of this constitution, we may well have expected the many forms of Marxism to have rejected the lack of meaning in people's struggles. Yet, with only a few exceptions, orthodox Marxism has stayed within this modern constitution. The masses of people have most often been treated as barely conscious of the meaning of their struggles, requiring them to be interpreted and organised by vanguard parties and elite intellectuals to be effective and meaningful. More recently, these same kinds of criticisms have been made of the cultural approaches of post-Marxists such as Jameson and Baudrillard on a variety of similar counts (Grossberg 1989: 174–175).

Related criticisms can also be made of the current 'regulation approach' in critical neo-Marxist theories, which has sought to describe the recent changes in the world of production. Though this approach is a more sophisticated objectivist analysis, it tends to concentrate upon the economic/state institutions which oversee the social relations of production and overlooks the struggles and transformations of human subjects in these changes (Gambino 1996: 44). This approach has therefore generally slipped into functionalism and a form of technological determinism in its theorisations (*Aufheben* 1994; Bonefeld and Holloway 1991).

Although the work of Foucault has initiated a new focus on resistances to power and disciplining technologies, even the avowedly Foucauldian-influenced 'actor-network theory' (ANT) in science studies, which has introduced a notion of non-human agency into social theory, tends to ignore the everyday resistant actions of the human masses (see Mort 1994). ANT, though a variable grouping of discourses itself, tends towards the same perspectives (and failings) as classical Marxism, viewing politics as about changing society, and seeing order as being produced, while also having an elitist concentration upon spokespersons for the masses of humans (Harbers and Koenig 1996: 14; Law 1994).[4] This leads to the struggles of people against domination being, on the whole, ignored in the analysis of network building, as well as notions of gender, class or race being bracketed out, as these are deemed unacceptable social explanatory devices. Yet little is given in such analyses to compensate for the supposed illegitimacy of such social explanatory devices, and this becomes something of an absence in many of their works.

However, since the 1960s a broad Marxist approach has developed which seeks to put the self-activity of the masses of people, or what was once called the working class, at its centre of analysis of capitalist society rather than the dominant powers of capital (Witheford 1994). This is the autonomist Marxist current arising in Italy in the 1960s and 1970s and being reworked and developed in the 1990s in differing ways by groups in Europe and North America. Briefly put, these autonomist, and what we may now (rather unfortunately) call post-autonomist, theories tend to invert the view of orthodox Marxism and other forms of analysis to see the struggles of the masses of people as an important dynamic of capitalist development (Witheford 1994: 89). Thus, it is not only that the resistances to capital impose limits upon its operations, but also that it is the activities of those resisting that at times can create the conditions of crisis. For example, technological development is not seen as simply an objective tendency, but as often being a weapon that capital resorts to against those struggling against domination (Witheford 1994: 90; Panzieri 1980). Following such a trajectory it can be argued that the current information-based restructuring of capital is seen to be a widespread attempt at re-establishing capitalist control in which all social activities, not just those of the workplace, are subordinated to capitalist logic through a process of commodification of ever-larger areas of experience (Witheford 1994: 95). All of society is now seen as being permeated with the rules of capitalist relations of production, a kind of 'social' or 'ecological factory', and it is a process which is seen as analogous with Deleuze's transition from a disciplinary society to a society of control (Hardt and Negri 1994: 259; Negri 1989; Deleuze 1992). Moreover, as capital restructures in periods of crisis it seeks to decompose forms of resistances to it, but in this process autonomists claim that there also occurs a recomposition of opponents, the creation of new social subjects with fresh tactics and organisational forms against this attempted reimposition of control. These sites of resistance and appropriation are necessarily plural, focused around such things as education, the environment, work, race, feminism and many other sites, but they are also viewed as having the capacity to come together in a lateral 'circulation of struggles'. Such a perspective also accepts that not all forms of resistances or reappropriations of technologies and spaces can be viewed as simply 'progressive' or not, proposing instead that such practices are complex, often contradictory and always open to recuperation.[5] Resistances and domination may therefore always be found to be complexly entangled.

This is, of course, only to give a very brief overview of autonomist approaches, and it is one which is certainly not immune from criticism, especially regarding overly optimistic interpretations of resistances as well as the playing down of older forms of struggles and divisions between and within groups (see Negri 1989, 1992; *Aufheben* 1994; Witheford 1994, 1995: 38).[6] We must also be careful, in asserting this capacity of resistance, not to allow some notion of a self-conscious capitalist strategy to dictate our thinking. Rather the 'strategy' of capital should be seen as resulting from conflicting and competing strategies of individual capitals and their various agencies: capital, as a subject, emerges out of such internal conflicts as well as from resistances to it (*Aufheben* 1994: 40). Such autonomist approaches do, however, seem to hold to a possibility of a collective capacity for self-creativity on a larger scale than the often individualistic sites focused on by many postmodern analyses, as well as arguing against the inevitability of people becoming an inert and semi-passive mass in the face of capital restructuring. In doing this they critique what we have termed the modern constitution from one aspect at least – that of the human. However, such autonomist approaches still lack a developed ecological dimension, and when it comes to wider non-human forms of agency this is obviously a serious lack.

Indeed, within the vast plurality of Marxist discourses the role of 'nature' has been a constant problem that has been debated in the twentieth century; however, it has rarely been one of the most important strands of most of these discourses.[7] Debates in the early twentieth century tended to waver between geographical determinism and promethean views which argued that within the capitalist mode of production (and in the state capitalism of the USSR) nature was so dominated that it was effectively neutralised as a significant factor (Bassin 1996). There have been relatively few forms of Marxist analyses which have embraced a notion of nature, or even animals, as active agents that have been non-determinist in their orientation, and this has been something of a failing as some Marxist writers have noted (Leff 1995).[8] The question of non-human agency is being increasingly reworked and debated, mainly through the work of environmentalists and the influence that they have had upon social theorists and environmental historians. The ways that this embracing of non-human agency is undertaken vary widely, ranging from semiotic approaches to more realist and naturalist positions. Although there are problems with some of these theories, I do not aim to go into these here. Instead, I want to discuss a political construction of non-human agency which many will find unacceptable for it seems guilty of all the 'isms' we are not supposed to engage in. Yet such political constructions of non-human animal agency draw attention to some wider debates about agency found around ecology, and the specific interest here is the question of whether non-human animals are capable of resistance in the sense that is implied by some eco-anarchist groups. Before this we need to say a little about relations between humans and non-human animals, particularly wild animals.

Human and non-human relations

Some time ago Lévi-Strauss argued that natural species are important not because they are good to eat, but because they are good to think with (Lévi-Strauss 1969: 162). This purpose of thinking with and through animals does not just have the purpose of classifying them, but rather, as the anthropologist Richard Tapper argues, cultural constructions of animals are used for moralising and socialising purposes – they are thus good to teach and learn with (Tapper 1994: 51). Yet in modern western societies, as various commentators have stated, animals have in many ways seemingly vanished from the lives of humans. Animals have become marginalised, particularly in their wild forms since the nineteenth century. Harbingers of this disappearance and marginalisation can be found in some nineteenth-century Romantic painting, and more especially in Grandville's ruth-less illustrated caricature of the 1840s depicting the bourgeois idealist attempt to subsume nature under its own subjective categories. In much of Grandville's work the world of nature and everyday objects seems to be precociously alive, but in a form where everything has been transformed into commodities, bringing to expression what Marx called the 'theological capers of commodities' (Buck-Morss 1989: 154). Indeed, all kinds of human artefacts, such as musical instruments or umbrellas, become animated, while animals are often humanised and given the characteristics of people. More disturbingly, several drawings create montages of different animals, while others mix animals and humans, becoming 'monsters', in ways which have peculiar resonances with current developments in genetic manipulation whereby the deeper levels of life become mere commodities to be mixed at will. Buck-Morss (1989: 158) argues that Grandville's caricatures 'mimic the hubris of a humanity so puffed up with its new achievements that it sees itself as the source of all creation' – and thus able to do with it what it pleases. For Berger one particular drawing by Grandville captures the myths of domination and commodification of nature and increasing disappearance of wild animals from human lives (see Figure 11.2). Here lines of anthropomorphised animals dressed as humans in the latest fashion styles depart off-stage into a steamship Ark, a departure that Berger terms a reversal of the first assembly of humans and animals in the Christian myth of Noah's Ark (Berger 1977b: 665).

However, animals remain present (if not often visibly so) and intimately involved in the lives of even the most urbanised westerners. But this is generally so in the kind of highly mediated and commodified forms to which Grandville made early allusions, through such things as zoos, packaged foods, sundry commodities,

Figure 11.2 'The new Ark' by J. J. Grandville
Source: J. J. Grandville (1969) *Grandville: Des Gesamte Werk*, 2 vols, Munich: Rogner und Bernhard, p. 1,345

television, and most obviously as domestic pets (Berger 1977a; Anderson 1995; Tuan 1984; Philo 1995). The different forms of these human–animal relations are varied and complex, but all such relations tend to involve high amounts of domination and exploitation, even when they also involve affection (Tuan 1984).

It can be argued that cultural constructions of animals are dependent upon people's familiarity with them, as well as upon human–animal relations of production and the availability of knowledge of alternative ways of living, among other factors (Tapper 1994: 51–52). Even as wild animals have retreated from many people's lives in some, but not all, ways (see Wolch *et al*. 1995), such animals are still used 'to think with' and 'to teach with', and this occurs increasingly in an imaginary sense imbued with political and moralising discourses. This can be seen, for example, in popular children's cartoons and stories, wildlife programmes, and the use of animals in advertising.

Tapper (1994) has identified two main senses in which animals can be viewed here. First, they can be regarded as Other, as the model of disorder and of how things should not be done, and second, certain animals are idealised as agents or characters with motives, values and morals, at times almost indistinguishable from people (Tapper 1994: 51). My argument is that with the rise of a range of environmental protests around animals in recent years both these senses of thinking with animals are used, but are also often inverted so that 'nature' or non-human animals sometimes become symbols of the good for eco-anarchist political groups which often term themselves 'primitivist'.

Human and non-human resistances

Philo (1995: 656) remarked that there are many difficult theoretical issues bound up in making statements about animals, particularly wild or semi-wild animals, as a social group having the potential for resistance or transgression of human orders; this is certainly the case. Although in his article Philo questions the notion of animals resisting, on the grounds that this is to impute human characterisations on to non-humans, he does acknowledge that certain animals can on occasions transgress the socio-spatial order created and policed by humans (Philo 1995: 656). Whether this transgression is simply by intruding into the ordered life of human homes, or escaping from captivity into cities or towns, animals can at times be seen to be out of place, undermining the established human order. However, empirical studies of 'primitivist' political groups can identify a certain sense of political playfulness around this problematic notion of animal resistances, a way of thinking with animals as a political ploy and identification with the supposed interests or needs of animals.

Specifically this identification of common human and animal interests centres on notions of wildness. Here the argument goes that the exclusion of wildness, in the form of wild animals and wild nature, has come about from viewing animals as exclusive Other to be dominated and appropriated. This has occurred through the domesticating processes of civilisation, a process of domination which has also been applied to humanity. Such groups tend to valorise indigenous peoples – or, as they often term them, primitive or primal peoples – as having complex mythic and metaphorical relations with nature. In such societies, it is argued, animals and plants are regarded as centres, metaphors and mentors of the different skills, traits and roles of people (Shepard 1992: 45). They are treated as agents and social beings with differing motives and values. It follows from this that humans need to rediscover their own internal wildness as well as wild nature, for 'we' need this wildness to develop as humans; it is a guide to human life.[9] Wildness here is closely equated with freedom. Thus, we are also seemingly asked to entertain the idea that those who have been domesticated, or repressed, in some sense by modern society can revert back to the wild, to be 'feral' beings at odds with authority and order, whether animal or human, and through this wildness, somehow, there will be a return to authenticity (Zerza 1995: 82).[10] What we see occurring in the more political forms of these arguments is then an inversion of value. Wildness, which it is argued has been viewed as suppressed Other in western societies, becomes valorised above domestication or civilisation as the primary reality. Industrial civilisation, which seeks to tame wild nature through developments that degrade wild lands and threaten wild species, is the political adversary, and in this we can see a strong neo-Luddite, anti-technological tendency among such groups. This anti-civilisation, anti-technological tendency can be found in various journals and magazines of eco-anarchist groups where some animals are viewed as being (in some sense) in alliance with eco-activists; animals are seen to be resisting domestication, domination and industrialisation just as some humans are. For example, in the British magazine *Do or Die – Voices from Earth First!* (1995 no. 5) we find a page headlined 'animal antics', simply repeating stories culled from press reports on some actions of animals and other non-humans above a picture of a bull attacking a bullfighter. Here are two examples of these stories:

> An elephant whose calf was knocked down by a locomotive in the Sythet region of Bangladesh on the 21st of February 1993 blocked the next train that passed an hour later and banged her forehead against the engine for 15 minutes until it could no longer run. She then walked off into the jungle, leaving over 200 passengers stranded for over 5 hours.

> Nature's war on man's design grinds on in Seville where the world's Expo, having been plagued by armies of rats, which gnawed through most of the optical and electrical cables, is now being consumed by mosquitoes. Soaring temperatures, fountains, abundant vegetation and strong night lights have attracted every insect within miles and no one has a clue what to do.

In such press stories we see a strong profession of what are perceived to be the disturbing qualities and the disorder which the wild can bring to urban civilised spaces, as well as the threat that 'civilisation's' technology faces when it moves into the wild – a kind of destructive healing. But in these reinterpretations wildness is deliberately invoked in a positive sense as the anti-self of society, and these future primitivists play on, and play up, the disturbing, destructive, qualities of 'wildness' which these news stories already are about. For, as Taussig (1986: 215–219) notes, it is a seemingly perpetual facet of a dominant class (or we may say species in this case) to impute both destructive and healing powers to those that are dominated, the wild, or primitive, others; in this case stories of wild animals are invoked by their human 'allies' to play upon such powers and fears.

Similar views can also be found in the North American magazine *Fifth Estate* (1993 no. 28), only here the actions of animals are depicted in a more celebratory fashion. As well as stories of animals escaping from zoos, we are also told of hundreds of bats disrupting court sessions in Texas, and cows escaping from slaughterhouses and other perceived places of domination. Just as 'sympathetic humans' are engaging in the sabotage of places that exploit animals, so we are told that animals are fighting back: 'fighting for their own liberations'. As an example of this we are shown pictures of elephants crushing a car, and the article ends by spelling out this idea of a common resistance between some humans and some animals:

> The destruction of the wild (out there and in our souls) proceeds at an ever-maddening pace. Let us hope that acts of defence and resistance by animals, fish, birds and their human brothers and sisters will increase.
>
> *(Fifth Estate*: 32)

But this eco-anarchist paper also uses a host of graphics and collages which capture the dream of a symbolic nature returning to wreak havoc on civilisation. So too do we find some of Grandville's illustrations being used, especially those that invert the positions of humans and animals, such as the one described earlier of animals 'returning' to a steamship Ark. Other illustrations invoke what Buck-Morss (1989: 154) describes as an active and rebellious nature taking its revenge on humans who would fetishise it as a commodity (see Figure 11.3).

This invocation of animals is of course selective and unequivocally anthropomorphic – but this seems to be the main point, it is an argument that is itself transgressive of most modern ways of discussing animal behaviour. As Berger (1977a) argues, anthropomorphism was itself a strong feature of human–animal relations up until the nineteenth century and beyond. It is a residue of the continuous use of animals as metaphors, and one which has by no means completely disappeared from mainstream culture (Berger 1977a). Such metaphoric and anthropomorphic uses of animals do have their more acceptable and unacceptable forms, however, and, as I have said, some of the descriptions used by primitivists work as literary devices genuinely to disturb 'civilised' responses. Perhaps, as Berger implies, the increasing disappearance of wild animals from the lives of most humans means that their mystery also increases, lending force, and a greater sense of unease, to such political and anthropomorphic accounts of animals.

Figure 11.3 'The April Fool' by J. J. Grandville
Source: J. J. Grandville (1974) *Bizarries and Fantasies of Grandville*, introduction and commentary by S. Applebaum, New York: Dover, p. 28

Of course these articles and stories are not always entirely serious, but are a semi-playful way of thinking through animals. Yet, they remain problematic in several ways. For the equivalence of 'wildness' and 'freedom' is simply a projection which does not hold water, because animal behaviour is not wild in a sense of being

248

boundless. Indeed, animals tend to behave in accordance with their own instincts and social limits (Zimmerman 1992: 267). In fact, such arguments tend to verge on some of the ancient and Renaissance forms of *theriophily* – meaning literally the love of wild beasts (Boas [1933] 1966). In some forms of this varied doctrine about wild animals the latter were treated as superior to humans precisely because they did not have 'faculties of reason' which corrupted them. At times this indeed seems to be the sense towards which eco-primitivists tend.[11] Yet, at the same time if we accept the basic stories of these press reports animals do indeed appear to act purposively, at least in some sense. Such actions are reinterpreted in a political way by primitivist groups, which would generally be seen as in opposition to 'normal' societal ways – thus for them it is not just transgression, but resistance that animals can sometimes be said to engage in.

Intentionality, accommodations and resistances

But exactly what kinds of questions about animals and agency does this open up if we are not simply to disregard such notions as the most extreme forms of the current vogue to see resistance in more or less every situation? We may ask whether it is possible to talk of animal resistance without engaging in the deliberate and exaggerated forms of anthropomorphism that many future primitivist accounts deploy. Cresswell (1996: 22–23) has argued that resistance is perhaps overused to describe many actions, and this is certainly the case. However, his argument for this does not necessarily preclude the imputation of animal resistances. For Cresswell argues that resistance relies upon a sense of intentionality, a conscious intentionality, of purposeful action directed against something disliked. But would we need to go so far as to impute conscious intentionality to animals in order to speak of purposive action – which would be to commit the kinds of anthropomorphisms that eco-anarchists are perhaps sometimes guilty of? I think we do not, and that two ways seem open to describe actions of non-humans and interrelations with humans which do not force us into attributing conscious intentionality to animals and may allow us to talk not just of transgressions but of animal resistances. The first could be thought of as taking an approach which seeks to argue that in some ways humans and non-human animals are perhaps not quite so different as often argued with regard to actions. The second emphasises more how agency is a relational effect, something which emerges out of interactions between heterogeneous materials.

First, we can talk of animals having intentional actions of a non-reflexive kind, that is not conscious intentionality but simply intentional action. This is to follow the anthropologist Ingold in separating animal consciousness from animal thinking. Ingold argues that at a practical, everyday level of being in the world animals act, much as humans often do, without reflexively thinking, but in ways directed by a practical consciousness; that is animal action is conscious and intentional but animals do not think about what they feel and do (Ingold 1994: 96). What separates humans from this level of activity is not just the processes whereby this practical consciousness is formed, but the human ability reflexively to attribute intentionality and to frame this within discourse. Thus, we could argue that when an animal intentionally acts to escape from the confines of a zoo it does not reflexively think and conceptualise its plan, it simply acts (though it may well learn from failed attempts), and in this sense it could be viewed as a form of resistance to human ordering; it acts intentionally against human goals to confine it. Such an approach may, in the last resort, be unprovable, although it is perhaps the general approach to agency taken in more analytical and science based accounts and one which seeks understanding through commonality.

The second view is actually to look at the process of formulating goals, or future states of affairs, from present states in a process of modelling as occurs, for example, in scientific practice (Pickering 1993: 578). The creation of models as a way of formulating goals is generally seen to be constitutive of such scientific practice and is always open-ended, with no necessarily fixed destination. In this reflexive process of modelling, according to Pickering, there is a complex process of *resistance* and *accommodation* in which the model works and then perhaps meets resistances to it, forcing changes to be made to the model in the form of accommodations to such resistances. These resistances can be material, a lack of skill, or may be a machine not working as desired. But resistance, or what Gooding (1992: 68) terms 'recalcitrance' can also be seen to emanate from what we term 'nature', perhaps from non-human animals or even viruses under study. These resistances are temporally emergent in the goal-directed activity of scientists and others, and according to Pickering (1993: 577) are always liminal, taking place at the intersection of human and non-human agency. Gooding (1992: 70) has gone as far as to state that: 'Recalcitrances shape and constrain the development of experimentation: they enable empirical constraint'. However, just as these resistances have generally been neglected from accounts of practice, so we should not move too far to see them as in some sense determining structures of knowledge. Just as it was noted earlier that the self-activity and resistances of antagonistic social subjects affect the way that capital is organised, but do not determine wholly how it is organised, so too is knowledge not wholly determined by lack of skills, pre-existing material constraints, or the emergent resistances and activities of animals or 'nature'. Rather, it is in the temporally emergent resistances to goal-oriented activity and accommodations to these resistances that both human and non-human material realms are interactively reconfigured at the same time. However, it may also be the case that this notion of affective agency in resistance is too narrow a notion of agency (Gingras 1997), just as it is too narrow to view human agency as occurring only in resistances or transgressions, or indeed to attribute agency only to strategic speakers.[12] Indeed, with regard to non-human animals and physical processes, this is an area which requires much more empirical work in order to examine such possible difficulties.

These two forms of being able to talk of non-human resistance are not without their own problems, but I put them forward as two possibilities of thinking towards non-human, and more specifically, animal resistances, arguably in a less playfully political sense than discussed earlier. Both of these approaches, Ingold's based on Gibson's ecological perception, and Pickering's post-phenomenological view, can be regarded as starting from a perspective where the world is viewed as an interrelational field of agency, or where every body has a capacity of affecting and being affected within certain maximum and minimum thresholds (Deleuze 1988: 123–124). Such a perspective offers much in being able to ascertain the affectivity of non-humans, as well as humans, in the world across particular contexts and situations.

Conclusion

While autonomist Marxism challenges what has been termed the modern constitution from the aspect of the possible meaningful activity of some humans, the imputations of agency to both humans and non-human animals of eco-anarchists would seem on the face of it to be more of a symmetrical challenge to this constitution. In fact the eco-anarchistic constructions and imputations of resistances to animals discussed here are simply one part of a complex range of ways of thinking and teaching with animals that can be found among western environmental movements. I am not arguing that how groups such as *Earth First!* or *Fifth Estate* talk of 'wildness' as a form of resisting is correct, but then I do not expect them to care much about that. Certainly, it seems that often such arguments miss their target, being overly simplistic, especially in blaming something as abstract as civilisation for current ecological and social problems. Moreover, by inverting the privileging of the primitive over civilisation, such groups remain within a system where an historical term is overlaid upon cultural differences and thereby denies the diverse peoples included in 'primitivism' their own cultural development and living histories.

Groups such as *Earth First!* represent, for many people, the most extreme form of a neo-Luddite tendency of resistance to so-called progress. This has partially involved a celebratory construction of resistance that involves animals and humans seemingly struggling for the same ends, the rediscovery of wildness and an opening up of political and cultural processes to explore and realise new potentialities. This move is only one among a multiplicity of sites of resistance which celebrate direct actions, and one which is itself becoming the focus of growing private and state security crackdowns in North America, the UK and many other areas of the world.

However, this invocation of human/non-human alliances is of course asymmetrical, because the non-humans do not really come together fighting a common cause. But this does not then invalidate questions of non-human resistances wholly. What I have sought tentatively to put forward are two approaches highlighting the possibility of non-human resistances which occur at the micro-levels of everyday life and which are thus often missed, ignored or are unattributed in social and cultural accounts that tend to attribute agency to strategic speakers. Yet, these other agencies are often not missed by some of these strategic speakers, such as scientists or others who work more openly in what we have termed a 'field of agency' (Pickering 1995). What these approaches may point to is that there are many different ways of acting within the world with other non-human beings and physical processes, which may be thought of as active in a variety of ways. From the perspective of how we live and interact, some of these practices are more dominating and others less so, just as many ecologists argue, and a sensitivity to such non-human potential for activity and affectivity may well help to develop better ways of living in which 'nature' is not simply approached as a factory for production, a resource for exploitation by capital. But any such approach to nature must include the social, for what we have found is that the social and the natural can no longer be effectively separated, if indeed they ever could.

Notes

1 This argument of course relies very heavily upon Shapin and Schaffer (1985). However, Latour (1991, 1993: 13–48) does critique this work for its lack of symmetry in its analysis of science and politics.

2 Much of Bloor's article in turn relies on the work of J.R. and M.G. Jacob in the early modern history of science.

3 There is a large variety of work that could be included here including the various different works on perception by both Gibson and Merleau-Ponty, as well as more recent theories such as Lovelock's Gaia theory.

4 However, Akrich and Latour (1992: 261) do introduce the relativised concept of de-inscription to describe 'the reaction of anticipated actants – humans and non-human – to what is prescribed or proscribed to them', according to their own 'antiprograms'. Such semiotic notions are rarely developed in their analyses and remain profoundly textual in their form.

5 For example, Bonnett (1996: 26–27) argues, through an example of young men creating a spatial logic of masculinity by crossing roads at improper places, that the connection between conservative identities and transgressive practices has rarely been understood, or wholly acknowledged, but demands to be integrated 'into the political interpretation of everyday spatial manoeuvres'.

6 Negri has become a somewhat unpopular figure among revolutionary groups in Italy due to his enforced exile, and, rather less forced, forgetting of his comrades who remained in jail in Italy after the crackdown on the autonomist movement and their false complicity in the terrorism enacted by the Red Brigades. However, in 1997 Negri voluntarily returned to Italy to draw attention to the plight of some of his

former comrades. He was immediately jailed and has seemingly been forgotten. See *Semiotext(e)* (1980) for a background to the violent crackdown on the revolutionary movement in Italy in the 1970s.

7 Exceptions could be said to be the work of the Frankfurt School.

8 Of course this is an overly simplistic statement of a very complex field. More recently, some critical-realist-influenced work leans in the direction of an active sense of nature (Bhaskar 1978; Benton 1989). Other work seeks to reassert a position of Marx's whereby nature is seen to be an active agent in production in the form of use values. For more on this see Leff (1995). Some work in environmental history can also be seen to take a weak Marxist or neo-Marxist approach, particularly that of Worster.

9 Similar arguments, of a less obviously political nature, can be found in the 'Biophilia hypothesis' of Wilson and others, where wild animals are regarded as necessary for human development (Wilson 1984; Kellert and Wilson 1993).

10 The exception to this ability of animals to revert to the wild for some 'primitivists' are some domesticated animals. For example, according to Shepard (1993): 'Domestic animals become numerous, docile, and flaccid, their brains diminished, their anatomy and physiology subject to disfunction, and their ethology abbreviated.' For Shepard domestic animals are the results of the first genetic engineering. However, he departs here from many other 'primitivists' when he argues that it is the sentimental concern for pets that has given rise to much of the 'righteous socialising' about the animal kingdom and its moralising arguments about animal rights, anti-hunting and vegetarianism.

11 See also my unpublished PhD thesis for a more extended discussion on this subject (Wilbert 1998).

12 Similarly, it can be argued that Pickering tries to take his theory of scientific practice too far, even arguing that it can be seen as a theory with application in almost every practical situation, a theory of everything. Gingras's (1997) critique of Pickering's approach as a 'new dialectic of nature', in the sense attempted by Engels, is probably accurate. However, this does not make the whole theorical approach invalid by any means, rather it calls for a limited and more careful application that requires further empirical working, as well as a little humility on Pickering's part.

References

Akrich, M. and Latour, B. (1992) 'A convenient vocabulary for the semiotics of human and nonhuman assemblies', in W. Bijker and J. Law (eds) *Shaping Technology/Building Society: Studies in Sociotechnical Change*, Cambridge, MA: MIT Press.

Anderson, K. (1995) 'Culture and nature at the Adelaide zoo: at the frontiers of "human" geography', *Transactions of the Institute of British Geographers* 20: 275–294.

Aufheben (1994) 'Decadence: the theory of decline or the decline of theory', *Aufheben* 3: 24–34.

Bassin, M. (1996) 'Nature, geopolitics and Marxism: ecological contestations in Weimar Germany', *Transactions of the Institute of British Geographers* 21: 315–341.

Benton, T. (1989) 'Marxism and natural limits', *New Left Review* 178: 51–86.

Berger, J. (1977a) 'Animals as metaphor', *New Society* 10 March: 504–505.

Berger, J. (1977b) 'Vanishing animals', *New Society* 31 March: 664–665.

Bhaskar, R. (1978) *A Realist Theory of Science*, Brighton: Harvester.

Bloor, D. (1982) 'Durkheim and Mauss revisited: classification and the sociology of knowledge', *Studies in the History and Philosophy of Science* 13(4): 267–297.

Boas, G. ([1933] 1966) *The Happy Beast in French Thought of the Seventeenth Century*, New York: Octagon.

Bonefeld, W. and Holloway, J. (eds) (1991) *Post-Fordism and Social Form*, London: Macmillan.

Bonnett, A. (1996) 'The transgressive geographies of daily life', *Transgressions* 2/3: 20–37.

Buck-Morss, S. (1989) *The Dialectics of Seeing: Walter Benjamin and the Arcades Project*, Cambridge, MA: MIT Press.

Cresswell, T. (1996) *In Place/Out of Place*, London: University of Minnesota Press.

Daston, L. (1995) 'How nature became the other: anthropomorphism and anthropocentrism in early natural philosophy', in S. Maasen, E. Mendelsohn and P. Weingart (eds) *Biology as Society, Society as Biology: Metaphors*, Dordrecht: Kluwer.

Deleuze, G. (1988) *Spinoza: Practical Philosophy*, trans. R. Hurley, San Francisco: City Lights.

Deleuze, G. (1992) 'Postscript on the societies of control', *October* 59: 3–7.

Gambino, F. (1996) 'A critique of the Fordism of the Regulation School', *Common Sense* 19: 42–64.

Gingras, Y. (1997) 'The new dialectics of nature', *Social Studies of Science* 27:

Gooding, D. (1992) 'Putting the agency back into experiment', in A. Pickering (ed.) *Science as Practice and Culture*, Chicago: University of Chicago Press.

Grossberg, L. (1989) 'Putting the pop back into post-modernism', in A. Ross (ed.) *Universal Abandon? The Politics of Postmodernism*, Edinburgh: Edinburgh University Press.

Harbers, H. and Koenig, S. (1996) 'The political egg in the chicken basket', *EASST Review* 15(1): 9–15.

Hardt, M. and Negri, A. (1994) *Labor of Dionysus: A Critique of the State-Form*, Minneapolis: University of Minnesota Press.

Ingold, T. (1994) 'The animal in the study of humanity', in T. Ingold (ed.) *What is an Animal?*, London: Routledge.

Kellert, S. and Wilson, E. O. (eds) (1993) *The Biophilia Hypothesis*, Washington, DC: Island Press.

Latour, B. (1991) 'The impact of science studies on political philosophy', *Science, Technology and Human Values* 16(1): 3–19.

Latour, B. (1993) *We Have Never Been Modern*, trans. C. Porter, Hemel Hempstead: Harvester Wheatsheaf.

Law, J. (1994) *Organizing Modernity*, Oxford: Blackwell.

Leff, E. (1995) *Green Production: Toward an Environmental Rationality*, trans. M. Villaneuva, New York: Guilford.

Lévi-Strauss, C. (1969) *Totemism*, trans. R. Needham, Harmondsworth: Pelican.

Mort, M. (1994) 'What about the workers?', *Social Studies of Science* 24: 596–606.

Negri, A. (1989) *The Politics of Subversion*, Cambridge: Polity.

Negri, A. (1992) 'Interpretation of the class situation today: methodological aspects', in W. Bonefeld, R. Gunn and K. Psychopedis (eds) *Open Marxism vol. 2: Theory and Practice*, London: Pluto.

Panzieri, R. (1980) 'The capitalist use of machinery: Marx versus the objectivists', in P. Slater (ed.) *Outlines of a Critique of Technology*, London: Ink Links.

Philo, C. (1995) 'Animals, geography and the city: notes on inclusions and exclusions', *Environment and Planning D: Society and Space* 13: 655–681.

Pickering, A. (1993) 'The mangle of practice: agency and emergence in the sociology of science', *American Journal of Sociology* 99(3): 559–589.

Pickering, A. (1995) *The Mangle of Practice*, Chicago: University of Chicago Press.

Semiotext(e) (1980) 'Italy: Autonomia, Post Political Politics', *Semiotext(e)* 3(1).

Shapin, S. and Schaffer, S. (1985) *Leviathan and the Air-pump: Hobbes, Boyle and the Experimental Life*, Princeton, NJ: Princeton University Press.

Shepard, P. (1992) 'A post-historic primitivism', in M. Oelschlaeger (ed.) *The Wilderness Condition*, San Francisco: Sierra Club.

Strathern, M. (1992) *After Nature: English Kinship in the Late Twentieth Century*, Cambridge: Cambridge University Press.

Tapper, R. (1994) 'Animality, humanity, morality, society', in T. Ingold (ed.) *What is an Animal?*, London: Routledge.

Taussig, M. (1986) *Shamanism, Colonialism, and the Wild Man: A Study in Terror and Healing*, Chicago: University of Chicago Press.

Tuan, Yi-Fu (1984) *Dominance and Affection: The Making of Pets*, New Haven, CT: Yale University Press.

Wilbert, C. (1998) 'The love of trees: concepts of place, origins, and roots in economies of society and nature from Linnaeus to ecological restoration', unpublished PhD dissertation, Anglia Polytechnic University, Chelmsford.

Wilson, E. O. (1984) *Biophilia*, New Haven, CT: Yale University Press.

Witheford, N. (1994) 'Autonomist Marxism and the information society', *Capital and Class* 52: 85–125.

Witheford, N. (1995) 'Cycles and circuits of struggle in high-technology capitalism', *Common Sense* 18: 34–80.

Wolch, J., West, K. and Gaines, T. E. (1995) 'Transspecies urban theory', *Environment and Planning D: Society and Space* 13: 735–760.

Zerzan, J. (1995) 'Feral', *Do or Die: Voices from Earth First!* 5: 82.

12

FALLING DOWN
Resistance as diagnostic

Tim Cresswell

In the film *Falling Down* (1992) we follow the progression of the hero/antihero known as D-Fens (played by Michael Douglas) as he abandons his car on a crowded Los Angeles freeway and makes his way, on foot, across the city-scape in an effort to reach somewhere he calls home. As he progresses across the city he inevitably transgresses the territories of a number of Los Angeles' social groups, including a convenience store owned by a Korean-American man, a graffiti-marked gang territory, an exclusive golf course and a barbed-wire-surrounded house with swimming pool. In each case he confronts the territory's owners becoming pro- gressively more violent in his desire to get home. In addition to his transgressions of territory, he also transgresses the other elements of Harvey's (1989) explanatory triad – time and money. In the convenience store he complains about being charged more than 50 cents for a can of Coke and proceeds to smash up the store while demanding that prices be returned to 1965 levels. In a fast food joint he arrives two minutes late for breakfast and holds the customers hostage in his quest for a breakfast at 11.02. In short the D-Fens character questions urban construc- tions of space, time and money in turn as he wanders across the city. Predictably such questioning brings down the wrath of discipline on him: he ends up being shot off the end of a pier in Venice Beach.

This film was almost universally condemned by liberal reviewers. It was seen as a piece of white, middle-class, male angst which glorified the violent actions of a white man as he threatens and kills women and people of colour. In contrast it provoked standing ovations by filmgoers throughout the conservative South who presumably saw it in much the same way. Indeed the film was marketed as the story of an 'ordinary man [read white and middle class] at war with the every- day world'. The director Joel Schumacher defended the violence in the film by stating that:

Movies reflect society, and there have been several movies in the US about anger in the street but they have all been by African-Americans. Well, they're not the only angry people in the United States.

(Davies 1995b: 145)

Here Schumacher lays claim to the ambiguous 'privilege of resistance' on behalf of white American men. The film is open to multiple readings emphasising various and often simultaneous forms of domination and resistance. Its obviously reactionary message forces us to ask how we might differentiate between forms of activity that might be described with the blanket term 'resistance'. In one reading Douglas's character is an out-of-work guy, made 'not economically viable' by the military industrial complex of the United States, resisting the dominant arrangements of space, time and money. The act of walking across Los Angeles can easily be romanticised as the act of a postmodern flâneur disrupting the functional structures of life in the city. As Harvey (1989) has put it:

Prices, the movements of the clock, rights to clearly marked spaces, form the frameworks within which we operate and to whose signals and significations we perforce respond as powers external to our individual consciousness and will . . . To challenge these norms and the concrete abstractions in which they are grounded is to challenge the central pinions of our social life.

(Harvey 1989: 188)

D-Fens challenged each of these pinions in turn and thus guaranteed his fate. His walk across the city can be read as 'resistance' to the very bedrock of modern capitalist America. But to make this reading would be to miss another, equally obvious, interpretation of his actions. The very same activity of walking across the city becomes self-evidently extremely oppressive to his ex-wife and the large number of people he kills and maims along the way.

The place he refers to as 'home' is, in fact, no longer his home but that of his estranged wife and child. As D-Fens gets closer and closer in his walk across the city, she becomes more and more confined to the house, scared for her life and the life of her child. One feminist reading (Davies 1995a, 1995b) points out the relative freedom implicit in the mobility of D-Fens in his masculine peripatetic odyssey. While one scene shows D-Fens strolling across a golf course and accosting golfers with the claim that the space would be better used as a children's play area, another shows his ex-wife and child holed up in the place called 'home' begging police to protect them from D-Fens' imminent, and potentially violent, arrival. While D-Fens is framed by public and semi-public spaces (the street, the freeway, shops, restaurants, a golf course) his ex-wife is progressively enveloped by the claustrophobia of her home with its locked doors and occasional police guard.

I have used this film in an urban geography course and students found themselves supporting both interpretations. Most sympathised with Douglas's evasion of the police, his refusal to sit in a traffic jam and his comic anger at being refused breakfast when he is two minutes too late. Students laughed at and applauded his actions. Yet when asked to rationalise this support, few would even try as D-Fens is far too objectionable a character to merit their sympathy. My students and myself were caught in a dilemma. For my part I had become used to looking for resistance in all cultural productions. When it was there I had become used to celebrating its existence and pointing it out as yet more evidence of the triumph of the human spirit over subtle and overwhelming power. I had tried over several years to pass this perspective on to students, encouraging them to look for resistance every-where. I realise that not everyone is guilty of the same simplifications, but I think it is fair to say that human geography, and cultural studies even more so, have been guilty of romanticising resistance (Ferguson and Golding 1997).

In our class we had just worked our way through several dense texts on the social construction of space in the capitalist city and here was a man clearly holding two fingers up to the main features of that construction. Even if that man was Michael Douglas, we looked at his character for signs of the heroic struggle against modern power. Sure enough we found them. Yet this was not someone to romanticise. The characteristic move of progressive social and cultural theorists has been to identify with and make moral investments in subaltern or subordinated groups. It is their actions that have been described as resistance – as contestations of dominant spatialised norms. Yet here was a rather nasty white middle-class (admittedly unemployed) man clearly contesting a series of disciplinary forces in ways that drew applause and laughter. Both my students and I had no language to describe these acts as resistance, because resistance seemed irrevocably heroic and infused with positive moral value. What I want to suggest here is that there is another language to use in relation to resistance – one which does not make quite such a moral investment in those doing the resisting. In short, what the film pointed to was the power of resistance not as a potent symbol of subaltern freedom, but as an indicator and diagnostic of power. It is this difference that I discuss in this chapter.

Social/cultural geography and the romance of resistance

In 1987, in a review of social geography, Jackson wrote:

> In virtually every arena of social life, the spatial strategies by which
> subordinate groups seek to contest their domination remain to be

investigated. The prospect of a geography of resistance is one that surely merits the most urgent consideration.

(Jackson 1987: 263)

At that juncture in the development of geography as a discipline this made a lot of sense. Jackson was drawing on developing work in cultural studies and funda- mental changes in theories of ideology and power to suggest that the object of resistance had been underemphasised almost to the point of invisibility. Things have certainly changed since then. While resistance may have been invisible in 1987, it was central to cultural and social geography in 1999. Few could get away with a paper on power which did not, in some way or other, deal with resistances to that power. Indeed, resistance may well be the central theme of contemporary social and cultural geography (Pile and Keith 1997). The new cultural geography has always been concerned to incorporate agency into debates about meaningful social life. Resistance is, of course, a particularly obvious form of agency which forms a central part of contemporary critiques of older superorganic forms of culture. To an outside observer such as Rowntree writing in 1988, the very heart of the new cultural geography was the foregrounding of 'the various strategies of resistance employed by subordinate groups to contest the hegemony of those in power' (Rowntree 1988: 580) .

What I would like to suggest is that resistance has taken centre stage in geographical, social and cultural analysis to the extent that there is a danger that no area of social life will not be described as resistance. We have had walking as resistance (Tester 1994), watching TV as resistance (Fiske 1987) and driving around America as resistance (Cresswell 1993). It is not unlikely that soon we shall have policing as resistance, conformity as resistance and perhaps domination as resistance. Resistance is in danger of becoming a meaningless and theoretically unhelpful term. Something that is applicable to everything is not a particularly useful tool in interrogating social and cultural life. The romance of resistance leads to a curious kind of inertia in which an apparently unitary, dominating power is seen to be challenged everywhere and thus by a curious magic remains un- challenged. Everybody is so busy resisting always, and already, that little more needs to be done. One problem is that an act such as an armed insurrection or a general strike is equated with the act of farting in public or telling jokes about the boss. The word resistance can apply to all of these and yet they are clearly more different than they are alike. When the geography of resistance that Jackson called for comes to be written, it will surely have to differentiate between resistance at different levels, visible and invisible, intentional and unintentional, active and passive.

It is no surprise that another one of the key concepts to re-emerge in recent cultural studies (and geography) is consumption (Shields 1992) . It is surely no coincidence that work referred to as the 'new economic geography' is primarily concerned with consumption rather than production. While production has been thoroughly wrapped up with various modes of domination such as Marxist theories of exploitation – consumption appears to be about choices and can quickly become conflated with resistance. As Gitlin (1997: 32) observed: 'Where once Marxists looked to factory organisation as the prefiguration of "a new society in the shell of the old", today they tend to look to sovereign culture consumers'. Productions seem regimented and dull while consumption is exciting – glamorous even. The consumption equals resistance equation is most evident in the work of cultural theorists such as Hebdige (1988) and Chambers (1994) who revel in the innovative use of motor scooters, listening to a Walkman while walking through the city or the inappropriate wearing of black bin-bags.

It seems that few who are writing today want to think that there are people, especially in the western world, who are completely without recourse to some form of resistance, however minor. Hence we are prone to seeing resistance in even the most mundane forms of human activity. In some ways it seems that the romance of resistance is tied up with the older sense of agency. Any self-respecting social scientist in a truly post-structuralist world wants to see and emphasise the limits to structuring. Recently any act that is not clearly the result of dominant structures has been described as resistance. Simply choosing to do something is resistance. Yet it could equally be the case that the vast majority of choices we make are far from resistant but rather serve to reproduce existing forms of domination.

I have to be careful here as it is probably the case that all of the activities I have mentioned can, in some way or another, be described as resistance. I have a lot of sympathy with the directions and impulses that have raised the issue. But I believe there is a distinct danger of looking hard for signs of human freedom and finding it in the shopping mall or the car boot sale – in the innovative use of motor scooters or the inappropriate wearing of black bin-bags. People making choices, con-suming, resisting. These will be seen as evidence for everyday heroism and the analysis will stop there.

In many ways I take my lead here from the work of Abu-Lughod (1990), who in her work on and with Bedouin women confronted this proliferation of resistance within anthropology. She confesses to an original involvement in a romance of resistance, believing that by studying Bedouin cultures she would be involved in revealing a proud form of romantic nomadic freedom pointing towards the impossibility of all inclusive power. Yet, on working in the field, the Bedouin women refused to live up to her rosy expectations. She also realised that her romantic view of the women led to a curious foreclosing of certain questions about the workings of power. By looking for resistance she was looking for the absence or incompleteness of power. It might be more fruitful, she decided, to use resistance instead to 'tell us about forms of power and how people are caught up in them' (Abu-Lughod 1990: 42).

Power and invisibility: the fish don't talk about the water

With this in mind let me turn to some thoughts concerning power. The big strawperson in any debate on power and domination is the orthodox or structuralist Marxist. It was theories of ideology arising from the work of the Frankfurt School – particularly Adorno, Horkheimer and Althusser – that allegedly posited the existence of a cultural apparatus that served to blind people to their 'real interests'. Almost all cultural productions from high religion to jazz music were at some time theorised as being tools to produce subservience and reinforce domination. Popular culture came out particularly badly in these theorisations, being seen as 'mere consumption'. The power in popular and mass culture was seen as a kind of power which blinded people and more or less forced them into subservience. It is notable that the word 'consumption' was key. To the Frankfurt School consumption was the opposite of agency and never came anywhere near 'resistance'. The transformation from a radical cultural studies which emphasised the power of culture over people to one which underlined the creative and resistant use of cultural products revolved around a reconfiguration of the both popular culture and consumption. As McGuigan (1997: 138) has argued, 'consumption was no longer to be seen as the "passive" moment in cultural circulation but, instead, "active" and nodal, involving popular appropriation of commodities and differential interpretation of texts'. Popular culture as consumption was thus transformed from a moment of acquiescence to a moment of resistant power and agency. Alongside this re-evaluation of the popular was a necessary re-evaluation of power. While the Frankfurt School subscribed to a fairly orthodox notion of power as something imposed from above, newer versions of cultural theory, including those in geography, began to see power as something of a two way process.

Lukes (1974) defined power as that which makes individuals act against their own interests, where interests are those which people would want and prefer were they able to make that choice. Both Weberian and Marxist accounts of power would have little problem with this. The pivot of traditional notions of power is the idea that power, through force or persuasion, diverts people from pursuing their 'real interests'. Modern conceptualisations of power tend to be less conspiratorial and judgmental. The obvious point of reference here is Foucault. In a post-Foucault world, power is no longer understood as a repressive force possessed by people and institutions at the top of a clearly demarcated social hierarchy (i.e. 'the state', 'the culture industry) to be overthrown by mass organised resistance. Rather, power is seen to be something which is exercised by everyone, that is potentially productive and at the heart of all social relations.

The observations of Foucault (1980, 1984) together with those of Bourdieu (1990, 1992) and others have suggested that power works through the subject's own sense of agency. That is to say that power is reproduced through the multitude of individual acts that conform to a pre-given mode of power. Bourdieu writes of people's common sense, or taken-for-granted sense of limits ,which unconsciously serves to reproduce the power relations that produce common sense. Foucault is slightly more tricky, being more directly concerned with power. Krips (1990) has referred to Foucault's concept of power as 'subjection'. The crucial point in Foucault's concept of modern power is that the process of subjection imposes the framework in terms of which a subject's interests are constituted. There are no such things as real interests. This marks his work as different from that of Lukes as well as Bourdieu, who implicitly invokes 'real interests' or 'objective conditions' as the conditions of judgement against which a subject's actions are measured.

Foucault and Bourdieu also invoke intentionality in different ways. In Bourdieu's work, power is reproduced by people's actions or 'practical philosophy' as Gramsci called it. The actions are intentional but the intentions are based on objective or real conditions which are historically constituted but misrecognised as natural. Again Foucault's conception of power is different. Power in Foucault rests on a distinction between micro and macro scales (Heller 1996; Krips 1990). At the microscale people's actions are intensely premeditated and intentional. These intentional microscale actions add up to a global strategy which is incomprehensible to the local actors. The difference between Bourdieu and Foucault can be seen by comparing two statements on intentionality. To quote Foucault, 'People know what they do; they frequently know why they do what they do; but what they don't know is what what they do does' (Foucault 1984: 95). So although each local act of power is intentional, the global product of these acts in unintentional and unorchestrated. Power is thus locally transparent and globally opaque. Bourdieu, meanwhile, has stated that 'it is because subjects do not, strictly speaking, know what they are doing that what they do has more meaning than they know' (Bourdieu 1977: 72). The key difference, of course, lies in the intentionality of the actor. In Foucault's world actors are completely aware of the reasons for their acts of power but unaware of the global consequences. In Bourdieu the actors are in some way mistaken about their acts and it is in this mistake that the unknowability of the consequences lies. So Foucault's conception of power is different from other views of power in that it does not rely on the notion that people are being forced directly or coercively to act against their interests. Also power in the form of a global strategy is not seen as an intentional form of oppression but as an unintended consequence of locally intentional actions.

There are also crucial similarities between these views of modern power. The most important of these is the observation that power is rarely direct, brutal repression but usually indirect, invisible and constituted through practice. People are not simply the inert targets of power but active agents in its articulation. Modern forms of power which do not act directly on the body are often impossible to point at and name, as the only names that we have are those given to us by the pre-existing structures. Power, in other words, is a taken-for-granted part of the atmosphere, as mundane and crucial as the air we breathe. This observation leads to particular problems when studying power and its effects. Risseeuw's (1988) excellent book on power and resistance among women in Sri Lanka describes the problem of investigating everyday forms of power with reference to the South Asian saying 'the fish don't talk about the water' – which points towards the problems that arise in trying to delineate and articulate these invisible forms of power. If power in general is often invisible, spatialised power in its everyday varieties is even more so. Clearly there is a paradox here. Socially constructed spaces are all around us and often take highly visible and material forms. In that sense they are not invisible, but these spaces are the contexts within which we lead out lives. Quite literally they are the water referred to in the South Asian saying. The very obviousness of these spaces give them the illusion of nature, of just being. Spatialised power in the forms of norms, prohibitions and expectancies is thus especially powerful.

As an aside I would like to point out that our search for subtle and everyday forms of power and resistance often has nothing or little to say about more simple forms of power that make up some people's everyday existence. The door that is broken in the early hours, the boots and fists meeting flesh, the condition of immigrants treated as slaves by their employers, or the not so subtle intrusions of the state into the private lives of consenting adults. All of these remain, for the most part, unchallenged by those seeking to fine tune Foucault or by the ironic postmodern shoppers deconstructing the shopping mall. Nonetheless, such moments are beyond the scope of this chapter and I would like now to return to the problem of rethinking everyday resistance.

Reconceptualising resistance

What follows from a Foucauldian conception of power are new ideas concerning resistance. One pessimistic derivation is that power is so all encompassing that resistance is either futile or non-existent. Indeed any attempt to resist power can only hope to institute new kinds of power with different, but equally oppressive characteristics. The second, optimistic, derivation is that resistance, like power, is everywhere. Indeed Foucault made this point himself when he wrote: 'where there is power there is resistance' (Foucault 1984: 95). While Foucault did not spend too much time on the subject of resistance, we can see what he might have said by holding a mirror up to his theory of power. Such a view of power necessarily requires a rethink of normal views on resistance.

The editors of this collection have insisted that domination/resistance form a dyad and that the two terms cannot be disentangled. This is somewhat in contrast to orthodox thinking about resistance which pits it against 'power'. If domination and resistance are indeed entangled, they are so under the concept of power. The orthodox assumption seems to be that resistance is against power and that effective resistance will eventually overturn power. A Foucauldian conception of power holds that power is a transformative capacity – the ability to transform the actions of others in order to achieve certain strategic goals. This transformative capacity cannot be destroyed as it exists at the heart of all possible social relations. Power is not something to be overthrown, but rather to be used and transformed. How then can we think about resistance in a way that is not opposed to power? One possibility is to think of resistance in terms of intentionality – as a motivation for the deployment of power. Cooper (1994) has made just this argument. She argues that subordinate groups, like anyone else, exercise power so it becomes futile to argue that their actions are against power. By what means could such groups resist, she asks, if not through and by power? Resistance, then, is not opposed to power but is a subset of it. Resistance becomes the deployment of power with the motivation of alleviating or transforming the conditions under which one lives. Resistance thus reconceptualised would not be romanticised as an indicator of power's absence, but rather seen as evidence for power's existence and an intervention that serves to delineate the mode of power in question.

Consider this observation with reference to 'carnival'. This is one of the areas to which cultural geographers and others have looked to find signs of heroic resistance. A heated debate has ensued as to whether carnival is a pointless letting off of steam or a truly revolutionary activity indicating the incompleteness of power. If we think about resistance as a diagnostic of power, the grounds of the debate are transformed. One of the key acts of carnival is laughter. Eco (1980) has provided some interesting, and I think relevant, insights into the causes and role of the comic. Laughter, he argues, comes from the violation of rules, the transgression of expectancies. We laugh, he suggests, when we allow ourselves the vicarious pleasure of a transgression of a rule that we have secretly wanted to violate, without the risk of sanctions. The key to laughter, in Eco's view, is in our awareness of the violated rule – an awareness that is rooted in common sense rather than discursive justification. After all, we all know that if a joke has to be explained, it is not funny. Typically comic laughter arises from the transgression of common-sense rules of the everyday – the pragmatic rules of symbolic interaction that constitute our environment. A custard pie in the face is funny because pies are eaten and not thrown. It is precisely because rules are so thoroughly absorbed that their violation becomes comic. What arises from this is the observation that comic laughter would be impossible in a world of absolute permissiveness because, as Eco (1980: 275) puts it, 'nobody would remember what was being called into question'.

Eco's analysis of comic laughter can be extended to the metaphysical world of the carnivalesque. While Bakhtin (1984) appears to believe that laughter is the liberatory tool of the people, Eco would suggest that it gives licence only to those who have so thoroughly absorbed the rule that it has become part of the atmosphere. Permanent carnival would not make sense: it makes sense only as a short space of time surrounded by a year of normality. So while it may not make sense to look at comic laughter as resistance outside of and against power, it would make perfect sense to think of it as an indicator of deeply held norms and expectations which we could then seek to delineate and transform.

The relationship between resistance and power is similar. Resistance presupposes the existence of power. By focusing on resistance, we allow dominant forms of power to achieve a certain kind of closure in that the groups or individuals who are resisting are only ever responding to the conditions which pre-exist the (powerful) acts of resistance.

Resistance (in space) as a diagnostic of power (in space)

What I want to suggest to end this chapter is that resistance can be used quite differently from the ways it has been used in social and cultural theory recently, not as an affirmation of agency and denial of overarching oppression, but as what Abu-Lughod (1990) calls a diagnostic of power. Foucault's famous dictum, that everywhere there is power there is also resistance, can of course be inverted to say that everywhere there is resistance there is also power. Studies of resistance clearly have implications for ideas concerning power in general. Yet by treating resistance as a heroic refusal – a gesture of the creativity of the human spirit – we overlook the differences between forms of resistance and divert our attention from the machinations of domination. By recognising processes of resistance in particular contexts, we can point to the existence of power and develop strategies for its study and perhaps for its transformation.

The geographies of power, domination and resistance are important parts of this process. As is indicated in Chapter 1, space constitutes the active medium within which power – both dominant and resistant – happens. In orthodox accounts of power versus resistance there is often an implied dualistic geography. Often, for instance, mobility is seen as a resistant act while boundedness and clear spatial boundaries are seen as inherently dominating. This is the clear suggestion in de Certeau's (1984) distinction between strategies of the powerful (based on a powerful space of the proper) and tactics of the weak (based on myriad movements through these spaces). In *Falling Down* it is also tempting to spatialise power and resistance in dualistic ways. The ability of D-Fens to walk and shoot his way through public space can be read as indicative of a particularly threatening form of masculine power in contrast to the private claustrophobia of his ex-wife's home. Alternatively his mobility can be read as resistance as he transgresses the spaces of the proper represented by the golf course, the guarded home and the fast-food joint. Geography, however, is rarely so simple. Once we see resistance as a deployment and diagnostic of power rather than its opposite, the spatiality of domination/resistance becomes more entangled (an appropriate metaphor if ever there was one). The mobility of D-Fens that the film is centred on, for instance, cannot simply be read as resistance or domination. The relation of his geographical act to a myriad of different forms of power is entirely contextual and dependent on the particular social relations to which we are referring. The power of his mobility makes visible the power of the territories that he transgresses, just as it makes clear the relative powerlessness of his ex-wife's private space. The movements and spaces are not essentially one thing nor the other, but always and already both.

It is important that we do not stop thinking about everyday forms of resistance, but equally important that we do not romanticise and essentialise them. Rather than telling us how people are free or partially free from forces of oppression inscribed in space, resistance can be used strategically to reveal how people are caught up in a multitude of often invisible modes of power. If it is the case that the predominant forms of modern power are invisible, then the observation that 'the fish don't talk about the water' is apt. But dominant power always has its moments of crisis, particularly in the face of acts which do not conform to expectations. The French historian Canguilhem (1989), in a brilliant discussion of the normal and the pathological in medical discourse, has suggested that while the pathological may be logically secondary to the normal, it is existentially primary. That is to say that people's experience of the normal may only surface to the level of consciousness in relation to experiences of the pathological. The same might be said of power and resistance. If the dictum, wherever there is resistance, there is power, holds – common sense would suggest that it should – then acts of resistance could be seen as indicators of the existence of various forms of power which are rarely present in consciousness.

Returning to *Falling Down*, the various transgressions of D-Fens as he traverses Los Angeles can be read in a way that does not fall into the trap of romanticising the anger of white middle-class males. Despite the obviously reactionary intentions of the director and the celebration of the film among conservative Americans, the wanderings of D-Fens can be seen as a form of resistance that articulates and makes visible the everyday forms of power that surround the residents of Los Angeles. If resistance is being read as a diagnostic of power, the identity of those who are resisting, be they real or imaginary, is not especially relevant to our projects. Such an analysis of power and resistance does not need to make any moral claims about the identity of the resister and the oppressor. A model of resistance as a diagnostic of power makes no investment whatsoever in the subject position of the agents – it simply uses their acts as evidence for various modes of power, including the power of resistance itself.

It should be no surprise that the film really helped as a pedagogical tool. Students strained to understand Harvey's paper on time, space and money. It is not that such concepts are in any way new and irrelevant to their lives, it is rather the opposite problem: these contexts are so thoroughly imbued in modern life that they are the realm of common sense. It is as if it could never be different. Douglas's transgressions foregrounded these elements of common sense in a fashion that made Harvey's triad seem both more real and more arbitrary. They enjoyed seeing the sense of limits questioned and this enjoyment, marked by nervous laughter, formed a useful mode of resistance that might, in turn, point towards the far trickier possibilities of truly transformative practice.

References

Abu-Lughod, L. (1990) 'The romance of resistance: tracing transformations of power through Bedouin women', *American Ethnologist* 17(1): 41–55.

Bakhtin, M. (1984) *Rabelais and his World*, Bloomington: Indiana University Press.

Bourdieu, P. (1977) *Outline of a Theory of Practice*, Cambridge: Cambridge University Press.

Bourdieu, P. (1990) *The Logic of Practice*, Stanford, CA: Stanford University Press.

Bourdieu, P. (1992) 'The practice of reflexive sociology', in P. Bourdieu and L. Wacquant (eds) *An Invitation to Reflexive Sociology*, Chicago: University of Chicago Press.

Canguilhem, G. (1989) *The Normal and the Pathological*, Cambridge, MA: Zone.

Chambers, I. (1994) *Migrancy, Culture, Identity*, London: Routledge.

Cooper, D. (1994) 'Productive, relational and everywhere? Conceptualising power and resistance within Foucauldian feminism', *Sociology* 28(2): 435–454.

Cresswell, T. (1993) 'Mobility as resistance: a geographical reading of Kerouac's "On the Road"', *Transactions of the Institute of British Geographers* 18(2): 249–262.

Davies, J. (1995a) 'Gender, ethnicity and cultural crisis in *Falling Down* and *Groundhog Day*', *Screen* 36(3): 214–232.

Davies, J. (1995b) '"I'm the bad guy?" *Falling Down* and white masculinity in 1990s Hollywood', *Journal of Gender Studies* 4(2): 145–152.

de Certeau, M. (1984) *The Practice of Everyday Life*, Berkeley: University of California Press.

Eco, U. (1980) 'The comic and the rule', in Eco, *Travels in Hyperreality*, New York: Harcourt, Brace, Jovanovich.

Ferguson, M. and Golding, P. (eds) (1997) *Cultural Studies in Question*, London: Sage.

Fiske, J. (1987) *Television Culture*, London: Methuen.

Foucault, M. (1980) *Power/Knowledge*, New York: Pantheon.

Foucault, M. (1984) *History of Sexuality*, vol. 1, Harmondsworth: Penguin.

Gitlin, T. (1997) 'The anti-political populism of cultural studies', in M. Ferguson and P. Golding (eds) *Cultural Studies in Question*, London: Sage.

Gramsci, A. (1971) *The Prison Notebooks*, London: Lawrence & Wishart.

Harvey, D. (1989) *The Urban Experience*, Baltimore, MD: Johns Hopkins University Press.

Hebdige, D. (1988) *Subculture: The Meaning of Style*, London: Routledge.

Heller, K. J. (1996) 'Power, subjectification and resistance in Foucault', *SubStance* 25(1): 78–110.

Jackson, P. (1987) 'Social geography: social struggles and spatial strategies', *Progress in Human Geography* 11(2): 266–269.

Krips, H. (1990) 'Power and resistance', *Philosophy of the Social Sciences* 20(2): 170–182.

Lukes, S. (1974) *Power: A Radical View*, London: Macmillan.

McGuigan, J. (1997) 'Cultural populism revisited', in M. Ferguson and P. Golding (eds) *Cultural Studies in Question*, London: Sage.

Pile, S. and Keith, M. (eds) (1997) *Geographies of Resistance*, London: Routledge.

Risseeuw, C. (1988) *The Fish Don't Talk About the Water: Gender Transformation, Power and Resistance among Women in Sri Lanka*, Leiden: E. J. Brill.

Rowntree, L. (1988) 'Orthodoxy and new directions: cultural/humanistic geography', *Progress in Human Geography* 12(4): 575–586.

Shields, R. (ed.) (1992) *Lifestyle Shopping: The Subject of Consumption*, London: Routledge.

Tester, K. (ed.) (1994) *The Flâneur*, London: Routledge.

13

ENTANGLEMENTS OF POWER

Shadows?

Nigel Thrift

Foucault has a lot to answer for. Though he embraced a positive notion of power, the fact is that his world view is not very positive. In Foucault country, it always seems to be raining. Foucault's work is illustrative of a more general tendency among intellectuals writing on power: to look for the bad news. For example, a writer like Bourdieu often seems to tend towards a view of the speech act as simply a rite of institution. Of course, both authors have been careful to distance themselves from overly negative interpretations of their work. Foucault is clear that power can be turned against itself 'to produce alternative modalities of power, to establish a kind of political contestation that is not a pure opposition, a transcendence of contemporary relations of power, but a difficult labour of forging a future from resources relatively impure' (Butler 1993: 241). Similarly, Bourdieu is clear that the habitus cannot be reduced to the practice of self-consciously following a rule and that every 'misfire or misapplication highlights the social conditions by which a performative operates, and gives us a way of activating these conditions' (Butler 1997: 151). Still, the overwhelming impression is, too often, of a world that has given up the ghost. And ghosts, I shall argue, are important.

It does not have to be like this. There are alternatives. Over the past few years, like a number of other authors, I have been sketching out a rather more optimistic view of the world, one in which power is undoubtedly present and hurts but is neither everywhere nor all-pervasive. In particular, this has meant that I have sought for those elements of life which continually and chronically undermine all forms of power. In this way, I hope to provide a corrective to accounts like those of Foucault and Bourdieu which too often seem to transmit a political quietism with which we know their authors would have had no truck.

In this brief afterword, I want to mention four of these elements. First, and most straightforwardly, all powerful systems are being constantly subverted. It is, one might argue, a condition of their existence. A particularly good example of this principle is provided by Jowitt's (1992) analysis of Leninist regimes. For Jowitt it is what he calls dissimulation, rather than legitimisation, which is the chief way in which the public and private are tied together in these societies. For Jowitt (1992: 80), dissimulation is a central feature of a 'ghetto political culture' in which people retreat from an official sphere which can mean only trouble for them.[1] Instead they attempt to obtain as much 'free time and easy life' as possible outside of politics, minimising the 'regime's interference in one's private life'. Thus:

> Jowitt managed to pinpoint the central social practice familiar to every Soviet citizen in the 1970s and 1980s: saying something while believing the opposite to be true; participating in social ritual, just for the sake of participation; even bringing up children telling them that a schism exists between what is and what ought to be, a schism not to be mentioned in public statements that should describe the world as if the ideal were real.
>
> (Kharkhordin 1995: 209)

I would argue that all powerful systems are constantly being undermined by the undertow of everyday practice, not least because they themselves consist of everyday practices. Though they can call on various forms of technology to make them more durable, they cannot transcend other ontological skills.

Second, power is constituted across many registers of experience. It is a commonplace of current geographical enquiry that we live in an ocularcentric, scopophilic world which privileges vision, but acknowledging this insight requires that we be aware of the existence of many other practices that constantly correct this vision. For example, in performance studies an interpretative crisis has been brought about by the realisation that performance cannot be read as text. Take dance:

> Too much is lost in the gap between performance and plot description. There is an important distinction between critical interpretations of plots and analyses of performances. Very often, exactly because choreographic texts are elusive, dance historians rely on literary plots to make their critical interpretations. Privileging plot descriptions over performance descriptions, however, overlooks the most crucial aspect of dance. It is, after all, a *live*, interpretative art. It is not fixed on the page, nor can all its meanings be accurately conveyed through verbal means. And bodies can impart different meanings – sometimes diametrically opposed meanings – than words suggest.
>
> (Banes 1998: 8)

This is an important issue in considering power: the Leninist empire of the text is subject to dissimulation by the performative body. Thus, a plot may describe a female character as weak and passive but the dancer may subvert the role with affective verve: 'even dances with misogynist narratives or patriarchal themes tend to depict women as active and vital' (Banes 1998: 9).[2]

Again, take opera. Abbate (1991, 1993) has argued against Clément's (1988) contention that operas are misogynist because the plots so often kill off women. Instead, she argues that the presence of women singing on stage creates performance meanings different from those of the plots. 'Clément neglected their triumph: the sound of their singing voices. This sound is unconquerable, it cannot be concealed by orchestras, by male singers or – in the end – by murderous plots' (Abbate 1991: ix). In addition, Abbate argues that, by focusing on bodily performance rather than on instrumental music in her analysis of opera, she opens up a space for two underrated aspects of music – its physical force, and the contributions of the performer and the audience as well as the composer (Cook 1990). Specifically contesting Mulvey's theory of the scopophilic ideology of the male gaze, by changing the sensory register, Abbate (1993) is able to sense different powers:

> Listening to the female singing voice is a . . . complicated phenomenon. Visually, the character singing is the passive object of our gaze. But, aurally, she is resonant; her musical speech drowns out everything in range, and we sit as passive objects, bolstered by that voice. As a voice she slips into the 'male/active/subject' position in other ways as well, since a singer, more than any other musical performer, enters into that Jacobin uprising inherent in the phenomenology of live performance and stands before us having wrested the composing voice away from the librettist and composer who wrote the score.
>
> (Abbate 1993: 254)

Third, there is a strong tendency in the literature on power to ignore creativity. The reason for this is, I believe, a wholesale underestimation of the power of the imagination (Abercrombie and Longhurst 1998). But there are writers who have been trying to incorporate the imagination into accounts of power.[3] Thus Castoriadis (1994, 1997), starting from his root philosophical principle (and aim) of autonomy – to give oneself one's own laws – went on to privilege self-creation which, in turn, meant confronting the mystery of creation. For Castoriadis, creation was more than a combination of pre-existing elements. Instead, 'the various strata of being create themselves' (Curtis 1997: xxi),[4] producing an upsurge of radical novelty, an unpredictable discontinuity:

> And, at the source of all creation there is the imaginary, the inventor of a world of forms and meanings, which in the individual is radical

271

imagination, and in society the instituting social imaginary. Imagination and creation are everywhere linked, including at the very source of thought.

In contrast to the dominant conceptions, for which the imaginary is merely an illusion of superstructure, Castoriadis reintroduces it at the root of our human reality, just as, in contrast with conceptions unable to grasp the notion of the subject, Castoriadis rediscovers the constituents of the subject (the 'for-itself', the fact that everyone creates his or her world and has the power of imagination). He stresses the radical importance of the emergence of the autonomous subject two thousand years ago in Athenian democracy. His thought . . . takes an epistemological form: nothing which is living can be exhaustively and systematically reduced to our classical logic, which he calls ensemblist – identitarian. Castoriadis sees in what he calls 'magma', a substance without form which is creative of forms, the genetic substrate of all creation.

(Morin 1998: 4)

Joas (1996) draws especially on the American pragmatist model of situated creativity – namely, that new variations of action are generated by the tension of problems contained in situations – to generate a new model of human action which, in its commitment to new wholes created out of parts, bears some similarity to Castoriadis's work:[5]

a third model of action should be added to the two predominant models of action, namely rational action and normatively-oriented action. What I have in mind is a model that emphasises the *creative* character of human action. Beyond that, I hope to show that this third model overarches both the others. I do not wish simply to draw attention to an additional type of action relatively neglected to date, but instead to assert that there is a creative dimension to all human action, a dimension which is only inadequately expressed in the models of rational and normatively-orientated action. Both these models ineluctably generate a residual category to which they then allocate the largest part of human action.

(Joas 1996: 4–5, original emphasis)

The fourth element relates to Joas's latter point.[6] I would argue that there are a whole series of classes of human life which, with the power of what Ford (1998) calls 'hindspite' and without an emphasis on creativity and imagination, it is all too easy to miss and which, in turn, makes it all too easy to reduce human societies to the play of power. One such class consists of those diverse practices which go under the heading of 'play'. The legacy of seminal writers like Bateson, Turner and

Schechner shows that play is 'a mode, not a distinctive behavioural category . . . play is viewed as an attitude or stance that can be adopted towards anything . . . [it] occurs at a logical level different from that it qualifies . . . play is functional because it teaches about contexts; it teaches about forces not being at the same level as the acts they contain' (Schwartzman 1978: 169). More specifically, play is a means of conjuring up a virtual reality. It is a kind of everyday dissimulation of the everyday. Thus:

> one may . . . speak of mundane reality and virtual reality, where writers need to write of the real and the unreal. The latter chiefly privileged the real over the unreal, making the unreal usually a mimicry of the former. Now that we recognise that human cultures are built out of imagination and fantasy, not just out of physical discoveries, the present duality of mundane and virtual reality is more appropriate. It concedes that the mundane and virtual are both real worlds but in different ways, without in general privileging one over the other.
>
> More centrally . . . the play actions 'play off' the mundane actions they in part model, mimic or mock. Every action . . . is an action about actions. It communicates by its own stylised character that it is a play reality, not an everyday reality, which is to say it is metacommunicative, as Bateson puts it. But underneath that, every action about actions is also a meta-action. Between the original action . . . and the ludic commentary we have a binary tension that, it can be argued, is the initial source of the dialectical engagement of play. Play is novelty, but it also is typically at the beginning of an incongruity. But as the play proceeds, one is impressed less with the initial incongruity and more with the enjoyable estab-lishment of the internal incongruities of this separate kingdom of play . . . Play proceeds, as it were, sometimes with increasing transformation, even with increasing nonsense, from play to playful. To discuss play in this way is to imply that at the centre of play's dynamism is a dialectical relationship between its enactments and their everyday references.
>
> (Sutton-Smith 1997: 145)

Intimately connected with classes of life like play are other related classes of life and, in particular, dreams, daydreams,[7] fantasies and other imaginative fabulations which are as much embodied practices as those we have normally regarded as such.[8] These practices of creative proliferation are now being studied by geographers, usually through psychoanalytic traditions (e.g. Pile 1997), although many other traditions of work are also available.[9]

My intention in this afterword has been to write against the grain of much current work on power which would have us believe that we live in a world of crushing systems that can stamp on all expectation. Yet, as I have tried very briefly to show, almost as a matter of routine people slide around these systems, both in the mundane and virtual worlds. Of course, there are some awful things afoot in the world. But, if you are sensitive to them, there are other things too (Haraway 1991). Currently we tend to figure these other things through the language of ghosts and apparitions and hauntings and shadows. But, perhaps, as we come to sense these phantasms more accurately, so they will help us in our search for the 'address of the present':

> To stretch toward and beyond a horizon requires a particular kind of perception where the transparent and the shadowy can haunt each other. As an ethnographic project, to write the history of the present requires grappling with the form ideological interpellation takes – 'we have already understood' – and with the difficulty of imagining beyond the limits of what is already understandable. To imagine beyond the limits of what is already understandable is our best hope of retaining what ideology critique traditionally offers while transforming its limitations into what, in older Marxist language, was called utopian possibility.
>
> (Gordon 1997: 195)

And space? Well, space is clearly the stuff of power. But, equally clearly, it is the stuff of creativity and imagination. Work by geographers has constantly documented this generative aspect of space (while too often shoehorning it back into negative theoretical formats as mere 'resistance'). What we need now are theoretical frameworks and empirical arenas that can acknowledge all the different 'species of spaces' (Perec 1997) that geographers have begun to document in their richness and possibility.

For example, I have been trying to achieve this goal through the frame of non-representational 'theory' and the example of dance (Thrift 1996, 1997, 1999). Non-representational theory is an attempt to develop a body of work that emphasises the development of sensitivities (*or disclosure*), rather than knowledge *per se*, toward all of the everyday practices that usually go unnoticed in the background of our lives, practices which take place in an intermediate and indeterminate in-between what we currently frame as behaviour or action. These are unreflective, lived, culturally specific, bodily reactions to events which cannot be explained by causal theories (accurate representations) or by hermeneutical means (interpretations):

> Indeed, to the extent that anything done by a living individual is done in spontaneous response to the others or otherness around him or her, we cannot think of them as wholly controlling the outcome of their actions.

Yet it has not been wholly brought about by any causes external to them either. Such spontaneously responsive reactions are in fact a complex mixture of many different kinds of influences; they have neither a fully orderly, nor a fully disorderly structure; a neither completely stable, nor a fully objective character . . .

It is as if each disclosive space in which we find ourselves had 'its own requirements', so to speak. Thus 'it' lives us as much as we 'live' it.

(Shotter 1998: 282)

In turn, dance acts as a means of focusing some of these concerns and taking them on into new *performative* territories.[10] For if 'thought is a matter of gesture, shape, movement' (Marks 1998: 11), then dance can help us to think in new performative ways.

In other words, my project is about moving towards a *disclosive politics* which is an attempt to both valorise and deepen particular 'intuitive' skills (Spinosa *et al.* 1997).[11] But there are, no doubt, many other ways of thinking the possibilities of the world and for changing our structure of involvements.[12] Curved perceptions, muddy paths.

Notes

1 This retreat was hastened by a system which was 'structured not as the Panopticon but as the Abbey of Thélème so picturesquely described by Rabelais – where everybody watched everybody' (Khakhordin 1995: 214).

2 Lest the example be thought too precious (it is not), let me take another example from dance which shows, I think, the importance of expressive bodily performance as a *political moment* and as the *politics of the moment* – in the most extreme and and impoverished of circumstances. This example can be found in work by Stuckey (1987, 1995) on dance and song in slave culture (and specifically the role of the Ring Shout). Though Stuckey might be accused of a certain romanticism, he is surely right that

dance was the principal means by which slaves, using its symbolism to evoke their spiritual view of the world, extended sacred observance through the week. In an environment hostile to African religion, that denied that the African had a real religion, slaves could rise in dance, give symbolic expression to their religious vision.

Dance was the most difficult of all art forms to erase from the slave's memory, in part because it could be practised in the silence of aloneness where motor habits could be initiated with enough speed to seem autonomous. In that lightning-fast process, the body very nearly was memory and helped the mind recall the form of dance to come.

(Stuckey 1995: 55)

Or as Douglass (cited in Stuckey 1995: 54) put it: 'no thanks to the slaveholder or to slavery that the vivacious captive sometimes dances in his chains: his very mirth in such circumstances stands before God as an accusing angel'. (See also Note 7.)

3　I could have chosen many other authors who have tried to deal with creativity. For example, within Marxism a number of attempts have been and still are being made by French and Italian Marxists to salvage the human creativity evidenced in Marx's theories. Again, much interest is now being shown in the expressivist anthropology of Herder (cf. Taylor 1991).

4　This can be seen as equivalent to the self-organisation beloved of complexity theorists (see Cilliers 1998).

5　I had thought at this point of including Deleuze, whose pragmatics can be seen as a celebration of creativity, but there is a rising tide of criticism of Deleuze's work (e.g. Lecercle 1990; Williams 1998; Badiou 1998; Marks 1998) for his evacuation of the social, his lack of historical sense and his devotion to a critique of binary categories which he simply smuggles back in again.

6　My main concern about Joas's work is its tendency to an unnecessary humanism.

7　Abercrombie and Longhurst (1998: 103) have made the case that daydreaming is a specifically modern form of the imagination: 'Performance might be a very general feature of social life but modernity involves a particular, and intense, form of it. The modern faculty of daydreaming means that people are able to imagine themselves performing in front of other people and also imagine the reactions that others will have'. Maybe.

8　Both of these classes of life point to the importance of the *expressive* powers of embodiment. As Radley (1996) puts it:

> embodiment involves a capacity to take up and to transform features of the mundane world in order to portray a 'way of being', an outlook, a style of life that shows itself in what it is. Like the painted picture in a frame, it has self-referential qualities that allude to something not easily specified. This is the totality . . . which cannot be isolated in a particular movement or word because it transcends these when taken as a fragment of the mundane (e.g. the physical body). At the same time, it does not exist beyond the particulars of the act because it is only through the specific engagements of embodied people together that such symbolic realms are made to appear.
>
> (Radley 1996: 561)

9　I remain sceptical of much psychoanalytic theory. However, encounters between Freud and Wittgenstein suggest new possibilities, as does the work of writers like Jessica Benjamin (1998). Again, the tension between 'the penetrative gaze and resolute being of [Foucault] and the contemplative, somewhat dreamy gaze and stylised gestures' (Weigel 1996: 33) of Walter Benjamin may bear fruit. But I think there are other roads to the fantasmatic, which I have hinted at in this piece.

10　However, it would be a naive mistake to believe that dance is automatically a more open and expressive cultural form. From the geometric dance of the sixteenth century (Franko 1993) to Enlightenment motifs of the kind found in the *Encyclopédie* (Roach 1995) to modern rave (Pini 1998), dance has been a means of imposing order as well as a means of generating possibility (of course, order and possibility are not necessarily opposed to one another!).

11　This is, then, a symptomatological project of the kind favoured by authors like

Gordon (1997) and, of course, Ginzburg (1980) whose paper 'Clues' was far ahead of its time.

12 In using this word, I am not, of course, suggesting the thickening of behaviour of living things by reference to something (a 'subject') which is co-present-at-hand (Glendinning 1998). Rather, I am trying to disrupt structures of familiarity by emphasising possibility.

References

Abbate, C. (1991) *Unsung Voices: Opera and Musical Narrative in the Nineteenth Century*, Princeton, NJ: Princeton University Press.

Abbate, C. (1993) 'Opera, or the envoicing of women', in R. A. Sollie (ed.) *Musicology and Difference: Gender and Sexuality in Music Scholarship*, Berkeley: University of California Press.

Abercrombie, N. and Longhurst, B. (1998) *Audiences*, London: Sage.

Badiou, A. (1998) *La Clameur de l'être*, Paris: Hachette.

Banes, S. (1998) *Dancing Women: Female Bodies on Stage*, London: Routledge.

Benjamin, J. (1998) *Shadow of the Other: Intersubjectivity and Gender in Psychoanalysis*, New York: Routledge.

Butler, J. (1993) *Bodies that Matter: On the Discursive Limits of 'Sex'*, New York: Routledge.

Butler, J. (1997) *Excitable Speech: A Politics of the Performative*, New York: Routledge.

Castoriadis, C. (1994) *Crossroads in the Labyrinth*, Hassocks: Harvester.

Castoriadis, C. (1997) *World in Fragments: Writings on Politics, Society, Psychoanalysis and the Imagination*, Stanford, CA: Stanford University Press.

Cilliers, P. (1998) *Complexity and Postmodernism: Understanding Complex Systems*, London: Routledge.

Clément, C. (1988) *Opera, Or the Undoing of Women*, trans. B. Wing, Minneapolis: University of Minnesota Press.

Cook, N. (1990) *Music, Imagination, and Culture*, Oxford: Oxford University Press.

Curtis, A. (1997) 'Introduction', in C. Castoriadis, *World in Fragments*, Stanford, CA: Stanford University Press.

Ford, R. (1998) *Women with Men*, London: Harvill.

Franko, M. (1993) *Dance as Text: Ideologies of the Baroque Body*, Cambridge: Cambridge University Press.

Ginzburg, C. (1980) 'Clues', *History Workshop Journal* 9: 5–36.

Glendinning, S. (1998) *On Being with Others: Heidegger – Derrida – Wittgenstein*, London: Routledge.

Gordon, A. F. (1997) *Ghostly Matters: Haunting and the Sociological Imagination*, Minneapolis: University of Minnesota Press.

Haraway, D. J. (1991) *Simians, Cyborgs and Women: The Reinvention of Nature*, London: Free Association.

Joas, H. (1996) *The Creativity of Action*, Cambridge: Polity.

Jowitt, K. (1992) *New World Disorder: The Leninist Extinction*, Berkeley: University of California Press.

Kharkhordin, D. (1995) 'The Soviet individual: genealogy of a dissimulating animal', in M. Featherstone, S. Lash and R. Robertson (eds) *Global Modernities*, London: Sage.

Lecercle, J. (1990) *Philosophy through the Looking Glass: Language, Nonsense, Desire*, La Salle, IL: Open Court.

Marks, J. (1998) *Gilles Deleuze: Vitalism and Multiplicity*, London: Pluto.

Morin, E. (1998) 'Cornelius Castoriadis, 1922–1997', *Radical Philosophy* 90: 3–5.

Perec, G. (1997) *Species of Spaces and Other Pieces*, Harmondsworth: Penguin.

Pile, S. (1997) *The Body and the City: Psychoanalysis, Space and Subjectivity*, London: Routledge.

Pini, M. (1998) 'Peak practices: the production and regulation of ecstatic bodies', in J. Wood (ed.) *The Virtual Embodied: Presence/Practice/Technology*, London: Routledge.

Radley, A. (1996) 'Displays and fragments: embodiment and the configuration of social worlds', *Theory and Psychology* 6: 559–576.

Roach, J. (1995) 'Bodies of doctrine: headshots, Jane Austen, and the Black Indians of Mardi Gras', in S. L. Foster (ed.) *Choreographing History*, Bloomington: Indiana University Press.

Schwartzman, H. B. (1978) *Transformations: The Anthropology of Children's Play*, New York: Plenum.

Shotter, J. (1998) 'Action research as history making', *Concepts and Transformation* 2: 279–286.

Spinosa, C., Dreyfus, H. and Flores, F. (1997) *Disclosing New Worlds: Entrepreneurship, Democratic Action, and the Cultivation of Solidarity*, Cambridge, MA: MIT Press.

Stuckey, P. S. (1987) *Slave Culture: Nationalist Theory and the Foundations of Black America*, New York: Oxford University Press.

Stuckey, P. S. (1995) 'Christian conversion and the challenge of dance', in S. L. Foster (ed.) *Choreographing History*, Bloomington: Indiana University Press.

Sutton-Smith, B. (1997) *The Ambiguity of Play*, Cambridge, MA: Harvard University Press.

Taylor, C. (1991) 'The importance of Herder', in E. Margalit and A. Margalit (eds) *Isaiah Berlin: A Celebration*, Chicago: University of Chicago Press.

Thrift, N. J. (1996) *Spatial Formations*, London: Sage.

Thrift, N. J. (1997) 'The still point: resistance, expressive embodiment and dance', in S. Pile and M. Keith (eds) *Geographies of Resistance*, London: Routledge.

Thrift, N. J. (1999) 'Steps to an ecology of place', in J. Allen and D. Massey (eds) *Human Geography Today*, Cambridge: Polity.

Weigel, S. (1996) *Body and Image – Space: Re-reading Walter Benjamin*, London: Routledge.

Williams, S. J. (1998) 'Bodily dys-order: desire, excess and the transgression of corporeal boundaries', *Body and Society* 4(2): 59–82.

14

ENTANGLEMENTS OF POWER

Reflections

Doreen Massey

The flier for the conference around which the chapters in this collection are based focused our attention on a number of questions. The first two put me in mind of those examination questions that give off a strong indication of the way in which it is expected they will be answered. The first in this pair of questions read: 'Are theories which suggest the opposition of processes of domination and resistance appropriate for our understanding of contemporary forms of power?' Strong hints here that the examinee is expected to answer in the negative, an impression re-inforced by the next question: 'Would a more nuanced approach that analyses the interconnections and interdependencies of power be more apt?' Does the examinee have the bravery to argue *against* a 'more nuanced' approach?!

The smile which this formulation produced was quickly followed, though, by thoughts of history, the history of theorising around these issues, and of how far the dominant perspectives have shifted. It was not an affront, nor startling, barely even a challenge, that the agenda was that we should answer 'no' to the first question and 'yes' to the second.

There has been a long history, which continues still, of wrestling with the notion of totality and of how, if society was to be thought in those terms, there could be any opportunity for alternatives, resistance or dislocation. In some formulations, change was deemed to come about through the production of internal contra-dictions (including the production of resistance) within 'the system' itself. In some of the heavier structuralist readings the all-encompassing nature of the totalities seemed to leave no room at all for anything that might seriously be called 'resistance'. (Some of the New Age versions of holism can leave one with a similar feeling of claustrophobia.) Moreover, and as so often happens, the attempt to struggle free from the worst aspects of the structuralist stranglehold has been marked by the very battle in which it was engaged.[1] This is true in very general ways: the language which is so often deployed to express the social form and spatiality of (western?) society today is utterly dominated by negations: fragmentation, *dis*location, *dis*ruption . . . Perhaps the positivities of which these words are a denial in fact refer less to changes in the organisation of society itself (I use this phrase with a suitable degree of wariness, but I shall use it nonetheless) than to shifts in the stories that we tell about it. We write 'dislocation' because once we used to think in terms of coherent structures. One step that might be taken is to characterise this new formulation in more positive terms. The adoption of the term 'entanglements' would seem in that light to be a very definite step forward.

The inheritance of structuralism, and of the long engagements around it, shows through in more precise and detailed ways as well, and in ways which relate directly to questions of the spatiality of power. The work of de Certeau (1984) is a classic case in point. This has become a key focus for theorists of space, power and the city, both in social and cultural studies generally and in geography in particular. Yet the model of power with which it operates is essentially one which opposes the strategies of the powerful to the tactics of the little people, those who resist. It is an engagement with structuralism – an attempt to find a way out of a system which in fact had no exit signs. What de Certeau does is to retain 'the structure' as his conceptual starting point, and then recruit guerrillas to attack it.[2]

This general approach is problematical philosophically in that it preserves the structure of big binaries which is now so widely recognised to channel thinking down particular, and not particularly helpful, pathways. It is also problematical politically, for it overestimates the coherence of 'the powerful' and the seam-lessness with which 'order' is produced. Moreover, I would argue, it reduces (while trying to do the opposite) the potential power of 'the resisters'. In fact, as is now generally recognised, the whole thing is far more fraught, unstable and contingent, as well as multiple, than that. This latter characterisation is the one broadly adopted in this collection, and it is one with which in general terms I agree.

Before moving on, however, it is perhaps worth pausing just to mark two issues (among many potential ones) with which this amended approach will itself have to deal. The first is a point often made, and I simply repeat it here. This is that a recognition that power is everywhere – and that we must pay attention to the micro-politics of power (which we must) – should not lead to a position where the real structural inequalities of power are lost, dissipated in a plethora of multiplicities. This is an argument which has been made against Foucault (Hartsock 1990) and also against other micro-relational readings of the world, such as actor-network theory (see e.g. Murdoch 1997).

This is connected to the second issue, which is that the question of 'totality' and of the possibility/impossibility of genuine alternatives and genuine resistance is by no means settled (as if such fundamental questions ever could be). The

Situationists, deeply opposed to the structuralists, nonetheless struggled with the same issue. Their analysis of the society of the spectacle raged against its tendency to be all-encompassing. It was a constant battle to maintain the possibility of a continued potential for demonstrating a different way of doing things. The Situationists' insistence that the system is always open enough to allow such opportunities was as much born of the necessity of political conviction as it was an outcome of their theoretical analysis of society.[3]

These issues are pertinent now. With globalisation told as a story of planetary inevitability, with neo-liberalism the only economics in town (though here some cracks may be beginning to show), with that claustrophobic constellation of family/car-ownership/supermarket-shopping, having won the west, now being exported everywhere else, with – perhaps most of all – every attempt at radical otherness being so quickly commercialised and sold or used to sell . . . With all of this, one might well ask what are, and where are, the possibilities for doing things differently? In spite of our more complex talk of entanglements, it is sometimes hard to see how resistance can be mounted without it subsequently, and so quickly, being recuperated. Perhaps this remains as much a struggle for us today as it was for the Situationists and their like. And, we must ask, are 'entanglements' necessarily any less totalising? If dominance and resistance are mutually constitutive, then is there anything 'other'? So, although the power of the dominant is indeed 'fraught, unstable and contingent' (my words) and though it does indeed have continually to be shored up, it is also at times despairingly difficult to imagine a form of refusal which will resonate, which will have effects, which might just overturn a few things.

Much of the rest of this chapter is concerned to ponder the implications of these questions, and indeed to do so with a (small) degree of optimism. Indeed there is no alternative to doing so; like the Situationists we have to believe that it is possible, and there is plenty of work to build on (Gibson-Graham 1996 comes to mind). But these reflections do lead also to a note of caution – about how we characterise, and how we (as academics) treat with, resistance. Certainly it is often tough to distinguish between transgression and the reproduction of the same old conventions and relations of power (one thinks of debates over punk or Madonna). But Meaghan Morris did once, somewhere, ironically express her profound relief that now – after new cultural analysis – cleaning her car on a Sunday afternoon could be read as subversive of the system. It is important in other words not to trivialise resistance, nor to underestimate what real resistance costs. Moreover, an academic romanticisation of resistance will not do either. Indeed, it could be part of the problem. While the extension of intellectual endeavour to those who do not conform to the 'straight white male etc etc' model is entirely to be welcomed, and welcomed as a victory for those many who fought for it, the current tendency for (often highly conventional) academics to pursue 'exotics' of various forms has its dangers too. If pursued as part of a strategy of being part of, or furthering one's career in, academe such moves can precisely be part of the process of recuperation, of normalisation. Such academics then precisely do become part of that 'hall of mirrors' of which the Situationists wrote with such disdain, and which makes real resistance so difficult.

Now, the shift away from thinking in terms of coherent totalising structures and/or big central blocks of power has a particular relevance to geographers as a result of the former model's complicity in a persistent connotational differentiation between space and time. From Lévi-Strauss, through certain forms of structural Marxism, to those still today engaged with the inheritance of structuralism, the structure itself – the synchrony – has been seen as spatial. In part this was a result of the structuralists' own struggle to escape the negative aspects of certain forms of narrative, and in part it was the result of an all-too-easy, but in my view incorrect, assumption that whatever is not time must be space. This is a long and complex story (for certain elements, see Massey 1992), but what emerged was an understanding of space as characterising the static block of power – the system. In opposition to this, time was read as the disjunctive, the disruptive. Most significantly, time is associated with politics and/or resistance. And so Laclau (1990) sees dislocation, the necessary condition for the existence of politics, as temporal (and as opposed to the spatial). De Certeau (1984) equates strategy, the proper, the system of power, with space; while tactics – the street politics of resistance – are equated with the temporal. It is, I would argue, a wholly misconceived imaginary. Most certainly, this envisaging of a big structure (spatial) opposed by the guerrilla tactics of the fleet-of-foot (temporal) is part of what has given 'the spatial', in disciplines outside of geography, its supposed character of 'the fixed, the dead'. It is little wonder that Foucault (1980), engaged in his own struggles with structuralism, was moved to reflect as he did.

Divesting ourselves of that inheritance, therefore, potentially releases 'the spatial' to be conceived as a realm of much more active engagement in the process of making history. Indeed, I think the point is in fact stronger than this: that thinking space as actively and continually practised social relations precisely gives us one of the sources of 'the system's' inability to close itself. The accidental and happenstance elements intrinsic to the continuous formation of the spatial (when imagined in this way) provide one aspect of that openness which leaves room for politics.

Whatever view one has of dominance/resistance, there is somehow an implied spatial imaginary. Thus the view (a) which envisages an opposition of domination and resistance might take the form of a central block under attack from smaller forces. In contrast, the view (b) which envisages entanglement may call to mind a ball of wool after the cat has been at it, in which the cross-over points, or knots or nodes, are connected by a multitude of relations variously of domination or resistance and some only ambiguously characterisable in those terms.

Certainly in its oppositional characterisation of the spatial and the temporal, the former view (a) made it difficult to think space actively and politically. The relational/entangled view (b) certainly seems to open up the potential for the forefronting of the spatial (that is, the spatial configuration of power-imbued social relations) as an active force in the formulation and operation of dominance/ resistance. But that still leaves questions about how we imagine – and how we as academics produce and promulgate an imagination of – the spatiality of those entanglements.

One of the most powerful geographical imaginaries around at the moment is that of globalisation. Moreover, and at the risk of some overgeneralisation, it seems to me that the imagination of globalisation which is all too frequently accepted and disseminated, especially in cultural studies and related fields, is gloriously and grossly overdrawn. It matters not whether an account of power is employed which leans on a notion of a central fortress (the Identity, against which everything else must be ranged) or whether an entangled approach is adopted. In either case, pictures can be drawn which are so totalising that they effectively deny the possibility of resistance.

In such cases the discourse of globalisation is functioning, not just as a description, but as a normative insistence. It gives succour to the neo-liberal mantra that there is no alternative. The International Monetary Fund, World Trade Organisation, World Bank and most governing elites, to say nothing of the world's multinational corporations, are deeply engaged in a double practice of describing as inevitable (and therefore you must adapt your policies, open your borders, accordingly) something which they are at the same time strenuously striving to produce. It is that tendentially totalising imagination of a necessary future which enables the imposition of strategies of structural adjustment and which, for instance, Tony Blair (himself an active producer of the very globalisation that he avers is inevitable) uses as an excuse for, say, the necessity for 'flexibility' in the labour market. It is of consummate importance that, as active producers of geographical imaginaries, we persistently question and strive to hold open such discourses on the spatiality of power.

The flier for the conference also posed a question about how new theorisations of power might enable forms of practical political action. It is perhaps over-estimating our influence to speak simply of enablement. But neither must we underestimate the impact that 'academic' work can have. The very shifts which we are discussing in the conceptualisation of power itself, and the recognition of the multiplicities and mutual implication of dominance and resistance, have come about in some part because of engaged politico-theoretical challenges. Because, indeed, of resistance. Politics beyond academe, and intellectual labour within post-colonialism, sexual politics and feminism, have confronted both the old material fact of white/male/straight/western dominance *and* the assumption/assertion that theirs was the one and only – universal – voice. Part of the point, and the strength, of such struggles is that they stand as critiques of, resistances to, both systems of rule and systems of power/knowledge (see, on post-colonialism, Hall 1996). Moreover – another twist – they are challenges not only to what had previously been accepted to be the central 'blocks of power', but also to (what had previously been accepted to be) the dominant form of resistance: the (implicitly white/male/straight etc) industrial working class.

Indeed, if one takes seriously the implications of entanglements of power then not only does the identity of 'the resisters' shift and multiply, but so too does the characterisation of identity itself. In this model of power it is the relationship which is the thing. The identities of those positioned within the entanglements are themselves constituted through the practice of interrelation. The immediate implication of this is that it leaves you without a politics of identity. No longer can preconstituted 'women', 'blacks', 'lesbians' or indeed 'the working class' go ready-made into battle. What is required instead is to think through what it might really mean to have a politics of interrelation: a politics which, rather than claiming rights for a rapidly multiplying set of identities, concerns itself more with challenging, and taking responsibility for, the form of the relationships through which those identities are constructed, in which we are individually and collectively positioned and through which society more broadly is constituted.

This in turn poses a further issue of spatiality. For if relationships are practices (they are processes, they are dynamic) so the spaces of domination/resistance are active spaces of action, continually being made. These are spaces which are themselves formed out of practices, meaning in turn that, if we refuse in any simplistic way to counterpose domination and resistance, then so also must we reject any simplistic spatial imagery of centre and margins. (Likewise notions of 'interstitial spaces' seem to me to resonate all too much with Laclau's image of dislocations – of a previously conceived coherent structure.)

The language of marginality has been widely adopted, and indeed it does catch at the relative powerlessness experienced by those who feel themselves pitted against stronger forces. Yet the spatial imagery may nonetheless be unfortunate. Not only does it implicitly set a coherent centre against already-constituted margins, but also it reinstalls notions of central blocks of power. What is politically at issue is not the expansion of some spaces (the margins) at the expense of others (the centre), nor 'resisting' through finding/creating a 'space' where dominance is less effective, but rather transforming, subverting, challenging the constitutive relations which construct spaces in the first place. That is, challenging and changing that practised space of dominance/resistance. What is at issue is not the support of the margins, but ways in which the spatialities of power can be reordered through practices which are more egalitarian, less exploitative, more mutually enabling.

This point has become even more important to stress because of the development in academe in recent years of what might be called a romance of the margins, indeed a romance of the resisters. The overwhelming assumption is that the margins/resisters are good. But are they? necessarily? In Nicaragua in 1985–1986 the reactionary forces of the US-backed Contra resisted the government of the Sandinistas. In Germany in the late 1990s the neo-fascist Deutsche Volksunion (DVU) conducted a guerrilla resistance of the streets against a wider set of relations which, even if they left much to be desired, were at least in public utterance not explicitly racist. Are the Contra and the DVU marginal and therefore to be supported? Bhabha has spoken of the problem of 'the piously marginal' and of 'the periphery as a kind of moral stance'.[4] We need a more rigorous theorising, and the construction of a more rigorous spatial imagery than this. What is at issue politically is not so much where the spaces are (centre or margins) but what kinds of spaces they are. What kinds of spatialities of power do we want to build?

The converse implication of this romance of the margins/resisters is equally significant. It is that eternally romancing the margins is also to evade the *responsibilities of* power. Maybe it is as a result of always siding with the marginals that 'power' can so easily slip into being characterised as completely negative. What is certainly true is that the question of the responsibilities of being in a position of relative powerful*ness* is much less often addressed. Maybe there is something to be gained from talking 'between generations' here. I am of the 1968 and 1970s generation and my imaginings of power come also from the excitements (and the disappointments) of Cuba, China, Mozambique, Angola, Vietnam, Nicaragua, the Greater London Council, the Socialist Republic of South Yorkshire . . . OK, we lost them (nearly) all. Maybe we blew them all. But it has left me with an assumption that it is necessary to think also about 'power' in its possibilities and its responsibilities. Indeed, part of the message implicit in the use of the term 'entanglements' is that almost none of us is a 'good marginal/resister' on every front. All of us (academics/intellectuals) are likely to be in a position of relative privilege/powerfulness on at least some of the axes of social relations within which we are positioned and constructed.

Perhaps we might also address, then, what forms and relations of the spatialities of power we might aim at; not as a utopian blueprint, but as an emplaced/situated reformulation of some currently oppressive entanglements. What might a more democratic/egalitarian map of power look like? If these things really *are* relational, then such questions should really anyway be implicit in any consideration of the relations which it is the business of 'resistance' to reformulate.

Notes

1 It is perhaps too easy these days to be simply negative about structuralism; it should be remembered that in its day it too was a response to what it saw as politically reactionary formulations. Althusser's famous 'There is no point of departure' remains one of the most revolutionary of thoughts.
2 I am aware that others see more complexity than this in de Certeau, but I think this formulation is essential to his approach.
3 Of course, such elements do frequently go together.
4 In a paper delivered to the 'Stuart Hall Conference', at the Open University, 15 May 1998.

References

de Certeau, M. (1984) *The Practice of Everyday Life*, Berkeley: University of California Press.

Foucault, M. (1980) 'Questions on geography', in C. Gordon (ed.) *Michel Foucault: Power/Knowledge – Selected Interviews and Other Writings, 1972–1977*, Brighton: Harvester.

Gibson-Graham, J. K. (1996) *The End of Capitalism (As We Know It): A Feminist Critique of Political Economy*, Oxford: Blackwell.

Hall, S. (1996) 'When was the post-colonial? Thinking at the limit', in I. Chambers and L. Curtis (eds) *The Post-Colonial Question*, London: Routledge.

Harstock, N. (1990) 'Foucault on power: a theory for women', in L. Nicholson (ed.) *Feminism/Postmodernism*, New York: Routledge.

Laclau, E. (1990) *New Reflections on the Revolution of Our Time*, London: Verso.

Massey, D. (1992) 'Politics and space/time', *New Left Review* 196: 65–84.

Murdoch, J. (1997) 'Towards a geography of heterogeneous association', *Progress in Human Geography* 21: 321–337.

AUTHOR INDEX

AUTHOR INDEX

SUBJECT INDEX

Acroma treaty 101
action: actor-network theory 233;
    collective 9, 10–11; and conduct 231;
    neighbourhood 138–42, 144; noisiness
    232–3; and power 229–30
actor-network theory (ANT) 222–3,
    240–1; action 233; embodiment
    230–1; Latour 25, 223, 229, 233, 235;
    organisations 205, 214–15; power
    227–8, 280
Adams, Gerry 28
advocacy planning 6–7
Africa: European imperialism 93,
    94–5; sport as process 149–50; see
    also Cyrenaica; Kenya; South
    Africa
agency 252; affective 250; animals 238,
    242–3, 249–51; as collective
    production 232; human 219, 221–2,
    238, 261–2; nature 239–40;
    non-human 238, 242–3; power 261–3;
    refigured 234; and structure 6
agents provocateurs 23, 29
al-Mukhtar, Omar 105, 110, 115
Algeria 32
'Ali al-Sanusi, al-Sayyid Muhammed bin
    101
Andean America 167, 173–4, 176–7
animals 243, 248; agency 238, 242–3,
    249–51; commodified 243–4;
    consciousness 249–50; as cultural
    construction 243; domesticated
    253n10; and humans 233, 243–5;
    intentionality 250; as Other 245;

resistance 245, 249, 251; transgression
    245, 249; wildness 245–6
anthropomorphism 247
anti-politics 21
anti-technology 246
apartheid 73–4
Arthur, Dr John 150
assembly line work 208, 213
assimilation 164, 167
athletes: black 156–7, 159–61; mobility
    158; nationality 158–9, 161; resistance
    148, 156
authenticity 175, 246
autonomist approach 241, 242, 251
avoidance protest 11

Badoglio, Pietro 110, 116
baseball 154
Baum, John 48
Bedouins: concentration camps 113–14,
    117; guerrilla tactics 104–5; Ottoman
    troops 103; resistance 94; sottomessi
    104, 107–9; space 111–12, 113;
    territory 111–12, 114; women 260
biopower 16, 17
black media production 192
Black Power 153
Blair, Tony 283
bodies, power 219, 220, 234, 235
bodily space and identity 170–1
body-cultures, indigenous 151, 154,
    156
body-subject 234
Brigaddo Rosso 23

SUBJECT INDEX

Printed and bound by CPI Group (UK) Ltd, Croydon, CR0 4YY

01/11/2024

01782633-0002